MICROBIAL RESPONSES TO LIGHT AND TIME

For the majority of organisms two key environmental components are light and time. Biologically, time has various scales: the annual cycle of seasons, the daily cycle of the sun or the potentially rapid cycle of cell division. What is apparently fundamental to many organisms is the ability to respond in a manner predictive of time. The time-dependent changes in physiology, morphology and ecology are responsive to various environmental cues, most prominent of which is light. A variety of microbial and plant species can harvest light for photosynthesis, but the ability to perceive various forms of light is fundamental to a variety of other biological responses in many if not all organisms. This volume, which concentrates primarily, but not exclusively, on micro-organisms both prokaryotic and eukaryotic, provides a series of chapters which explore various aspects of the underlying biology, providing a wide-ranging and up-to-date review of this fundamental area of biology that is currently advancing at an exhilarating rate.

Mark X. Caddick is a lecturer in the School of Biological Sciences, University of Liverpool, UK

Simon Baumberg is a Professor in the School of Biological Sciences, University of Leeds, UK

David A. Hodgson is a lecturer in the Department of Biological Sciences, University of Warwick, UK

Mary K. Phillips-Jones is a lecturer in the School of Biological Sciences, University of Leeds, UK

SYMPOSIA OF THE
SOCIETY FOR GENERAL MICROBIOLOGY

Series editor (1996–2001): Dr D. Roberts, Zoology Department, The Natural History Museum, London
Volumes currently available:

MICROBIAL RESPONSES TO LIGHT AND TIME

EDITED BY

M. X. CADDICK, S. BAUMBERG, D. A. HODGSON AND M. K. PHILLIPS-JONES

FIFTY-SIXTH SYMPOSIUM OF THE
SOCIETY FOR GENERAL MICROBIOLOGY
HELD AT THE UNIVERSITY OF NOTTINGHAM
MARCH 1998

Published for the Society for General Microbiology

PUBLISHED BY THE PRESS SYNDICATE OF THE UNIVERSITY OF CAMBRIDGE
The Pitt Building, Trumpington Street, Cambridge CB2 1RP, United Kingdom

CAMBRIDGE UNIVERSITY PRESS
The Edinburgh Building, Cambridge CB2 2RU, UK http://www.cup.cam.ac.uk
40 West 20th Street, New York, NY 10011-4211, USA http://www.cup.org
10 Stamford Road, Oakleigh, Melbourne 3166, Australia

First published 1998

Printed in the United Kingdom at the University Press, Cambridge

Typeset in Monotype Times 10/12pt, in 3B2

A catalogue record for this book is available from the British Library

ISBN 0 521 62286 7 hardback

Coventry University

CONTENTS

CONTRIBUTORS

ALLEN, J. F. Department of Plant Cell Biology, University of Lund, Box 7007, S-220 07 Lund, Sweden

ARMITAGE, J. P. Microbiology Unit, Department of Biochemistry, University of Oxford, South Parks Road, Oxford OX1 3QU, UK

ARPAIA, G. Istituto Pasteur, Fondazione Cenci Bolognetti, Dipartmento di Biotecnologie Cellulari, Sezione di Genetica Molecolare, Università di Roma 'La Sapienza', Viale Regina Elena, 324, 00161 Roma, Italy

BALLARIO, P. Dipartimento di Genetica e Biologia Molecolare, Centro di Studio per gli Acidi Nucleici, Piazzale A. Moro 5, Università di Roma 'La Sapienza', 00185 Roma, Italy

BELL-PEDERSON, D. Department of Biochemistry, Dartmouth Medical School, Hanover, NH 03755-3844, USA

BERRY, A. E. Department of Biological Sciences, University of Warwick, Coventry CV4 7AL, UK

COGDELL, R. J. Department of Biochemistry and Molecular Biology, University of Glasgow, Glasgow G12, 8QQ, UK

COLE, A. B. Department of Biochemistry, Dartmouth Medical School, Hanover, NH 03755-3844, USA

COLLETT, M. Department of Biochemistry, Dartmouth Medical School, Hanover, NH 03755-3844, USA

COVE, D. J. School of Biology, University of Leeds, Leeds LS2 9JT, UK

CRIELAARD, W. E. C. Slater Institute Plantage, Muidergracht 12, 1018 TV Amsterdam, The Netherlands

CROSTHWAITE, S. K. Department of Biochemistry, Dartmouth Medical School, Hanover, NH 03755-3844, USA

DUNLAP, J. C. Department of Biochemistry, Dartmouth Medical School, Hanover, NH 03755-3844, USA

FRASER, N. Department of Biochemistry and Molecular Biology, University of Glasgow, Glasgow G12 8QQ, UK

FREER, A. Department of Chemistry, University of Glasgow, Glasgow G12 8QQ, UK

FYFE, P. Department of Biochemistry and Molecular Biology, University of Glasgow, Glasgow G12 8QQ, UK

GARCEAU, N. Department of Biochemistry, Dartmouth Medical School, Hanover, NH 03755-3844, USA

GILBERT, D. A. 11 Mornington Avenue, Ilford, Essex IG1 3QT, UK

GOLDEN, S. S. Department of Biology, Texas A&M University, College Station, TX 77843-3258, USA

HARZ, H. Botanisches Institut der Ludwig-Maximilians, Universität München, Menzinger Straße 67, 80638 München, Germany

HEGEMANN, P. Institut für Biochemie 1, Universitätstrasse 31, 93040 Regensberg, Germany

HEINTZEN, C. Department of Biochemistry, Dartmouth Medical School, Hanover, NH 03755-3844, USA

HELLINGWERF, K. J. E. C. Slater Institute Plantage, Muidergracht 12, 1018 TV Amsterdam, The Netherlands

HIRSCHIE JOHNSON, C. Department of Biology, Vanderbilt University, Nashville, TN 37235, USA

HODGSON, D. A. Department of Biological Sciences, University of Warwick, Coventry CV4 7AL, UK

HOWARD, T. Department of Biochemistry and Molecular Biology, University of Glasgow, Glasgow G12 8QQ, UK

ISAACS, N. Department of Chemistry, University of Glasgow, Glasgow G12 8QQ, UK

ISHIURA, M. Division of Biological Science, Graduate School of Science, Nagoya University, Chikusa, Nagoya, 464-01, Japan

KONDO, T. Division of Biological Science, Graduate School of Science, Nagoya University, Chikusa, Nagoya, 464-01, Japan

KORT, R. E. C. Slater Institute Plantage, Muidergracht 12, 1018 TV Amsterdam, The Netherlands

KYRIACOU, C. P. Department of Genetics, University of Leicester, Leicester LE1 7RH, UK

LAMPARTER, T. Institut für PflanzenPhysiologie, Free University Berlin, Königin Luise Strasse, 12–16, D-14195, Berlin, Germany

LAW, C. Department of Biochemistry and Molecular Biology, University of Glasgow, Glasgow G12 8QQ, UK

LINDEN, H. Istituto Pasteur, Fondazione Cenci Bolonetti, Dipartimento di Biotecnologie Cellulari, Sezione di Genetica Molecolare, Università di Roma 'La Sapienza', Viale Regina Elena, 324, 00161 Roma, Italy

LIU, Y. Department of Biochemistry, Dartmouth Medical School, Hanover, NH 03755-3844, USA

LLOYD, D. Microbiology Group (PABIO), University of Wales, Cardiff, PO Box 915, Cardiff CF1 3TL, UK

LOROS, J. J. Department of Biochemistry, Dartmouth Medical School, Hanover, NH 03755-3844, USA

LUO, C. Department of Biochemistry, Dartmouth Medical School, Hanover, NH 03755-3844, USA

McLUSKEY, K. Department of Chemistry, University of Glasgow, Glasgow G12 8QQ, UK

MACINO, G. Istituto Pasteur, Fondazione Cenci Bolognetti, Dipartimento di Biotecnologie Cellulari, Sezione di Genetica Molecolare, Università di Roma 'La Sapienza', Viale Regina Elena, 324, 00161 Roma, Italy

PHILLIPS-JONES, M. K. Department of Microbiology, University of Leeds, Woodhouse Lane, Leeds LS2 9JT, UK

PICCIN, A. Department of Genetics, University of Leicester, Leicester LE1 7RH, UK

PRINCE, S. Department of Chemistry, University of Glasgow, Glasgow G12 8QQ, UK

REHMAN, J. Institute for Medical Psychology, Chronobiology Group, Ludwig-Maximilians University, Goethestrasse 31, D-80336 München 2, Germany

ROENNEBERG, T. Institute for Medical Psychology, Chronobiology Group, Ludwig-Maximilians University, Goethestrasse 31, D-80336 München 2, Germany

ROSATO, E. Department of Genetics, University of Leicester, Leicester LE1 7RH, UK

SHINOHARA, M. Department of Biochemistry, Dartmouth Medical School, Hanover, NH 03755-3844, USA

SPUDICH, J. L. Department of Microbiology and Molecular Genetics, The University of Texas-Houston Health Science Center, Medical School, Houston, TX 77030, USA

WALSBY, A. E. Department of Botany, School of Biological Sciences, Woodland Road, Bristol BS8 1UG, UK

EDITORS' PREFACE

Light and time are usually considered to be the playground of the physicist. However, the work presented in this volume clearly places them as central to the living world. The link between light and time is also striking, even to the point where, at the molecular level, the perception of light and the mechanism of the biological clock are inseparable (Macino *et al.* and Dunlap *et al.*, this volume).

Life as we know it is impossible to envisage without the sun as our primary source of energy, and photosynthetic organisms to harvest this. In parallel with the evolution of the photosynthetic machinery, a range of intriguing responses to light have developed, allowing organisms to respond in subtle and even pre-emptive ways to changes in the light. The diurnal rhythm has also been adopted as a central component to many organisms which cannot harvest light energy, light being a source of information relating to many other aspects of the environment as well as a key agent responsible for damaging macromolecules. Not surprisingly therefore, light is directly involved in setting of the circadian clock which, in turn, affects a wide range of biological phenomena in organisms as diverse as cyanobacteria and human beings.

In the opening chapter, John Allen has presented an overview lucidly demonstrating the breadth of the field, which extends over a time scale ranging from femtoseconds to gigayears (see also Lloyd and Gilbert, this volume). The physics of light absorption, its concomitant dangers and benefits, the transfer of the energy, and its final demands and consequences on the cell's physiology and development are all addressed. The predictive nature of the circadian clock and questions relating to evolution, and the horizontal transfer of some of these key biological systems are also considered.

Light is an exploitable energy source with the consequence that, in the most common form of photosynthesis, oxygen is generated. However, oxygen and light are a lethal combination. The biological response has been to evolve a variety of behavioural and regulatory strategies. The principal behavioural response is phototaxis, with positive phototaxis towards light of a usable wavelength and negative phototaxis from wavelengths that are harmful. The regulatory approach includes the synthesis of photoprotective molecules such as carotenoids when appropriate (Hodgson and Berry, this volume) and the synthesis of photoharvesting assemblies only in the absence of oxygen, as in the purple bacteria (Cogdell *et al.* and Phillips-Jones, this volume) and halobacteria (Spudich, this volume). In addition to this regulatory response, both these bacteria also employ the behavioural option.

In the photosynthetic bacteria, described by Judith Armitage, it appears that behaviour and photosynthesis are linked via the photosynthetic/respiratory electron transfer chains of these organisms, or possibly a protometer (Δp), both of which are also influenced by other factors such as redox, and which therefore also influence behaviour. The ability of the organism to sense, signal and respond to several components within its environment simultaneously demands co-ordination and this, in turn, seems to have produced mechanical links to integrate the various systems involved.

However, simple strategies are also available such as the self-regulating buoyancy mechanism employed by some cyanobacteria involving gas vesicles (Walsby, this volume). In the case of some micro-organisms there is an apparent ability to determine the direction of a light source, such as in the flagellated alga *Chlamydomonas* (Hegemann and Harz, this volume). The eye, which is only $1 \mu m$ across, is able to detect the intensity of the light, and its variation with respect to the rotational movement of the organism also allows orientation. The life cycle and development of moss is dependent on light as a cue at various stages (Cove and Lamparter, this volume). One of the most intriguing aspects of this plant's responses to light is seen in the regeneration of protoplasts, which requires high levels of light; most interestingly, they display polarity of growth, which is principally determined by blue light.

The evolutionary link between the proteins responsible for harvesting light for energy transduction and those involved in signalling has been explored at the molecular level in *Halobacterium* by John Spudich. The four principal proteins are structurally related but have diverged, depending on their role in energy or signal transduction. A detailed structural analysis of the light harvesting centres from several organisms (the antenna complexes) reveals a diverse array of structures (Cogdell *et al.*, this volume). The available molecular detail, rather than defining a clear model, demonstrates variety which almost certainly reflects the ecological niches of the respective organisms, where the system has to be adapted to the appropriate wavelengths and intensities of light to maximally convert the available energy without damage. This careful balancing act requires precise regulation of the light harvesting machinery, which is considered in detail for both the pigment proteins in *Rhodobacter* (Phillips-Jones, this volume) and carotenoid synthesis in *Myxococcus xanthus* (Hodgson and Berry, this volume). Parallels can be drawn between these prokaryotic signalling pathways and those of the eukaryotes. The filamentous fungus *Neurospora crassa* is sensitive to blue light, which acts as a signal for developmental processes, including carotenoid biosynthesis, and the circadian clock (Macino *et al.*, this volume). The molecular characterization of the signal transduction pathway has recently led to the identification of two transcription factors. Interestingly, these have motifs common to bacterial proteins involved in redox sensing (the PAS domains) and DNA binding domains shared with the principal erythroid transcription factors of vertebrates. This suggests a convoluted evolutionary history as well as an intriguing correlation of light and oxygen.

The remaining five chapters are all concerned with time, and with the way in which light and other environmental components are crucial in setting the biological clock. Susan Golden and co-workers (this volume) have demonstrated that, in at least some prokaryotes (the cyanobacteria), the features that define the clock share properties common to eukaryotic organisms. The cyanobacteria are presented with the problem that the oxygen produced by photosynthesis is not compatible with the key enzyme in nitrogen fixation, nitrogenase. Thus the processes of photosynthesis and nitrogen fixation have to be separated temporally as well as spatially. This temporal requirement is achieved with a molecular clock which can be entrained by both light and oxygen.

Perhaps the most striking combination of circadian phenomena described in this volume is that of the marine dinoflagellate *Gonyaulax polyedra* (Roenneberg and Rehman). The circadian rhythm defines a periodic rise and fall of the organism within its marine environment of up to 30 metres, the timing of mitosis, photosynthesis and two distinct forms of bioluminescence. The signals involved in altering the clock are equally complex, with both red and blue light acting separately, combined with a role for nitrogen sources. Mitosis and the cell cycle, as aspects of biological time, are explored further in the review by Lloyd and Gilbert (this volume), who catalogue the wide range of time domains observed within living systems.

Perhaps the most interesting work relating to the molecular basis of the biological clock has come from the dissection of the molecular basis of the circadian clock in *Neurospora crassa* (Dunlap *et al.*, this volume). From this work we have our first detailed overview of the clock mechanism. One of the central components, the gene frequency, having been cloned and characterized, has been shown to integrate the known components of the circadian clock, the light response, a temperature compensation, and a periodic rise and fall in the level of the active protein product within the cell, which appears to define the circadian rhythm. This involves regulation at the level of transcription, translation and post-translation. When extending studies from microbial systems to insects, vertebrates (Kyriacou, Piccin and Rosato, this volume) and plants, it would appear that the basic components and organization of biological clocks are similar, with light as a principal signal, temperature compensation built in, and a variety of clock responsive genes integral to a wide range of processes from basic metabolism to development. Whether these parallels continue to be strengthened with the rapidly accruing molecular detail remains to be seen. The recent finding that the microbial PAS dimerization domain is involved, even in animal systems, suggests that microbial clocks are excellent models with broad relevance. The rapid pace of research in this field will mean that many of the remaining fundamental questions will not be unanswered for long.

The authors have presented an excellent range of current reviews and include a number of different scientific approaches, extending from the ecological to the biophysical, to describe and dissect a variety of phenomena which are pertinent to most, if not all, biological systems. We thank all the contributors for their considerable efforts, which have resulted in this wide-ranging and comprehensive account, and our hope as editors is that they prove both illuminating and timely.

M. X. Caddick, S. Baumberg, D. A. Hodgson
and M. K. Phillips-Jones, December 1997

LIGHT, TIME AND MICRO-ORGANISMS

JOHN F. ALLEN

Department of Plant Cell Biology, Lund University, Box 7007, S-220 07
Lund, Sweden

TIME DOMAINS IN BIOLOGY

Direct, individual human experience of the natural world can be measured in milliseconds or in decades. Our evolving picture of the world depends upon comparison of our individual experience with those of other people, which extends the time-scale, upwards, by perhaps a factor of a thousand. Thus rare astronomical and geophysical events, which most people never witness, are important ingredients of an overall picture, and now influence even our understanding of living organisms and their evolution.

Of course, experimental science incorporates indirect experience, too. This enormously extends, in both directions, the time domains with which we must feel comfortable in order to describe important events. A device for presenting time on an intelligible, logarithmic scale, and designed to be relevant to early events in photosynthesis, was introduced many years ago by the physicist Martin Kamen (Kamen, 1963). This device has been widely adopted (Gregory, 1989; Whitmarsh & Govindjee, 1995). By analogy with pH (the negative logarithm, to the base ten, of hydrogen ion concentration), Kamen introduced the pt_s scale, where the lifetime t of a process produces a value for pt_s for that process given by:

$$pt_s = -\log_{10}t(s)$$

where $t(s)$ denotes time, measured in seconds.

In an attempt to provide an overview of microbial responses to light and time, Kamen's pt_s scale is extended here into time domains of which a photophysicist might be expected to disapprove, but with the same objective of making a reference axis upon which mutually related, if temporally distant, events may be described and compared (Fig. 1).

Figure 1 is arranged in four columns. The left-hand column is the pt_s scale itself, with $t(s)$ values also given. $t(s)$ will be more familiar to, for example, certain kinds of photosynthesis researcher, for whom millisecond events are at the fuzzy, qualitative, and descriptive end of the scale, and are there seen in rather the same way as a molecular microbiologist tends to view topics such as epidemiology.

The second column of Fig. 1 places important processes along the pt_s scale at positions corresponding to their lifetimes, where these are known, or can

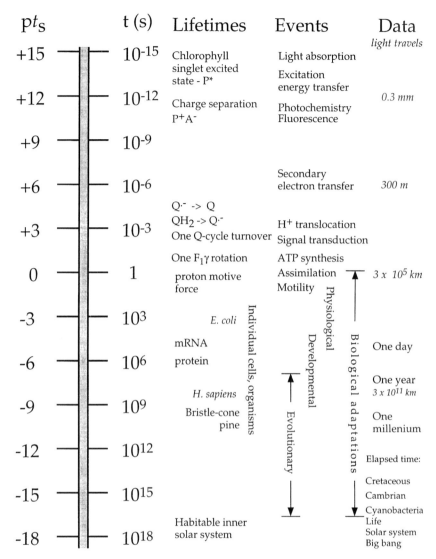

Fig. 1. The pt_s scale. pt_s = $-\log_{10}$t(s) where t(s) is time, measured in seconds. As a supplement to this overview, the reader may find it useful to consult the internet site http://plantcell.lu.se/ltm/, where the pt_s scale and some of the structures and events included here are presented dynamically and interactively, and include hypertext links to primary or related sources of further information.

be guessed with reasonable confidence. The third column is a similar arrangement of the events to be discussed in this chapter. Columns two ('Lifetimes') and three ('Events') are central to the topics discussed in this overview, and encompass the events to which reference will inevitably be made elsewhere in this symposium.

The right-hand column of Fig. 1, 'Data', is intended to provide points of reference. At positive values of pt_s, these are given as the distance travelled by light during the corresponding time interval. For example, during the initial charge separation of the primary photochemistry of purple bacterial photosynthesis, light travels approximately 1.2 mm. The velocity of light is 3×10^8 m s^{-1}, and is not entirely remote from direct experience, at least for those who may initially be puzzled by the delay of 245 ms between terrestrial and satellite reception of a 'live' BBC radio broadcast. At larger, negative values of pt_s, the 'Data' column includes selected historical events whose names are placed on the scale next to the time that has elapsed since they took place, and 'pt_s' then, strictly speaking, means 'Δpt_s'. In these time domains the distance travelled by light is normally expressed in light-years (that is, in multiples of 3×10^{11} km).

One curious feature of Fig. 1, to which the reader's attention is now drawn but which will not be discussed further, is that the seconds, minutes, and hours of direct human experience, experimental biology, and so on, stand midway between the fastest and slowest events of which we can conceive, in a manner analogous to the 'anthropic principle' where the log scale is not one of time, but of distance (Barrow & Tipler, 1986).

This overview is divided into the present introduction, a concluding section, and, in-between, a description of selected events occurring within each of the five time domains that each have $\Delta pt_s = 6$ and which, together, cover almost the whole pt_s scale of Fig. 1. Actually, the whole scale has the value $\Delta pt_s = 33$; the last three units encompass events that will not be discussed in any sort of detail, and these will be subsumed into the fifth domain. No overview from this distance could be generally accepted as balanced, still less complete: the selection of events reflects the theme of the symposium, recent progress, and the author's current interests.

$$15 < pt_s > 9$$

FEMTOSECONDS TO NANOSECONDS: THE FIRST TIME DOMAIN

Light absorption

The use of pumped lasers and synchrotron radiation pulses as actinic light sources has now extended studies of the interaction of light with matter ('spectroscopy') into the sub-nanosecond ('ultrafast') time domain. The term 'femtochemistry' describes the study of the fastest of these events. Photosynthesis is well known as the process by which light is absorbed and converted into a useful form in the biosphere, and as the source of the global redox disequilibrium that appears to be the earth's unique planetary signature of life. Besides this primary and, for us, indispensable interaction of light with living matter, light as an environmental signal must also depend upon principles governing its absorption and the chemical changes that may then take place.

Figure 2 shows a simplified 'Jablonski' diagram, in which the energy of an electron in an atom or molecule can be described from quantum mechanics as possessing one of a relatively small number of discrete values. Absorption of a single quantum of light depends on the availability of an electron whose permitted energy change falls within the range determined by the energy of the light quantum that induces it. The energy of the quantum is proportional to the frequency of the radiation, and inversely proportional to its wavelength – a Jablonski diagram turned through 90° yields an absorption spectrum.

Light absorption and the movement of the electron between energy levels occur on a femtosecond time scale. Internal conversion may occur between one energy level and a lower one, and is accompanied by release of energy as heat. There are four processes that may then take place. The first is thermal, or non-radiative, de-excitation. This is the usual route for atoms and molecules absorbing specific energies without further events of direct biological relevance. The remaining three processes that follow absorption can occur at different rates, and all are important properties of photosynthetic systems. Fluorescence is important for what it tells us. Fluorescence is the re-emission of a quantum of light, and occurs in picoseconds to seconds. Slow fluorescence, or 'luminescence' may be induced by changes in the physical or chemical state of the molecule containing the electron. Chlorophyll and bacteriochlorophyll have an inherently high, and variable, yield of fluorescence. Fluorescence emission from chlorophyll occurs with a lifetime typically measured in picoseconds, and the fluorescence lifetime of chlorophyll depends on its physical environment as well as on the rate of the competing processes of energy transfer and photochemistry. The next route for the falling electron (Fig. 2) is energy transfer, which is a sub-picosecond phenomenon. This is the fate of excitation of most chlorophyll molecules bound to protein in photosynthetic membranes as well as that of the water-soluble phycobilins that perform a similar, light-harvesting, function in cyanobacteria and red algae.

Light harvesting

The function of light-harvesting pigments is to collect light energy over a far larger area than would be possible if each molecule were required itself to participate in storage of that energy in chemical form. The reason for this constraint is the relatively long time required for regeneration of the ground state by the final route for de-excitation, namely photochemistry. Although the chemical structure of the molecules involved may be identical, the

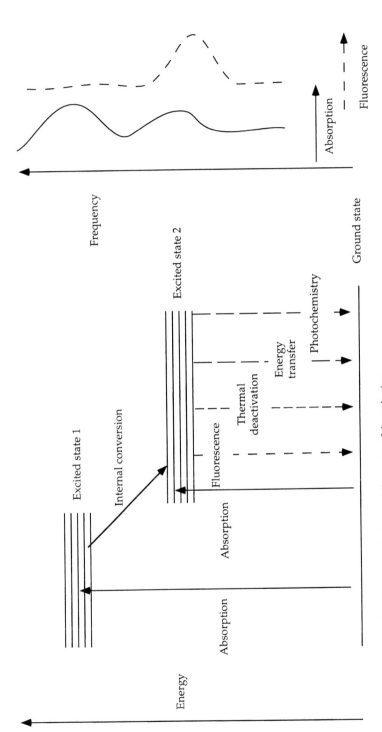

Fig. 2. Energy states of an electron, excitation, and the routes of de-excitation.

division of function between light-harvesting pigments (carrying out energy transfer) and reaction centre pigments (carrying out photochemical charge separation) is a fundamental feature of all photosynthetic systems. This division is imposed by the physical nature of light and the disparity between the time taken for its absorption and for its conversion into chemical form.

Our understanding of the early events in the biosphere's harnessing of light energy has increased dramatically in the last two decades, and it happens that the progress has been made almost entirely with photosynthetic bacteria. Some events have certainly been clarified for the chloroplasts of multi-cellular, green plants, but a rigorous description of the fundamental events of photosynthesis depends absolutely on an atomic resolution of the distances involved in energy and electron transfer. These were first secured for a membrane-intrinsic light-harvesting complex by Cogdell and co-workers with their structure of the LH II of the purple, non-sulphur photosynthetic bacterium *Rhodopseudomonas acidophila* (Freer *et al.*, 1996). For the photochemical reaction centre, the equivalent step was made a decade earlier by Michel and co-workers with their extraordinary structure for the reaction centre of *Rhodopseudomonas viridis* (Deisenhofer *et al.*, 1985) (Fig. 3(a), see colour plates). The broadly confirmatory structures of light-harvesting complexes (see Freer *et al.*, 1996) and reaction centres (Yeates *et al.*, 1987) from other species were also obtained with purple bacteria. There are also the high-resolution structures of Huber and associates of the more specialized light-harvesting structures of phycobiliproteins (Duerring, Schmidt & Huber, 1991), and the partial structure of the chloroplast light-harvesting complex II, at slightly lower resolution, of Kühlbrandt and co-workers (Kühlbrandt, Wang & Fujiyoshi, 1994).

The structure and function of the intrinsic membrane pigment–protein complexes that function in light-harvesting in purple, non-sulphur photosynthetic bacteria (the Rhodospirillaceae) are discussed by Cogdell (this volume), and any further description of this breakthrough is, at this point, superfluous.

Reaction centres

In reaction centres, the chlorophyll or bacteriochlorophyll molecules (P) involved in the primary photochemical reaction of photosynthesis receive excitation energy from their chemically identical light-harvesting antenna pigment molecules, but the reaction centre pigments themselves are held by histidine ligands in an environment close to an electron donor (D) and acceptor (A). The excited state (P*) returns to the ground state (P) via an oxidized species (P^+), the electron being lost to the acceptor, thus:

$$DPA \rightarrow DP*A \rightarrow DP^+A^-$$

In the third state (DP^+A^-) the excitation energy is said to have been 'trapped' by photochemistry. The rise-time of the absorption change (a photochemical bleaching) that reports on the generation of P^+ has been timed at 4 ps $(pt_s = 11.4)$ for reaction centres of purple bacteria, and is synchronous with the reduction of the acceptor, bacteriophaeophytin (Youvan & Marrs, 1987). Subsequent events are again determined by the kinetics of a number of competing reactions, but the 'useful' reaction is forward electron transfer from A^- to a secondary electron acceptor, a quinone, which takes 200 ps $(pt_s = 9.7)$. Re-reduction of P^+ to restore the ground state by the donor (in purple bacteria, a c-type cytochrome), together with movement of the electron from the first to a second quinone, takes about 200 ms $(pt_s = 3.7)$. The second quinone accepts a second electron by the same route, and moves on to provide electrons to the Q-cycle, discussed here in time domain two. The sequence of events is shown in Fig. 3(b) (see colour plates).

Around 2×10^9 years ago $(\Delta pt_s = -16.8)$ some photosynthetic bacteria appear now to have developed the singularly useful trick of supplying the electron to P^+ from a tyrosine side chain, generating tyrosine cation radicals that are capable of sequential abstraction of hydrogen atoms from water, producing molecular oxygen. The oxygen-evolving donor side of photosystem II in modern cyanobacteria and green plants is thus now viewed as a member of the growing class of free-radical proteins, along with ribonucleotide reductase and galactose oxidase (Hoganson & Babcock, 1992). There are strong arguments, supported by conservation of the histidine chlorophyll ligands, that the reaction centres of oxygenic photosystem II are homologous with those of anoxygenic purple bacteria (Deisenhofer et al., 1985). The best photosystem II structures to date, for example, that of Rhee et al. (1997), support this idea. An intriguing evolutionary consequence of this view is that oxygenic photosynthesis, which requires the coupling, in series, of two distinct types of reaction centre (photosytems I and II), must have depended upon lateral transfer of genes between the evolutionary precursors of the modern green, sulphur bacteria (whose single reaction centre resembles photosystem I) and those of the purple bacteria (Blankenship, 1994).

$$9 < pt_s > 3$$

NANOSECONDS TO MILLISECONDS: THE SECOND TIME DOMAIN

The second time domain is, for the purposes of this overview, one of secondary electron transfer, photoisomerization, ligand binding, catalysis, and protein structural change. Figure 3(b) already extends the primary events of photosynthesis well into this region. To this can be added the photocycle of bacteriorhodopsin (Subramaniam et al., 1993).

In reality, the early events of signal transduction must also fall into this domain, but most lie outside the range of conventional biochemistry – even the best stopped-flow apparatus has a mixing time of a few milliseconds. Light signal transduction (for example, Hoff, Jung & Spudich, 1997; Spudich, this volume), like photosynthesis, is an exception, since the substrate can be added in pulses. Long before the use of lasers, conventional flashlamps with μs flashes took photobiology and photochemistry well into this time domain. In photosynthesis, light itself has already done its job within the first time domain, and the 'light reactions' of photosynthesis are largely passive, thermodynamically 'downhill' electron transfers that are initiated by the much faster events of primary photochemistry.

Nevertheless, there is a crucial set of secondary electron and proton transfers that occur in bioenergetic membranes, including photosynthetic membranes if the donor happens to be a reaction centre. One intermediate in these reactions is particularly short-lived, but may provide a key to understanding quite distant biological responses to light and time.

The Q-cycle

The Q-cycle (Mitchell, 1975) is an indispensable component of energy coupling in most photosynthetic and respiratory systems, and appears to be the primary function of cytochrome bc_1 (or 'b-f') complexes. The Q-cycle takes two electrons from the bulk quinone pool in the membrane, passes one on to an iron–sulphur protein and hence to cytochrome c, and recycles the other back into the pool by means of its transfer between two cytochrome b haems, arranged across the membrane. At the 'Q_o' site ('o' for 'outer') each of the two electron transfers releases one proton. At the 'Q_i' ('inner') site a proton is bound when quinone is re-reduced by the electron from cytochrome b, and another when an electron is supplied from the original donor to the pool. The overall result is that movement of one electron from the donor, via the quinone pool, to cytochrome c drives translocation of two protons from the inner to the outer aqueous phase. This resolves a stoichiometric anomaly apparent in earlier formulations of the chemiosmotic hypothesis. The overall process belongs in the third time domain, since, in general, one complete turnover of the Q-cycle is complete in about 60 ms: $pt_s = 1.2$. However, a recent proposal for the component reactions is now outlined. These lie in the second time domain and, arguably, are necessary for understanding events in all subsequent time domains.

One inexplicable feature of the Q-cycle has, until recently, been the requirement for bifurcation of the electron transport chain at the Q_o site: by what means may the quinol be forbidden from donating both of the electrons it carries in the thermodynamically favoured direction, that is, to the Fe–S centre? Even after a single electron transfer, the intermediate semiquinone should be a good donor to the iron–sulphur protein. In other words, why does recycling through cytochrome b occur at all?

Recent structures from X-ray crystallography for the iron–sulphur and for the intrinsic membrane domain of the bc_1 complex have provided Crofts et al. (1998) with the structural basis for an ingenious solution to this problem. Figure 4 (see colour plates) presents the fundamentals of Crofts's idea. The quinol in its Q_o site is initially much closer in space to the iron–sulphur centre than it is to the low-potential b-haem, and electron transfer to the iron–sulphur centre is then kinetically favoured. After the first electron transfer, the semiquinone ring moves closer to the low-potential b-haem, permitting the second electron transfer to occur in a different direction. After accepting the second electron, the iron–sulphur centre moves away from the Qo site and towards cytochrome c as a result of rotation of the mobile head-group of the iron–sulphur protein.

The kinetics of the component reactions are known in some detail, and are consistent with Crofts's proposal (1998). One way of considering the Qo site mechanism in functional terms is to view the semiquinone anion radical as indispensable but dangerous. This dilemma may provide an insight into the persistence, in evolution, of extra-nuclear genetic systems in eukaryotes, as discussed in time domain five. Thus the steady-state concentration of the semiquinone is maintained at the lowest possible value by the first electron transfer, from the quinol to the iron–sulphur centre, being around ten times slower than the second electron transfer, from the semiquinone to the low-potential b-haem. Quinol oxidation to semiquinone occurs in 600 μs: $pt_s = 3.2$. the semiquinone is oxidised to quinol in 60 μs: $pt_s = 4.2$. The electron transfer between the low-potential and high-potential b-haems takes around 200 μs: $pt_s = 3.7$.

$$3 < pt_s > -3$$

MILLISECONDS TO KILOSECONDS: THE THIRD TIME DOMAIN

ATP synthesis

Mitchell's 'chemiosmotic' mechanism for the coupling of electron transport to ATP synthesis (Mitchell, 1961) falls clearly into the early part of this domain. Recent progress in understanding the elements of chemiosmosis has been remarkable, and has again come from structural biology, as illustrated by the primary route for generation of the proton motive force that was discussed in the previous section.

F-ATPases

The proton motive force, a transmembrane electrochemical gradient of hydrogen ion concentration, may drive any one of a number of endergonic reactions, chief among which is synthesis of ATP. The means by which ATP synthesis results from movement of protons back across the coupling membrane can be inferred clearly from the structure of the extrinsic

F_1-ATPase of bovine heart mitochondria. The structure, from Walker and co-workers (Abrahams *et al.*, 1994), shows a radial, 3-fold symmetry, with each $120°$ sector, although composed of an common $\alpha–\beta$ heterodimer, containing a different ligand binding site associated with the catalytic subunit, β. The shape of each ligand binding site appears to be determined by the asymmetry of the single γ subunit, which forms a spindle-like structure inserted through the central core of the roughly cylindrical $\alpha_3\beta_3$ domain, where the points of interaction between γ and $\alpha_3\beta_3$ are large, hydrophobic amino acid side chains. From the structure alone (Fig. 5, see colour plates) the irresistible conclusion is that F_1 is a bearing. Rotation of γ within the central axis of F_1 would obviously induce sequential changes in the conformation of each α-β heterodimer, a structural basis for Boyer's binding-change mechanism. Each heterodimer binds ADP and phosphate loosely; then ADP and phosphate tightly; and, finally, ATP, which is seen *in situ* in the crystal structure in the form of a non-hydrolysable ATP analogue.

The idea that the γ subunit acts as a camshaft is currently supported by two quite independent lines of evidence. A direct and visually compelling demonstration has been provided by Yoshida and co-workers (Noyi *et al.*, 1997), who successfully tethered the γ subunit to an actin filament and the $\alpha_3\beta_3$ headgroup to an inert, metal surface: upon addition of ATP, some actin filaments were observed, in a light microscope, to rotate. During ATP hydrolysis, the angular velocity of the actin filament depended on its length, but all rotations were anticlockwise. Junge and co-workers (Sabbert, Engelbrecht & Junge, 1996) used a fluorescence tag (eosin) on the γ subunit of an immobilized chloroplast (C)F_1 in order to study movement of the tag by polarized absorption relaxation after photobleaching. The conclusion is that ATP induces rotation of γ relative to the hexagonal $\alpha_3\beta_3$ array. One complete ATP-induced rotation takes 100 ms: $pt_s = 1.0$.

In the reverse process, ATP synthesis, what causes the rotation of the γ subunit of F-ATPase? Here Junge *et al.* (1997) have made the suggestion that the core of the membrane-intrinsic F_o is linked mechanically to the γ subunit, and that rotation of the core of F_o is driven by the inward movement of protons. According to this hypothesis, protonation of regularly spaced, acidic amino acid side chains occurs from the outer, aqueous phase. Provided these side chains are in their protonated, uncharged, form, they are able to enter the hydrophobic environment of the membrane. Their deprotonation occurs into the inner aqueous phase, as favoured by the electrochemical gradient of proton concentration, but is sterically possible only after rotation of the core of F_o. This proton-driven stepping motor (Fig. 6) may have been adapted, in evolution, not only to use monovalent cations other than H^+, but to provide different ring sizes of F_o, in effect giving different gear ratios, that is, H^+/rotation and therefore H^+/ATP stoichiometries, according to the free energy available from the respiratory or photosynthetic chain that the

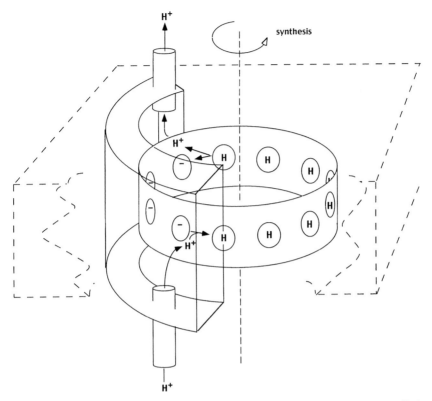

Fig. 6. F_o-ATPase as a proton-driven, rotary stepping motor, as proposed by Junge, Lill & Engelbrecht (1997). The inner ring, with 12 proton or hydrogen atom binding sites, is coupled mechanically to F_1-γ (Fig. 5). The sense of rotation of the inner ring of F_o, viewed from the top, is anticlockwise during ATP synthesis (as shown), when protons move inwards, down the gradient of electrochemical potential. If F_o remains coupled to F_1-γ during ATP hydrolysis, outward proton translocation will be driven by clockwise rotation of the inner ring. Drawing after Junge *et al.* (1997).

cell is able to deploy. This mechanism also suggests the tantalizing possibility that the gear ratio may be selected according to physiological circumstances. In the author's laboratory it has recently been found that plant mito-chondrial F_1-ε and Fo-b subunits are phosphorylated under specific con-ditions (A. Struglics, K. Fredlund, I. M. Møller and J. F. Allen, unpublished data). Fo-b is proposed by Junge *et al.* (1997) as the stator that prevents rotation of the hexagonal $\alpha_3\beta_3$ array whilst the γ subunit rotates within it. If not a gear lever, phosphorylation of subunits of ATPase may plausibly suggest the existence of a molecular clutch. If the automotive analogy can be extended, a clutch, or coupling control, may be a necessary pre-requisite for any physiological changes in H^+/ATP stoichiometry that maintain an optimal balance of effort and load without energy transfer inhibition, or stalling.

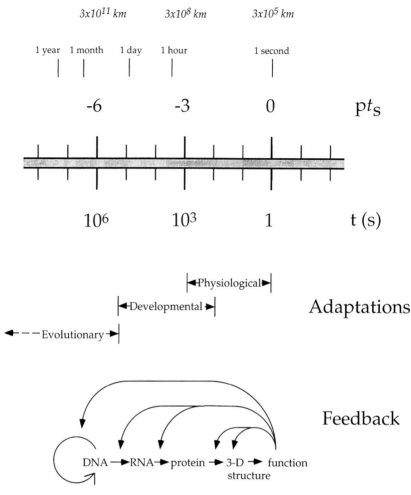

Fig. 7. The time domains of biological adaptations and gene expression. This diagram represents an expanded and supplemented part of the scale in Fig. 1, but with adaptations more specifically defined as having $p(t)_s$ ranges appropriate for prokaryotes.

Adaptation

In Fig. 1 it is seen that adaptations begin in the third time domain. This is true even if one includes responses such as nerve signal transmission and the perception of environmental signals, though for vision the initial events probably occur in domain two. Figure 7 magnifies the part of the pt_s scale of Fig. 1 corresponding to the boundary between domains three and four. Figure 7 adds the reference points of light-distance and more familiar units of time above the scale, and presents a correlation of different categories of adaptations with an extended 'central dogma' diagram denoting various stages of gene expression. Although it anticipates part of the section devoted to events in time domain three, there must logically now follow a discussion of biological adaptations generally.

Homeostasis, the maintenance of a constant internal environment in the face of external changes, is a necessary and fundamental feature of all living systems. I propose to divide the adaptations involved into three categories, here termed 'physiological'; 'developmental'; and 'evolutionary'. Other authors have wished to reserve the term 'adaptation' for only one of these three kinds of response to the environment. Maynard Smith (1966), for example, reserves 'adapted' for 'genetically adapted', which is equivalent to 'evolutionarily adapted', and, by way of distinction, uses the terms 'physiologically versatile or tolerant' and 'developmentally flexible'. It is suggested here that the last three adjectives imply passive properties of systems, and that 'adaptation', which suggests an active process, is preferable. Furthermore, in plant physiology, a careful distinction is sometimes made between 'adaptation', which may be either physiological or evolutionary, and 'acclimation', which refers to what a microbiologist would perhaps recognize as a developmental process.

Apart from the idealized arrangement of the three sorts of adaptation along the pt_s axis, there are further, non-trivial distinctions to be made between them. First, only physiological and developmental adaptations occur within the lifetime of an individual organism. The latter depends, of course, very much on the sort of organism being considered. For most microorganisms (especially when grown in a laboratory culture) both physiology and development occur well within the time domain $9 < pt_s > 3$. For all living things it is also possible to arrange the three sorts of adaptation according to the level of gene expression at which feedback is exerted as a result of any given environmental change (Fig. 7).

Thus physiology concerns largely a feedback control of structure, function, and molecular recognition of macromolecules – for the large part, proteins – that are already in place. Physiological adaptation is a fine-tuning of the machinery you already have. Where such processes as motility are involved, physiological adaptation takes the form of behavioural adaptation, for example, phototaxis (for example, Sprenger et al., 1993).

Development then becomes a matter of influencing transcription, mRNA processing and stability, and translation, in such a way that the environmental signal affects the final composition of the system that interacts with the new environment. Developmental adaptation is usually a matter of choosing a new set of components with which to work, and sometimes one of assembling new components from the old set in a different way. The borderline between physiology and development is perhaps less clear in microbiology than in other fields, but here I propose a line of demarcation, for convenience: any adaptation that occurs within the lifetime of an individual cell or organism can be described as 'developmental' if it requires protein synthesis *de novo*, and 'physiological' if it does not. The question of whether protein assembly counts as physiology or development then depends on one's definition of a protein, and becomes a matter of semantics.

Evolutionary adaptation, which comes into time domains four and five, arises from the question of whether the new environment can be used at all to make a second set of instructions, however these may be interpreted by development and physiology. The feedback loop from function to DNA replication in Fig. 7 is intended to represent this crude, qualitative line of information flow, and not the more subtle, developmental effects seen, for example, in control of the cell cycle. Evolution is the gradual departure of each succeeding set of instructions from an arbitrarily-chosen original. Clear thinkers long ago abandoned teleology in evolution (Dawkins, 1986, 1996). Attempts to rescue it now seem largely incoherent (Teilhard de Chardin, 1966). Nevertheless, the full implications of the view that evolution proceeds by a blind mechanism have perhaps not been thoroughly assimilated by our culture and society. Figure 7 presents what some may regard as an equally bleak overview of physiology and development, in which the wonders of development might be thought to be reduced to the operation of a few hierarchically arranged feedback loops. In the final section I shall return to this question, and suggest that it is not quite so simple. To forestall concern, it must be added that teleology does not come into it, and that the intention is, in fact, to provide an explanation of what might otherwise seem to be purposeful behaviour in physiology and development. Moreover, the supporting evidence is recent, central to this symposium, and comes first from microbiology; the area of the life sciences where one is, perhaps, least likely to become emotionally involved with the subject matter.

Protein phosphorylation

If physiology is fine-tuning the structure, function, and interactions of pre-existing proteins, a clear example is post-translational, covalent modification. The most versatile and widely deployed case appears to be protein phosphorylation. In chloroplasts, protein phosphorylation has an accepted, central role in the mechanism by which the light-harvesting complex II becomes redistributed between photosystem I and photosystem II (Allen, 1992). Protein phosphorylation is implicated in physiological control of excitation energy transfer in cyanobacteria and purple photosynthetic bacteria, too, despite the absence, in the latter, of the problem of redistribution of light-harvesting function between reaction centres of two different kinds. In chloroplasts, it is now known that phosphorylation of a sub-population of light-harvesting complex II polypeptides at the periphery of photosystem II induces a structural change at their amino-terminus (Nilsson

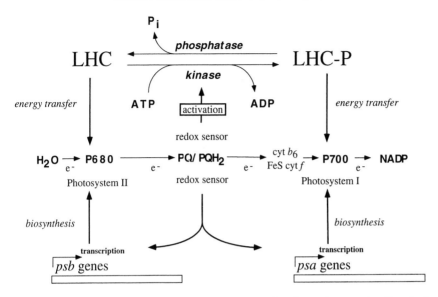

Fig. 8. Redox control of phosphorylation of chloroplast light-harvesting complex II and of reaction centre gene expression. Control of photosystem stoichiometry is assumed to result from redox effects on expression of genes for components of the two photosystems. In this scheme, photosystem stoichiometry adjustment and fine-tuning of light-harvesting by protein phosphorylation are complementary adjustments to the same redox signal. 'LHC' is LHC II in chloroplasts, and a putative phycobilisome component in cyanobacteria and red algal plastids.

et al., 1997). The structural change initiates a series of events, most likely involving dissociation of LHC II trimers, that lead to the eventual association of phospho-LHC II as a functional part of the light-harvesting antenna of photosystem I (Fig. 8). In an attempt to isolate components of the protein kinase–phosphatase system that must be involved in this process, a technique of time-resolved fluorescence imaging has been developed in the author's laboratory. This technique allows the screening of a population of cells, colonies, or small plants for mutants whose phenotype is an inability to carry out redistribution of excitation energy between photosystem I and photosystem II in synchrony with the control, or wild-type (Allen, Dubé and Davison, 1995). Whatever the enzymology behind this physiological adaptation, it is clear that light controls the ability of the photosynthetic apparatus to collect light energy efficiently by means of its effect on photosynthetic electron transport: the LHC II kinase is under redox control (Allen, 1992; Allen *et al.*, 1981) (Fig. 8). The light-induced phosphorylation of LHC II results from activation of its kinase by either chemical or photochemical reduction of the quinone pool (plastoquinone in chloroplasts and cyanobacteria) to which reference was made in the context of the Q-cycle. The half-time of light-induced LHC II phosphorylation in isolated chloroplasts is four minutes (Telfer *et al.*, 1983): $pt_s = -2.4$. In cyanobacteria, the phenomenology is essentially the same, but the lateral movement of LHC II is replaced by movement of its analogue, the extrinsic phycobilisome, between photosystem I and II (Allen, Sanders & Holmes, 1985; Mullineaux, Tobin & Jones, 1997). Although protein phosphorylation has still not been directly demonstrated to be involved, there is good circumstantial evidence that this is the case, and the half-time of the movement of the phycobilisome to photosystem I from photosystem II is 45 seconds (Mullineaux & Allen, 1988): $pt_s = -1.7$.

$$-3 < pt_s > -9$$

KILOSECONDS TO GIGASECONDS: THE FOURTH TIME DOMAIN

Although the border of the fourth time domain corresponds fairly closely with the doubling time of *Escherichia coli* in log phase, the fastest biological events to be discussed in this context are predominantly developmental. The first example can be considered as an extension of the physiological adaptation described at the end of the previous section.

Developmental adaptations: photosystem stoichiometry adjustment and complementary chromatic adaptation

The imbalance in distribution of absorbed excitation energy that initiates the physiological, and relatively rapid redox response of LHC II phosphorylation has a long-term, developmental counterpart in adjustment of photosystem stoichiometry. The response is to the same signal, and the adjustment in both cases is one of greater balance in light capture by the two photosystems in order that their rates of electron transfer may remain equal, a requirement of their being coupled electrochemically in series (Hill & Bendall, 1960). Photosystem stoichiometry adjustment (Allen, 1995) differs from LHC II phosphorylation in the time-scale over which the response occurs. In the cyanobacterium *Synechococcus* 6301 in one laboratory culture (Allen *et al.*, 1989), and for the higher plants mustard and pea (T. Pfannschmidt *et al.*, unpublished observations), the half-time of the increase in the ratio of photosystem I to photosystem II after transfer to photosystem II-specific light is approximately 16 hours: $pt_s = -4.5$.

Recent results from the author's laboratory support the idea that redox signals from the plastoquinone pool of chloroplasts initiate changes in photosystem stoichiometry in parallel to their effects in regulation of phosphorylation of LHC II in pea (Tullberg *et al.*, unpublished observations) and mustard (T. Pfannschmidt *et al.*, unpublished observations). For cyanobacteria, Fujita and co-workers (1994) have provided a number of independent lines of evidence that the redox state of plastoquinone is, again,

Fig. 3. Primary photochemistry of photosynthesis. (*a*) Structure of the reaction centre of *Rhodopseudomonas viridis*. On the left is the whole protein, with individual chains represented as individually coloured ribbons, and the heterogeneous atoms of the cofactors as red sticks. On the right the cofactors alone are shown as ball-and-stick models with cpk colours. (*b*) The sequence of primary electron transfer events. This presentation is based on one originally published, but using the *Rhodobacter sphaeroides* structure (Yeates *et al.*, 1987), by Youvan and Marrs (1987). Constructed from Brookhaven protein databank coordinate file 1prc using the program RASMOL (Sayle & Milner-White, 1995).

Fig. 4. The Crofts model for quinol oxidation at the Qo site of the cytochrome bc_1 complex: bifurcated electron transfer. The cytochrome bc_1 complex contains an iron–sulphur protein, a cytochrome b, and a cytochrome c_1 (the latter not shown). At the Q_o site of the complex is the iron-sulphur centre itself, and the two haems of the b-type cytochrome. On entering its binding site, the quinol passes an electron to the iron–sulphur centre and releases a proton. The semiquinone then moves in space, from the iron–sulphur protein to the low potential b-haem, passing an electron to the haem and releasing a second proton. Electron transfer occurs from the low-potential b-haem to the high potential b-haem, and the quinone leaves its binding site. Subsequently, the mobile head of the iron–sulphur protein rotates, carrying the iron–sulphur centre away from the Q_o site and towards the haem of cytochrome c_1, to which it donates the electron. The overall result is that single electron has moved from the quinone pool to the cytochrome c_1, but two proteins have been released into the outer, aqueous phase. The figure is adapted from graphics kindly provided by A. R. Crofts (Crofts *et al.*, 1998).

Fig. 5. Structure of the F1-ATPase. *Left*, the complete hexagonal array, viewed parallel to the membrane plane, with the polypeptide chain as individually coloured ribbons and the adenine nucleotides as space-filling models. *Right*, a view normal to the membrane plane (as if from the inner, cytoplasmic aqueous phase) with a 57% Z-slab in order to reveal the binding sites for the ATP analogue and ADP. The rotation of the γ-subunit, which is mechanically coupled to F_o (Fig. 6), is predicted to be anticlockwise during ATP synthesis, and one complete rotation will release three molecules of ATP. Constructed from Brookhaven protein databank coordinate file 1bmf using the program RASMOL (Sayle & Milner-White 1995).

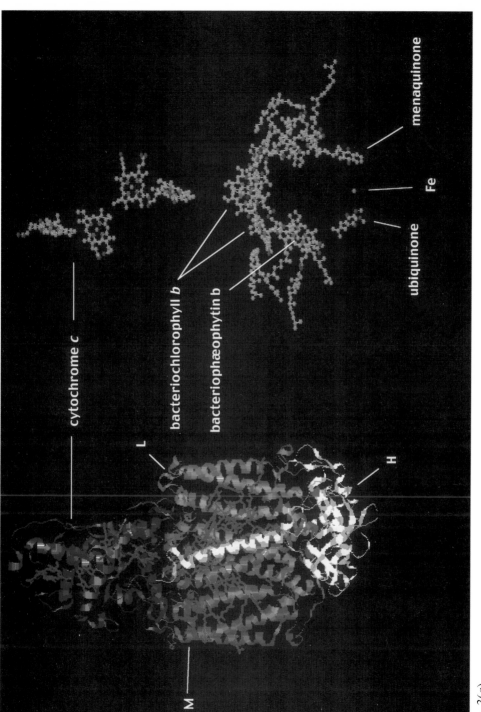

cytochrome *c*

bacteriochlorophyll *b*

bacteriophæophytin b

menaquinone

Fe

ubiquinone

L

H

M

Fig. 3(*a*)

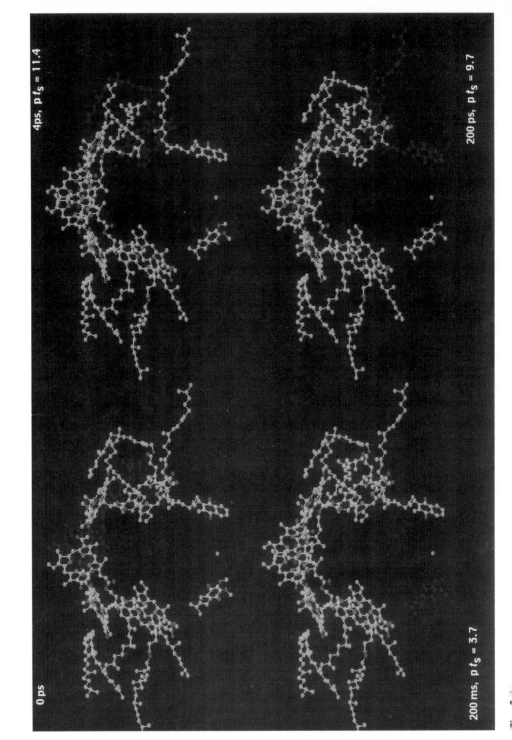

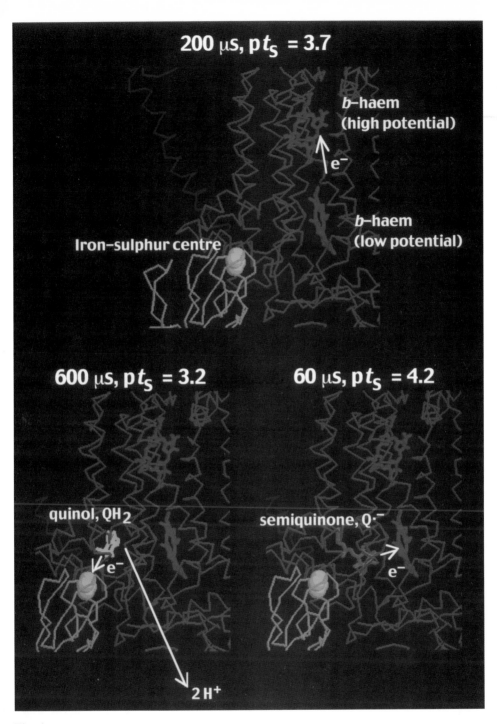

Fig. 4

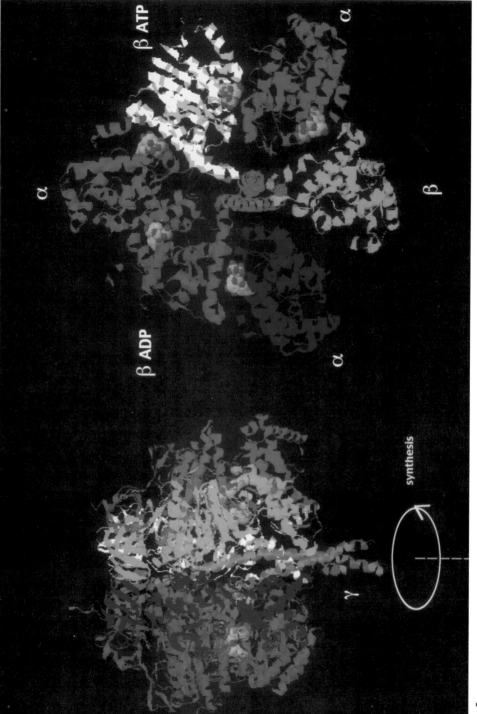

Fig. 5

the predominant factor determining the relative rates of assembly of the two photosystems. Results from the author's own laboratory tend to suggest that redox control of transcription of chloroplast genes is the primary response, at least in higher plants. The direction of the control is such that reduction of the plastoquinone pool by surplus photosystem II light or by chemical inhibition of the electron transport chain increases transcription of genes encoding subunits of photosystem I. Conversely, oxidation of plastoquinone by photosystem I light or by the action of inhibitors decreases photosystem I gene transcription and increases that of photosystem II. The conclusion is that expected, both teleologically and through the precedent of the scheme in Fig. 8: any imbalance in excitation energy distribution causes changes in gene expression, through redox control, that tend to correct the imbalance itself. Redox homeostasis is clearly important enough to require multiple pathways of feedback control, acting at different levels of gene expression (Fig. 7).

My aim in stressing here the action of light on gene expression through photosynthesis and redox control is not to detract from the importance of light itself as an environmental signal, which it obviously is, even in animals, in non-photosynthetic developmental stages in plants, and in some chemo-trophic bacteria. Progress has been made recently in understanding the photoreceptors and the associated signal transduction pathways of comple-mentary chromatic adaptation. Complementary chromatic adaptation is a long-established phenomenon by which some cyanobacteria switch on transcription of light-harvesting phycoerythrin genes in green light, and of phycocyanin genes in red light. The cells thereby assume a colour that is complementary to that of the light in which they grow. The signal transduc-tion pathway is seen, from genetic transformation that gives complementa-tion of mutant phenotypes, to contain a two-component system (Kehoe & Grossman, 1996), and the likely sensor shows some sequence similarities to phytochrome (for review see Allen & Matthijs, 1997), a major photoreceptor in plants.

Is a decision made about the appropriate level of adaptational response?

If physiological and developmental responses that achieve broadly the same effect can be distinguished and, at the same time, are initiated by a single environmental signal, how is the appropriate level of response selected? To take a homely analogy, one does not install new central heating every time the weather becomes overcast, nor does an overcoat provide a permanent solution to the discomfort of life in the sub-arctic. It is suggested here that measurement of the duration of the stimulus may play some part in the cell's decision. This is not to argue that cells possess foresight, although something analogous will be considered in the final section of this overview. Equally, one could argue that developmental events come into play only when

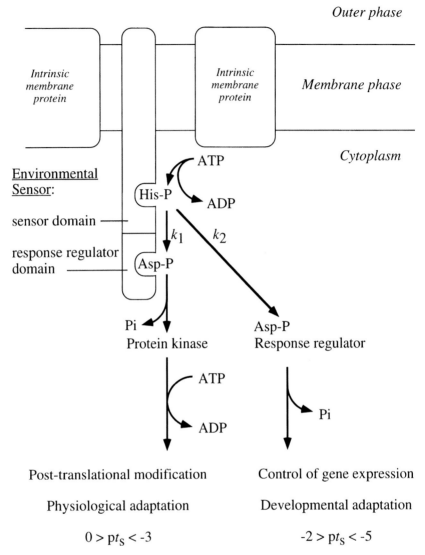

Fig. 9. A proposal for control of post-translational covalent modification by phosphorylation and, in parallel, of gene expression: bifurcated signal transduction. If the putative redox sensor contains both sensor and response regulator domains, and if the kinetically favoured pathway for phosphoryl transfer (k_1) is intramolecular, then phosphorylation of the sensor's own aspartate would initiate a physiological response, for example, by activation of a protein kinase. If full activation of the kinase then still fails to restore redox poise, phosphoryl transfer (k_2) may occur to the aspartate of the response regulator that regulates gene expression. An assumption of the proposed mechanism is that intramolecular phosphoryl group transfer is kinetically favoured over intermolecular transfer: $k_1 > k_2$. In Fig. 11, an additional possibility is suggested, which is that biological clocks provide an input that arbitrates beween physiological and developmental responses. The input could achieve this effect by acting on the sensor and response regulators in such a way as to alter the ratio $k_1{:}k_2$.

physiological responses that may restore the *status quo* have failed to do so. As regards the operation, in parallel, of control of protein phosphorylation (physiology) and of gene expression (development), a possible scenario is outlined in Fig. 9.

Two-component systems contain two components: a sensor and a response regulator. Response regulators are proteins that become phosphorylated on an aspartate on a surface-exposed loop. The phosphate group is transferred to the aspartate from a phosphohistidine side chain of the response regulator's cognate sensor (Stock, Ninfa & Stock, 1989). The histidine of the sensor becomes phosphorylated in response to the relevant environmental signal. The sensor is thus a response regulator kinase, mediating transfer of phosphate from ATP to the response regulator if, and only if, environmental conditions are right. Certain sensors, notably ArcB of *E. coli* (Iuchi & Lin, 1993), contain both sensor and response regulator domains, and the assumption can be made that a sensor's response regulator domain serves to damp the signal passed to the authentic, discrete response regulator. The suggestion outlined in Fig. 9 is that the branch-point of a bifurcated signal transduction pathway that leads either to a physiological or to a developmental response may be the transfer of the phosphate group of the sensor domain to an aspartate, either of its own response regulator domain, or to that of a separate response regulator.

The basis of the 'decision' taken by a system about the appropriate level of response to an environmental signal, would, according to this proposal (Fig. 9), be a simple consequence of success or failure of the most rapid response in restoration of the preferred internal environment. If the sensor contains both sensor and response regulator domains, and if the kinetically favoured pathway for phosphoryl transfer is intramolecular, then phosphorylation of the sensor's own aspartate (k_1) would produce a relatively rapid, physiological response, which, in this example, may be activation of a protein kinase. If full activation of this kinase then fails to restore homeostasis, it will also fail to switch off the stimulus initiating phosphorylation of the sensor's histidine. At the same time, the kinetically preferred intramolecular phosphoryl transfer from histidine is then prohibited by the acceptor (aspartate) being already in its phosphorylated form. The kinetically less favoured, intermolecular phosphoryl transfer (k_2) will then predominate. This second phosphoryl transfer is to the aspartate of the response regulator that controls gene expression. Figure 9 provides an outline of one mechanism by which developmental adaptation may be initiated when the stimulus caused by disparity in external and internal conditions persists even after physiological adaptation has reached the limit of its response. This general mechanism (Fig. 9) is built on a specific proposal for redox signalling in photosynthesis, where the sensor is a redox sensor and the two responses are phosphorylation of light-harvesting proteins and adjustment of photosystem stoichiometry (Allen & Nilsson, 1997).

$$-9 < \mathrm{p}t_\mathrm{s} > -15$$

GIGASECONDS TO KILOTERASECONDS: THE FIFTH TIME DOMAIN

The major transitions in evolution

Living organisms evolve, and their evolution has clearly been accompanied, in a general sense, by increasing complexity. It is nevertheless a simple fallacy to conclude that a more complex organism is in some sense more highly evolved, or 'higher' than a simpler one. Parasites are usually simpler that the free-living species from which they evolved. Parasites aside, many people still cling, intuitively, to a crude, anthropocentric idea that all other living things are failed attempts along the path to the eventual emergence of human beings. The terms 'lower eukaryotes' and 'higher eukaryotes' are nevertheless still in general currency. A cynical view is that these are just fancy ways of saying 'yeast' and 'myself', respectively. A good way of exposing the contradictions inherent in these terms is to ask their user where on the scale one should place an oak tree, or an octopus. One then often finds that the progressivist fallacy attaches great significance to such characteristics as being a chemoheterotroph, living on the land, and to having a nervous system or an immune response. An articulate cyanobacterium might regard each of these steps as retrograde, limiting future evolutionary possibilities to narrow and highly specialised physical environments. Microbiologists are mostly free of progressivist, anthropocentric tendencies, and will recognize that their favourite species, far from being 'primitive', has probably had longer to evolve in something like its present form. One might then wonder why vertebrates, for example, are thought to be more highly evolved than pseudomonads or purple photosynthetic bacteria. The latter, in particular, have retained the ability to do almost anything.

Maynard Smith and Szathmáry (1995) have considered the major transitions in evolution, and propose a solution to the apparent paradox of increasing complexity without there being any rational basis for the idea of evolutionary progress. The solution is the existence of transitions in levels of organisation, particularly where these involve information transfer. Thus the largest step, second to the origin of life itself, was probably the separation of information coding from its translation – the emergence of the specialised roles of DNA, RNA and protein from the 'RNA world' in which chemically similar macromolecules performed both catalysis and replication. The general theme of such major transitions seems to be division of labour, in which each new level of organization incorporates specialized components which, on their own, become less versatile, but which, acting in a way that complements other specialized components, create new possibilities for the environments that can successfully be exploited by the whole. According to this view, the evolution of human language ($\Delta \mathrm{p}t_\mathrm{s} \sim -13$) is then the most recent major evolutionary transition. Separate sexes (Allen, 1996) are another clear example.

The persistence of extra-nuclear genetic systems in eukaryotes

A further example of a major evolutionary transition, and one that intuitively fits with the general description, is the appearance of eukaryotes from prokaryotes. The idea that eukaryotes arose by endosymbiosis (championed, for example, by Margulis 1981) has now become orthodox. Among the divisions of labour that endosymbiosis allowed was the separation of energy-coupling membranes from those that formed the interface of the cell with the extracellular milieu. Following such specialization into subcellular compartments, or organelles, a large-scale copying of genetic information from the endosymbiont to the nucleus of the host cell must have occurred.

I have made a suggestion about why this copying of information has been incomplete in the case of chloroplasts and mitochondria (Allen, 1993). The argument depends on the interaction of some of the events discussed in this overview, and, at least for chloroplasts, upon the adaptations to light proposed for time domain four. A key feature of adaptations to changing light quality and quantity is that the primary influence of such changes on photosynthesis is felt through changes in the redox state of components of electron transport chains. In time domain three the precedent of redox-controlled protein phosphorylation in chloroplasts was considered, and it was pointed out that a parallel, developmental redox response also occurs, in the form of photosystem stoichiometry adjustment. The selective value of maintenance of redox homeostasis may be twofold. First, in order to function efficiently, redox poise must be maintained so that the input of light energy matches the capacity of the reaction centres to utilize it. Secondly, the redox chemistry initiated at photosynthetic reaction centres is an inherently hazardous process, producing the most indiscriminately reactive chemical species that occur in living cells. Single electron transfers at very high and low redox potential have a high probability of generating free radicals, especially of oxygen, such as superoxide, singlet oxygen, and the hydroxyl radical. Even at the moderate potentials of the proton-motive Q-cycle, the semiquinone anion radical has a crucial role (Fig. 4), and yet it is an effective generator of superoxide. It follows that adaptations that compensate for light-induced redox changes within photosynthetic systems may be essential in order to safeguard the cell from self-destruction. The assumption of the evolutionary proposal is that chloroplasts today encode those proteins whose redox functions require that their genes be retained within the same cellular compartment as that in which the redox signals originate (Allen, 1993).

A complementary proposal, that of Raven *et al.* (1994), is that the selective pressure tending to move genes from the endosymbiont to the nucleus is a decreased mutation frequency. The reason for the inherently high mutation frequency in organelles may be precisely the same generation of oxygen free radicals that, according to Allen's proposal (Allen, 1993), is minimized, though never eliminated, by redox control of gene expression. Genes are safer, and perhaps more capable of being appropriately deployed, in the nucleus. The chloroplast or mitochondrion is the wrong place to keep a genetic system. According to this synthesis of independently derived ideas (Allen & Raven, 1996), the proximity of genes for the key redox components of photosynthesis and respiration to the electron transport chains in which their gene products participate is a necessary price to be paid, and serves the same end as removal of other genes to the nucleus.

Thus there are powerful selective forces operating selectively on different endosymbiont, and, today, organellar, genes, which have thus segregated rather strictly between the nucleus, the chloroplast, and the mitochondrion. This view reinforces the idea that there must be an over-riding reason for the evolutionary retention of extra-nuclear genetic systems. The transition envisaged (Allen, 1993; Allen & Raven, 1996) for the evolution of chloroplast and mitochondrial genomes is depicted in Fig. 10. With small variations between major eukaryotic kingdoms and phyla, the present-day list of chloroplast and mitochondrial 'structural' genes is precisely a list of the components of photosynthesis and respiration that would seem to be most crucial for the regulation of organelle redox homeostasis (Allen, 1993). The chloroplast genome always encodes reaction centre proteins (time domain 2), for example, and most mitochondria likewise retain genes for components operating at the highest and lowest redox potentials of the respiratory electron transport chain. In addition, perhaps because of the involvement of the semiquinone anion radical in the Q-cycle (time domain 2; Fig. 4), cytochrome b is encoded in both chloroplasts and mitochondria. To this list of essential structural genes must then be added only genes for components of the minimal organellar genetic system that is required for the structural genes to be expressed. The components of the redox regulatory system itself (as depicted, for example, in Fig. 8, and in more general terms in Fig. 9) do not belong either to the category of key redox elements of electron transport chains or to the category of genetic system genes, and are therefore predicted to be nuclearly-encoded (Allen, 1993).

Other evolutionary effects of light

Micro-organisms may carry out photosynthesis, and may exhibit phototaxis, phototropism, and photonasty (photocontrol of development in which the change induced is not influenced by the direction of the source of light). True vision involves production of a focused image of the external world, and the optical requirements for an eye probably cannot be satisfied by micro-organisms, requiring true multicellularity with cell specialisation and division of labour. Nevertheless, the evolutionary origin of vision (Nilsson, 1996) in the late pre-Cambrian ($\Delta \mathrm{p}t_s = -16.5$) is only half as distant as the origin ($\Delta \mathrm{p}t_s = -16.8$) of oxygenic photosynthesis (discussed in time domain 2) and

endosymbiosis

redundancy

division of labour

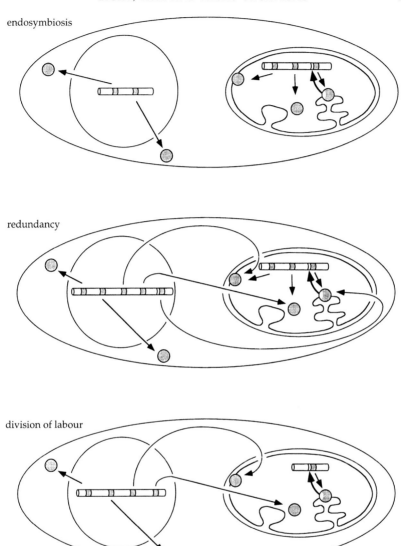

Fig. 10. Possible stages in the evolution of division of labour between nuclear and cytoplasmic (chloroplast and mitochondrial) genetic systems. Cylinders are sets of genes that may or may not be linked; shaded areas are individual alleles. Circles are proteins (gene products). Unshaded segments are promoter regions of genes or sensor domains of proteins. It is assumed that any gene may be copied between the endosymbiont (to the right of the schematic cell) and the host cell nucleus (to the left). Following endosymbiosis, most endosymbiont genes are subject to free radical mutagenesis, which they evade upon transfer to the nucleus: the endosymbiont copy is lost. In contrast, genes for certain key redox components of bioenergetic organelles remain *in situ*, since their expression must be subject to redox regulation in order to respond directly to environmentally induced changes in electron transport. Loss of the symbiont or organellar copy of such genes would uncouple redox control, and hence increase free radical production within the organelle and the cell as a whole. The nuclear copy of each of this subset of genes is therefore redundant, and is lost. Adapted from Allen (1993) and Allen & Raven (1996).

is likely to have depended on the free energy made available by increasing atmospheric concentrations of oxygen. Vision requires, and then places an extra selective value on, both size and motility, since it permits visually guided predation. This premium on size and motility may be expected to have enforced the dilemma of the mitochondrion: increased oxidative phosphorylation for motility must increase free radical production, and thereby make mitochondria less able accurately to replicate their own DNA. Mitochondrial replication is therefore in competition with mitochondrial function. Elsewhere I have suggested (Allen, 1996) a solution to the mitochondrial dilemma, which is entirely consistent with the global view of evolutionary transitions described earlier (Maynard Smith & Szathmáry, 1995). The proposal is that mitochondria replicate and synthesize ATP only because of a division of labour, and that an individual mitochondrion may carry out either ATP synthesis or replication, but not both – not even at different times. The division of labour is thus a specialization between one sex (female) that carries a subset of mitochondria which never function in ATP synthesis but instead act as a template for all other mitochondria, and another (male) in which mitochondrial replication, and thus mitochondrial inheritance, is abandoned altogether. Oxygenic photosynthesis was, according to this viewpoint, a necessary precursor of both vision and heterogametic sex, whose origins must therefore have been roughly contemporaneous. Furthermore, the adaptive mechanisms that may have evolved to allow prokaryotes to respond to changing light environments were the same as those that trapped genetic information in cytoplasmic, bioenergetic compartments, as a compromise in what might otherwise have been a complete separation of bioenergetic and genetic functions within the eukaryotic cell.

The pre-Cambrian explosion of varieties of multicellular animals (Gould, 1991), a few of which laid the foundations of all modern animal phyla, obviously produced immense evolutionary possibilities. In the context of this overview, it may be enough to point out that the acquisition, by the first cyanobacteria, of the ability to use the well-developed photochemistry of photosynthetic reaction centres as a thermodynamic sink for water oxidation was a revolution for more than geochemistry: it provided the conditions for an evolutionary route out of a purely microbial world. It is not argued here that this was progress. It is argued here that this was a major transition in evolution, and that subsequent transitions in levels of organization, including the one in which we now participate, would have been impossible without it.

CLOCKS AND THE EXPECTATION OF ENVIRONMENTAL CHANGE

Of the five familiar time intervals used for reference in Fig. 7, three correspond to the periods of a single astronomical rotation: the earth about its axis; the moon around the earth; and the earth around the sun. All three are used by at least some living organisms as environmental cues for developmental adaptation, crossing time domains three and four, and usually signalling the moment for progression from one developmental stage to another. It would be interesting to know if the lunar cycle, for example, is used as a cue only in multicellular animals that possess true vision. Although an eye may, to us, be helpful in distinguishing the phases of the moon, the accompanying changes in the intensity of the reflected sunlight are quite within the dynamic range to which living cells can and do respond. Circadian clocks in cyanobacteria should perhaps alert us to the additional possibility of 28-day cycles in plants and micro-organisms.

The discovery by Golden and co-workers (Kondo *et al.*, 1994; Golden, this volume) of a cyanobacterial circadian rhythm has far-reaching consequences. First, cloning and sequencing have allowed the comparison of 'clock' genes from widely differing types of organism and reveal clear indications of homology (Kay, 1997). The occurrence of a biological clock in a prokaryote gets completely away from the idea that such devices require a nervous system. The sequence similarities of clock components with those of known photoreceptors is interesting and exciting: it suggests that the photocontrol of development in processes such as complementary chromatic adaptation may have been the starting point for the elaboration, perhaps initially by autofeedback, of a developmental control that ran freely in the absence of light signals, but which retained the ability to respond to light in order to be regulated, or synchronized.

What is currently known of the molecular mechanisms underlying circadian rhythms will be described in a number of contributions to this symposium. At the time of writing this overview, the author hopes to learn, among other things, the frequency of the oscillator, which ought to be in place in Fig. 1 and Fig. 7 as a fundamental feature of living cells, and one that quite clearly must integrate cellular responses to light through time. In this attempted synopsis, finally, an observation will be made about the potential philosophical impact of any molecular or biochemical mechanism for measurement of time that can be demonstrated to operate even in single cells – and 'even' prokaryotes.

In the discussion of the interrelation of kinds of adaptation in time domain three, it was suggested that the 'blind' operation of negative feedback loops such as those depicted in Fig. 7 might be insufficient to account for adaptations, particularly developmental ones. In time domain four, the question was also raised of whether it is possible to imagine that cells can make a decision about the appropriate level of adaptive response, without invoking an intelligent or purposeful guiding hand. I suggest that the existence of molecular clocks in living cells may provide the basis for answering both these points.

The upper half of Fig. 11 shows a model of Sir Karl Popper (Popper, 1972). The arrows between boxes are intended to indicate the direction of information transfer. Popper was concerned with providing a conceptual and logical framework for scientific knowledge, without the 'progressivist' assumption that the development of science can be measured as the extent to which it approaches 'the truth' – whatever that may be. Thus, in Fig. 11, there is no direct connection between human knowledge ('World III') and the real world ('World I'). Nevertheless, World III may be modified by means of comparison of experience ('World II') with the expectations, derived from World III, that determine our perceptions. Conversely, there is a weaker sense in which the external world may be modified as a result of our perception of it: the arrow from World II to World I may indicate the selection of particular experiences, since all observation is guided expectation. The arrow from World War II to World War I could also be interpreted as changes that we make to our environment as a result of our knowledge of it.

The lower half of Fig. 11 show a simple translation of Popper's model into terms more specifically relevant to the topic of this symposium. Again, the arrows indicate direction of information transfer. In place of World I, the lower diagram has changes in the light environment. Adaptations then consist not merely of 'blind' responses to external changes, but of responses that arise in some way from the outcome of a comparison of the external signal with the internal reference provided by the oscillator or clock. Transient shading might thus be distinguishable from sunset, for example, and progressive changes in daylength might be identifiable as such, and correlated with seasonal changes in, for example, temperature. In the lower diagram, the arrow rising from left to right indicates the cell's or organism's ability to carry out changes – adaptations – that tend to restore its immediate physical environment and eliminate the need for further adaptation. If a physiological adaptation fails to do this, the internal time reference might arbitrate in the deployment of an equivalent developmental adaptation: trees do not shed their leaves at dusk. In a general sense, the box that replaces World III in the lower diagram could be labelled the sum of the cell's inherited predispositions to act in certain ways, as outlined by Popper himself (1972). The major evolutionary transition of human language then fits, again, into the definition of Maynard Smith and Szathmáry (1995): only in the upper diagram does the lower, right-hand box become separated from the continued existence of the individual organism that carries the reference: we may, if we choose, change our minds. In Popper's (1972) words: 'The difference between the amoeba and Einstein is that the amoeba must die with its theories'.

In conclusion, I propose that a major theme in biological evolution has been light, time, and the interplay between external signals that cells select as significant by means of reference to their internal clocks. Light is, and always has been, at least as important as any other factor in the physical environment. Because of their reliance on vision, this is still true in organisms that have long abandoned phototrophy. The existence of a circadian oscillator in

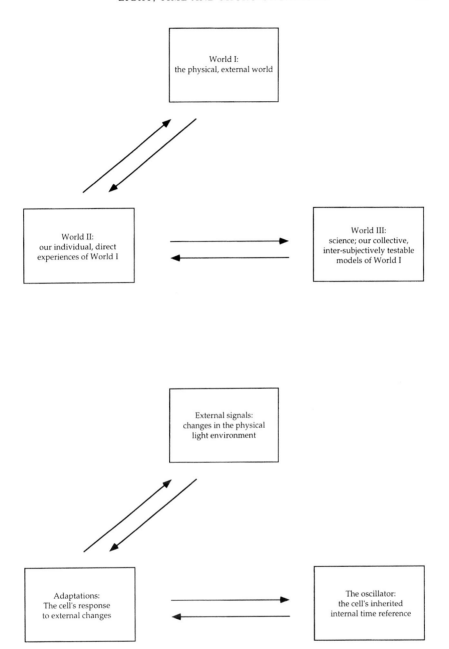

Fig. 11. An analogy between Popper's hypothesis of 'World III' (Popper, 1972) and biological responses to light that may be selected with reference to time. The oscillator or internal time reference arbitrates between levels of response to changes in the physical light environment. A possible mechanism for such arbitration is outlined in Fig. 9.

prokaryotes, particularly, is delightful, and underlines the central importance of life's responses to light and time. The possible origin of molecular clocks from photoreceptors, and their subsequent evolution and adaptive radiation can perhaps be seen as a new paradigm in biology. Biological clocks enlarge the meaning carried by Dawkins' striking metaphor for evolution – 'the blind watchmaker' (Dawkins, 1986). Besides producing intricate structures that have function without purpose, evolution has itself come up with time-reference, and one that uses light as a mechanism of calibration. Evolution is thus, in a literal sense, a watchmaker: only in a figurative sense is it blind.

ACKNOWLEDGEMENTS

Dr Nicholas Tsinoremas (Kondo *et al.*, 1994) gave me the basis of the final section with a simple, conversational remark that circadian rhythms represented an expectation of environmental change: I gratefully acknowledge this idea as his. I also thank Mrs. Carol Allen for discussions that led to a number of the ideas presented, specifically Fig. 9, and for comments on the manuscript; Professor Dan-E. Nilsson for his views on vision; and Professors Antony R. Crofts and Wolfgang Junge and Dr C. W. Mullineaux for copies of manuscripts prior to publication. This overview would not have been possible without support to the author for related experimental work described in some of the more recent references, and provided chiefly by the Swedish Natural Science Research Council (NFR) and the Swedish Council for Cooperation and Planning in Research (FRN).

REFERENCES

Abrahams, J. P., Leslie, A. G. W., Lutter, R. & Walker, J. E. (1994). Structure at 2.8 Å resolution of F1-ATPase from bovine heart mitochondria. *Nature*, **370**, 621–8.

Allen, J. F. (1992). Protein phosphorylation in regulation of photosynthesis. *Biochimica et Biophysica Acta*, **1098**, 273–335.

Allen, J. F. (1993). Control of gene expression by redox potential and the requirement for chloroplast and mitochondrial genomes. *Journal of Theoretical Biology*, **165**, 609–31.

Allen, J. F. (1995). Thylakoid protein phosphorylation, state 1-state 2 transitions, and photosystem stoichiometry adjustment: redox control at multiple levels of gene expression. *Physiologia Plantarum*, **93**, 196–205.

Allen, J. F. (1996). Separate sexes and the mitochondrial theory of ageing. *Journal of Theoretical Biology*, **180**, 135–40.

Allen, J. F. & Matthijs, H. C. P. (1997). Complementary adaptations, photosynthesis and phytochrome. *Trends in Plant Science*, **2**, 41–3.

Allen, J. F. & Nilsson, A. (1997). Redox signalling and the structural basis of regulation of photosynthesis by protein phosphorylation. *Physiologia Plantarum*, **100**, 863–8.

Allen. J. F. & Raven, J. A. (1996). Free-radical-induced mutation vs. redox regulation: costs and benefits of genes in organelles. *Journal of Molecular Evolution*, **42**, 482–92.

Allen, J. F., Bennett, J., Steinback, K. E. & Arntzen, C. J. (1981). Chloroplast protein phosphorylation couples plastoquinone redox state to distribution of excitation energy between photosystems. *Nature*, **291**, 25–9.

Allen, J. F., Dubé, S. L. & Davison, P. A. (1995). Screening for mutants deficient in state transitions using time-resolved imaging spectroscopy of chlorophyll fluorescence. In *Photosynthesis: From Light to Biosphere*, ed. P. Mathis, Vol. III, pp. 679–82. Dordrecht: Kluwer.

Allen, J. F., Sanders, C. E. & Holmes, N. G. (1985). Correlation of membrane protein phosphorylation with excitation energy distribution in the cyanobacterium *Synechococcus* 6301. *Federation of European Biochemical Societies Letters*, **193**, 271–5.

Allen, J. F., Mullineaux, C. W., Sanders, C. E. & Melis, A. (1989). State transitions, photosystem stoichiometry adjustment and non-photochemical quenching in cyanobacterial cells acclimated to light absorbed by photosystem I or photosystem II. *Photosynthesis Research*, **22**, 157–66.

Barrow, J. D. & Tipler. F. J. (1986). *The Anthropic Cosmological Principle*. Oxford: Oxford University Press.

Blankenship, R. E. (1994). Protein structure, electron transfer and evolution of prokaryotic photosynthetic reaction centers. *Antonie Van Leeuwenhoek*, **65**, 311–29.

Crofts, A. R., Barquera, B., Gennis R. B., Kuras R., Guergova-Kuras, M, & Berry, E. A. (1998). Mechanistic aspects of the Qo-site of the bc1-complex as revealed by mutagenesis studies, and the crystallographic structure. In *The Phototrophic Prokaryotes*, eds G. A. Peschek, W. Loeffelhardt & G. Schmetterer. New York: Plenum Publishing Corporation.

Dawkins, R. (1986). *The Blind Watchmaker*. Harlow: Longman.

Dawkins, R. (1996). *Climbing Mount Improbable*. London: Viking.

Deisenhofer, J., Epp, O., Miki, K., Huber, R. & Michel, H. (1985). Structure of the protein subunits in the photosynthetic reaction centre of *Rhodopseudomonas viridis* at 3Å resolution. *Nature*, **318**, 618–24.

Duerring, M., Schmidt, G. B. & Huber, R. (1991). Isolation, crystallization, crystal structure analysis and refinement of constitutive C-phycocyanin from the chromatically adapting cyanobacterium *Fremyella diplosiphon* at 1.66 Å resolution. *Journal of Molecular Biology*, **217**, 577–92.

Freer, A., Prince, S., Sauer, K., Papiz, M., Hawthornthwaite-Lawless, A., McDermott, G., Cogdell, R. & Isaacs, N. W. (1996). Pigment–pigment interactions and energy transfer in the antenna complex of the photosynthetic bacterium *Rhodopseudomonas acidophila*. *Structure*, **4**, 449–62.

Fujita, Y., Murakami, A., Aizawa, K. & Ohki, K. (1994). Short-term and long-term adaptation of the photosynthetic apparatus: homeostatic properties of thylakoids. In *The Molecular Biology of Cyanobacteria*, ed. D. A. Bryant, pp. 677–92. Dordrecht: Kluwer Academic Publishers.

Gould, S. J. (1991). *Wonderful Life. The Burgess Shale and the Nature of History*. London: Penguin Books.

Gregory, R. P. F. (1989). *Biochemistry of Photosynthesis*. Chichester: Wiley & Sons Ltd.

Hill, R. & Bendall, F. (1960). Function of the two cytochrome components in chloroplasts, a working hypothesis. *Nature*, **186**, 136–7.

Hoff ,W. D., Jung, K. H. & Spudich, J. L. (1997). Molecular mechanism of photosignaling by archaeal sensory rhodopsins. *Annual Reviews of Biophysics and Biomolecular Structure*, **26**, 223–58.

Hoganson, C. W. & Babcock, G. T. (1992). Protein–tyrosyl radical interactions in photosystem II studied by electron spin resonance and electron nuclear double resonance spectroscopy: comparison with ribonucleotide reductase and *in vitro* tyrosine. *Biochemistry*, **31**, 11 874–8.

Iuchi, S. & Lin, E. C. C. (1993). Adaptation of *Escherichia coli* to redox environments by gene expression. *Molecular Microbiology*, **9**, 9–15.

Junge, W., Lill, H. & Engelbrecht, S. (1997). ATP synthase: an electrochemical transducer with rotary mechanics. *Trends in Biochemical Science*, **22**, 420–3.

Kamen, M. D. (1963). *Primary Processes in Photosynthesis*. New York: Academic Press.

Kay, S. A. (1997). PAS, present, and future: clues to the origins of circadian clocks. *Science*, **276**, 753–4.

Kehoe, D. M. & Grossman, A. R. (1996). Similarity of chromatic adaptation sensor to phytochrome and ethylene receptors. *Science*, **273**, 1409–12.

Kondo, T., Tsinoremas, N. F., Golden, S. S., Johnson, C. H., Kutsuna, S. & Ishiura, M. (1994). Circadian clock mutants of cyanobacteria. *Science*, **266**, 1233–6.

Kühlbrandt, W., Wang, D. N. & Fujiyoshi, Y. (1994). Atomic model of plant light-harvesting complex by electron crystallography. *Nature*, **367**, 614–21.

Margulis, L. (1981). *Symbiosis in Cell Evolution*. New York: W. H. Freeman.

Maynard Smith, J. (1966). *The Theory of Evolution*. London: Penguin Books Ltd.

Maynard Smith, J. & Szathmáry, E. (1995). *The Major Transitions in Evolution*. Oxford: W. H. Freeman.

Mitchell, P. (1961). Coupling of phosphorylation to electron and hydrogen transfer by a chemi-osmotic type mechanism. *Nature* 191, 144–8.

Mitchell, P. (1975). Protonmotive redox mechanism of the cytochrome b-$c1$ complex in the respiratory chain: protonmotive ubiquinone cycle. *Federation of European Biochemical Societies Letters*, **56**, 1–6.

Mullineaux, C. W. & Allen, J. F. (1988). Fluorescence induction transients indicate dissociation of photosystem II from the phycobilisome during the state-2 transition in the cyanobacterium *Synechococcus* 6301. *Biochimica et Biophysica Acta*, **934**, 96–107.

Mullineaux, C. W., Sanders, C. E. & Melis, A. (1989). State transitions, photosystem stoichiometry adjustment and non-photochemical quenching in cyanobacterial cells acclimated to light absorbed by photosystem I or photosystem II. *Photosynthesis Research*, **22**, 157–66.

Mullineaux, C. W., Tobin, M. J. & Jones, G. R. (1997). Mobility of photosynthetic complexes in thylakoid membranes. *Nature*, **390**, 421–4.

Nilsson, A., Stys, D., Drakenberg, T., Spangfort, M. Forsén, S. & Allen, J. F. (1997). Phosphorylation controls the three-dimensional structure of plant light-harvesting complex II. *Journal of Biological Chemistry*, **272**, 18 350–7.

Nilsson, D-E. (1996). Eye ancestry: old genes for new eyes. *Current Biology*, **6**, 39–42.

Noji, H., Yasuda, R., Yoshida, M. & Kinosita, K. (1997). Direct observation of the rotation of F_1-ATPase. *Nature*, **386**, 299–302.

Popper, K. R. (1972). *Objective Knowledge. An Evolutionary Approach*. Oxford: Clarendon Press.

Raven, J. A., Johnston, A. M., Parsons R. & Kübler, J. E. (1994). The influence of natural and experimental high O_2 concentrations on O_2-evolving phototrophs. *Biological Reviews*, **69**, 61–94.

Rhee, K-H., Morris, E. P., Zheleva, D., Hankamer, B., Kühlbrandt, W. & Barber, J. (1997). Two-dimensional structure of plant photosystem II at 8-Å resolution. *Nature*, **389**, 522–6.

Sabbert, D., Engelbrecht, S. & Junge, W. (1996). Intersubunit rotation in active F-ATPase. *Nature*, **381**, 623–5.

Sayle, R. A. & Milner-White, E. J. (1995). RASMOL: biomolecular graphics for all. *Trends in Biochemical Science*, **20**, 374–6.

Sprenger, W. W., Hoff, W. D., Armitage J. P. & Hellingwerf, K. J. (1993). The eubacterium *Ectothiorhodospira halophila* is negatively phototactic, with a wavelength dependence that fits the absorption spectrum of the photoactive yellow protein. *Journal of Bacteriology*, **10**, 3096–104.

Stock, J. B., Ninfa, A. J. & Stock, A. M. (1989). Protein phosphorylation and regulation of adaptive responses in bacteria. *Microbiological Reviews*, **53**, 450–90.

Struglics, A., Fredlund, K., Møller, I. M. & Allen, J. F. (Submitted). Two subunits of the F_oF_1-ATP synthase are phosphorylated in the inner mitochondrial membrane.

Subramaniam, S., Gerstein, M., Oesterhelt, D. & Henderson, R. (1993). Electron diffraction analysis of structural changes in the photocycle of bacteriorhodopsin. *European Molecular Biology Organisation Journal*, **1**, 1–8.

Teilhard de Chardin, P. (1966). *Man's Place in Nature*. London: Collins.

Telfer, A., Allen, J. F., Barber, J. & Bennett, J. (1983). Thylakoid protein phosphorylation during state 1-state 2 transitions in osmotically shocked pea chloroplasts. *Biochimica et Biophysica Acta*, **722**, 176–81.

Whitmarsh, J. & Govindjee (1995). Photosynthesis. *Encyclopedia for Applied Physics*, Vol. **13**, pp. 513–32. Weinheim: VCH Publishers.

Yeates, T. O., Komiya, H., Rees, D. C., Allen, J. P. & Feher, G. (1987). Structure of the reaction center from *Rhodobacter sphaeroides* R-26: membrane–protein interactions. *Proceedings of the National Academy of Sciences, USA*, **84**, 6438–42.

Youvan, D. C. & Marrs, B. L. (1987). Molecular mechanisms of photosynthesis. *Scientific American*, **256**, 42–8.

MOTILITY RESPONSES TOWARDS LIGHT SHOWN BY PHOTOTROPHIC BACTERIA

JUDITH P. ARMITAGE

Microbiology Unit, Department of Biochemistry, University of Oxford, Oxford OX1 3QU, UK

> ...they are only considered to be the most simple of all existing organisms because they are the smallest, which we happen to know. But small and simple is not the same thing...
>
> Engelmann, 1881

HISTORY

Some of the earliest documented studies of bacterial behaviour were of photosynthetic species, probably because they are common and easily isolated from rivers and ponds. The first recorded observation of flagella *in vivo* were probably on a *Chromatium* species. A great German traveller and observer, Christian Ehrenberg, described the first observations of 'wave-shaped' flagella, which he thought were probably required for motility (Ehrenberg, 1883). Another German scientist, Thomas W. Engelmann, isolated bacteria from the Rhine outside his laboratory window. He carried out a range of experiments of what appear to have been a number of different species, and identified aerotactic and chemotactic behaviour, documenting the changes in swimming behaviour of cells near oxygen and near carbonic acid and suggesting that bacteria could sense the difference between oxygen and carbon dioxide in the same way as animals, and using this to support the idea of the 'unity of organic nature'. In 1881 he showed that 'bacterium photometricum', probably again a species of *Chromatium*, could accumulate in a light spot created on a darkened, sealed slide under certain conditions (Engelmann, 1881, 1883). His descriptions of *Chromatium* swimming and accumulation are so perfect that it is almost certain that this was the species he was studying. He described them swimming at 20–40 μm s^{-1} and rotating around their long axis at about 3–6 revolutions s^{-1}. The cells swam primarily in one direction but occasionally briefly reversed. He showed that these bacteria would accumulate around an oxygen bubble if incubated in the dark, but move away if illuminated (Engelmann, 1883). This was the first indication that oxygen and light sensing may somehow integrate.

Shining a spectrum of intense light from a gas light through a layer of the same bacteria he found that they accumulated in specific regions of the spectrum. A major accumulation was outside the visible region at 850 nm, while a second weaker accumulation was visible between 810 and 570 nm and a third in the region 550–510 nm. The blue and the visible red to orange regions became relatively empty. Watching the bacteria swimming, he saw that the bacteria appeared 'frightened' back when they swam over either the light/dark boundaries of the light spot, or the yellow/red or far-red/red boundaries, while moving in the opposite direction caused no response. The accumulations, particularly in the far-red, fascinated Engelmann. He found that the pattern of accumulation matched the absorption pattern of the pigmented cells. He had no facilities that would allow him to measure absorption in the infra-red, but he inferred from the coincidence of the accumulation and absorption in the visible regions that these bacteria must also absorb in this region and therefore probably used this region for photosynthesis. This was arguably the first description of the absorption spectrum of bacteriochlorophyll.

At this time it was still thought that oxygen was a by-product of bacterial photosynthesis, but Engelmann decided it was not oxygen causing the response to light in bacteria (he did, however, use the aerotactic response of other bacterial species to demonstrate oxygen evolution by chloroplasts of the green alga *Spirogyra*). He saw that bacteria incubated in the dark overnight stopped swimming, but started to swim again after 5–10 minutes in the light the following morning, and after a period in the light it took several hours for motility to be completely lost again when put back into the dark. He therefore decided that they were probably accumulating something like starch in the light and using its slow decay to power swimming after the light was removed. His observations on bacterial behaviour led Engelmann to write '... they are only considered to be the most simple of all existing organisms because they are the smallest, which we happen to know. But small and simple is not the same thing ...' How true!

After this spectacular start to the study of behaviour in photosynthetic bacteria, things basically stood still until the work of Roderick Clayton in the 1950s (1953*a*,*b*,*c*, 1958). Clayton was to go on to make great contributions to the world of photosynthesis, but his career started with the first quantitative study of photoresponses in the phototroph, *Rhodospirillum rubrum*. Manten had re-examined photoresponses in this species a little earlier and confirmed that the photosynthetic spectrum and that for the behavioural responses were coincident, but Clayton took this study further and provided the first quantitative data for the 'step-down' response, showing that it depended not only on the strength of the signal, but also the background light intensity. He examined the interaction between chemotactic responses and photoresponses and showed that the strength of the response could depend on the carbon source. As Engelmann had decided motility and responses were linked to the identified energy source of the day, starch, so Clayton decided the responses were controlled by the level of ATP.

Again the field remained more or less static for the next 25 years. While knowledge of the molecular basis of chemotaxis in *Escherichia coli* reached the point where it is now probably the best understood behavioural system in biology, the study of bacterial photoresponses basically stood still, with some descriptive work on the behaviour of certain species in response to light, but no genetic or quantitative studies. Why was this the case, and why is a great deal more known about photoresponses of an archaeon, *Halobacterium salinarium*, than of the much more common group of phototrophic bacteria? The problem probably lies in the intimate connection between swimming and the responses.

Bacteria swim by rotating semi-rigid helical flagella, using the electrochemical proton gradient (Δp) across the cytoplasmic membrane to drive that rotation (for recent reviews see Jones & Aizawa, 1991; Blair, 1995). As became apparent from the very early studies on the photoresponses, photosynthesis is probably involved in controlling the responses. Any change in photosynthesis also changes the Δp. Separating a response resulting simply from a change in Δp altering motor rotation rates from a true sensory signal dependent on photosynthetic activity is difficult. In addition, early attempts to isolate mutants in photoresponses which were still photosynthetic failed, all phototactic mutants turned out to also be photosynthesis mutants. In fact, the loss of photoresponses was one mechanism used in early experiments to isolate photosynthesis mutants. The failure to isolate specific photosensory mutants, we now know, probably reflects the choice of species studied, as some species appear to have much more complex sensory pathways than others, as will become apparent later, and this made the direct isolation of behavioural mutants impossible.

TERMINOLOGY

There has been a good deal of controversy surrounding what bacterial photoresponses should be called. In the strictest sense 'phototaxis' should describe the orientation of cells in the direction of a light source and then their movement towards or away from that light source (see Häder, 1987; Nultsch & Häder, 1988 for discussion). Free-swimming bacteria are probably too small to sense the direction of light (this chapter, p. 37), and therefore their responses are to a step-up or step-down in light intensity. Strictly speaking this is a photophobic response, a negative photophobic response or a positive photophobic response, depending on whether it is to a step-up or down in intensity. Gest and Bauer and their coworkers (Ragatz *et al.*, 1995) have argued that, because most phototrophic bacteria accumulate in the light and their response is to a step-down, bacteria are trying to stay out of the dark, and the response should be renamed scotophobia (fear of the dark). This does seem a reasonable term, however, under some circumstances the phototroph *Rhodobacter sphaeroides* can accumulate just outside of a light

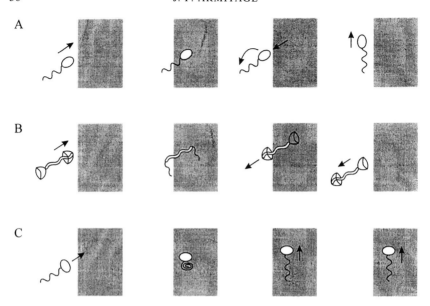

Fig. 1. Behaviour of different photosynthetic species when presented with a step-down in light. (A) *Chromatium*, which swims with a polar bundle of flagella, transiently reverses when passing over a light/dark boundary. (B) *Rhodospirillum rubrum* with a bipolar tuft of flagella reverses when passing over a light/dark boundary. (C) *Rhodobacter sphaeroides* stops swimming when passing over a light/dark boundary and the flagellum relaxes into a coiled form. On reformation of a functional helix, it may swim back into the light, but equally it may not.

spot. The large, and influential, world of bacterial chemosensing shows no inhibitions about the use of positive and negative chemotaxis to describe the behaviour of bacteria in a chemical gradient, although they are equally unable to sense the direction of a gradient. This review of the area of bacterial responses to light will be restricted to the (relatively) safe line of positive and negative photoresponses. This should not offend anyone.

PATTERNS OF BEHAVIOUR

Bacteria swim by rotating semi-rigid helical flagella, using the Δp to drive that rotation, which can be at rates of up to 300 Hz. This can drive swimming at speeds of up to 100 μm s^{-1}, although the average is closer to 30 μm s^{-1} (Mitchell *et al.*, 1991). Photosynthetic bacteria can have single polar flagella, single sub-polar flagella, or polar tufts. Few species identified are peritrichous when free swimming. Most studies have been on the response of specific species when they swim over a light/dark boundary and the majority reverse briefly, back into the light (Pfennig, 1968; Clayton, 1953a) (Fig. 1). When they swim over a dark/light boundary there is no obvious response, the cells just keep on swimming as normal. *R. sphaeroides* has a single sub-polar flagellum and swims rather differently from many other photosynthetic species (Armitage & Macnab, 1987). Rather than reversing to change direction, *R. sphaeroides* stops rotating its flagellum. The stopped filament relaxes its helical conformation from the distal end of the filament to form a short wavelength, large amplitude coil against the cell body. Slow rotation of the filament combined with Brownian motion reorient the cell so when the flagellar motor rotates rapidly again and the functional helix reforms the cell swims off in a new direction. The stops occur although the bacterium still has a Δp well above that needed to saturate motor rotation. When *R. sphaeroides* swims over a light/dark boundary or is subjected to a step-down in light intensity it therefore stops, rather than reverses (Packer, Gauden & Armitage, 1996).

This simple reversal response when passing over a light/dark boundary would result in cells becoming trapped in a region of light. However, the response must be a little more complex than this as, when a population of *R. rubrum* are examined microscopically after a step-down in light intensity the cells do not simply reverse once, but reverse repeatedly for a few seconds before adapting to the new light intensity and swimming as normal (Clayton, 1953c). *R. sphaeroides* stops when given a step-down, but again starts swimming normally after a few seconds, adapting to the change. This suggests that, although the behaviour appears simple, it may have some features in common with chemotaxis. In chemotaxis bacteria use a chemosensory phosphorelay system to control the flagellar activity (Amsler & Matsumura, 1995; Stock *et al.*, 1995). Bacteria are, as stated above, unable to sense a gradient directly, i.e. spatially, but sample their environment temporally, comparing the concentration now with a few seconds before. The phosphorelay system then controls whether the cell tumbles to change direction or not. If a bacterium is swimming in a positive direction it tumbles less frequently, biasing a usual random swimming pattern towards an optimum environment. To be able to sense a gradient, a bacterium must be able to sense a change when moving up or down the gradient. The sensory signal must therefore be terminated to stop the cell responding if the environment becomes stable, and the receptor system must be reset. The mechanisms involved in chemosensory adaptation will be discussed later, but the behaviour of bacteria to a step-down in light suggests that adaptation mechanisms must also be involved in these responses.

CAN A BACTERIUM SENSE A LIGHT GRADIENT?

The answer is probably not if free swimming, but it may be different for some species and in dense microbial mats or sediments. Many photosynthetic species swim very quickly, up to $100 \, \mu m \, s^{-1}$, but the swimming path is often not linear, because of the small size of bacteria relative to their local environment, and therefore the actual distance moved is unlikely to be great. The wavelength most effective at eliciting a response is about 850 nm, and given the nature of light in water the chances of a bacterium being able to swim far enough in open water to 'see' a change in intensity of 1% or greater in a few seconds, the optimum sensing time for chemotaxis, is unlikely.

This has recently been tested (Sackett *et al.*, 1997). A suspension of free-swimming *R. sphaeroides* or *Rhodospirillum centenum* was incubated in a light beam on a microscope slide. The light beam presented the cells with a gradual reduction in light intensity along the length of the beam, and a sharp step-down at the edges of the beam. The light scattering within the beam was followed over tens of minutes to measure whether cells accumulated in the beam, and could sense the direction of the light source. *R. centenum* accumulated rapidly in the source of the light beam, reaching a maximum concentration within a few minutes as the swimming cells in the regions around the light beam swam in and became trapped in the light. There was no movement of the cells towards the light beam. Unexpectedly, high light-grown *R. sphaeroides* slowly accumulated outside the light beam, as they swam over the light/dark boundary and stopped. When they started swimming again, they had as much chance of heading away from the light as swimming back into the light. These data indicate that free swimming cells are unlikely to sense the direction of light.

The case is different for colonies of *R. centenum* grown on an agar surface (Ragatz *et al.*, 1994, 1995). Liquid-grown *R. centenum* has a single sheathed flagellum, whose rotational direction reverses to change swimming direction. When grown on a surface, however, *R. centenum* becomes highly peritrichously flagellate, and if the agar surface is dry enough, the colony of growing *R. centenum* will move slowly over the agar surface. The colony shows both positive and negative phototaxis in a true sense, with the colony moving in a direct line towards the source of red light, around 800 to 850 nm, and away from light of around 550–600 nm. The observation that regions involved in positive and negative responses include wavelengths absorbed by bacteriochlorophyll suggest that, in this case, the sensory system may be more complex than in many other species, as absorption in both regions would result in increased photosynthetic activity. The colony response certainly looks like a true 'phototactic' response, as described for the oriented movement of some gliding cyanobacteria, and when confronted with two attractant beams of light at 90° to each other the colonies moved in a direction 45° between the sources, suggesting that the directional signals were being integrated. What is driving the movement is unknown as observations at the edge of the colony suggest that the cells inside the moving colony are not themselves directed but swirling around within the colony. The colony movement occurs under aerobic conditions as this species of photosynthetic bacterium has a full pigment complement under both aerobic and anaerobic conditions. Recent work has shown that there is an oxygen gradient formed within the colony and whether the cells are moving independently, balancing the responses to oxygen and light, and this drives the colony forward, is currently unknown (Romagnoli et al., 1997). However, if cells are isolated from a moving phototactic colony and resuspended the individual cells do not show any sign of sensing the direction of a light source; only responding to the step-down in light intensity in the same way as liquid grown cells.

WHAT IS THE PRIMARY SIGNAL FOR PHOTORESPONSES?

All the early data suggested that photosynthesis was somehow linked to the photoresponses. The signal could, however, come either from the activity of photosynthesis or a photosynthetic pigment acting as a receptor, in the same way as, e.g. sensory rhodopsin is a receptor (Spudich, this volume). Experiments in the 1970s added weight to the idea that the activity of photosynthesis was somehow involved in generating the signal. Mutants will full photosynthetic pigment complements, but no reaction centre proteins, showed chemotaxis but not photoresponses (Armitage & Evans, 1981). Inhibitors of electron transfer chain activity inhibited changes in membrane potential ($\Delta\Psi$) and photoresponses (Harayama, 1977). This suggested photosynthesis itself was the initial signal. The experiments had, however, been carried out on strains and mutants whose motility and phenotype had not been characterized and therefore the experiments were repeated and extended recently.

The photosynthetic electron transport chain of R. sphaeroides was inhibited, by using specific inhibitors under conditions where the cells were shown to remain fully motile. R. sphaeroides has a major advantage when it comes to measuring some bioenergetic parameters. The membrane-bound light harvesting carotenoid pigments change their absorption spectrum in the μs time scale when the membrane potential ($\Delta\Psi$) across the membrane changes. As the Δp at pH values of 7.2 and over is almost all $\Delta\Psi$, the change in carotenoid absorption can be used as a non-invasive measurement of changes in Δp (Clark & Jackson, 1981). The addition of antimycin A reduced the photoresponse to a step-down in light intensity, but did not completely inhibit it. Measurement of the $\Delta\Psi$ showed that photosynthesis had not been completely inhibited by antimycin A and the small size of the response correlated with the small amount of Δp sustained by the light. However, myxothiazol and stigmatellin independently both completely inhibited photosynthetic activity and they both inhibited any photoresponse to a step-down in light intensity (Grishanin et al., 1997).

That the primary signal comes from photosynthetic activity, and is not a measure of the actual change in light, is supported by the responses of high light- and low light-grown cells to an identical reduction in light intensity (Fig. 2). High light-grown cells have far fewer light harvesting pigments or reaction centres than low light-grown cells, which have highly invaginated

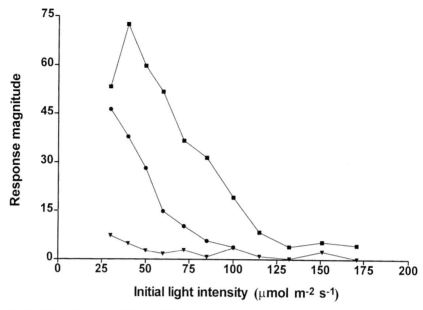

Fig. 2. Effect of growth at different light intensities on the photoresponse to a 97% step-down in light. The response was measured as the percentage of cells stopping and the length of the stop. The initial light intensity represents the intensity before the step-down by 97%. ■ = cells grown at high intensity, ● = cells grown at moderate light intensities, ▼ = cells grown at very low intensities. Note that only cells grown at high light respond to reduction in light from a wide range of starting intensities (from Grishanin, Gauden & Armitage, 1997, with permission).

membranes full of light harvesting complexes and higher numbers of reaction centres. The turnover kinetics of the components are also different. Photosynthetic electron transport therefore, saturates at much lower light intensities in low light grown cells, that can harvest all the light falling on them, than high light-grown cells that will 'miss' many photons. When subjected to identical 97% reductions in light intensity, the high light-grown cells responded to the step-down in light intensity over a much wider range of starting intensities, as the final intensity was below that required to saturate photosynthesis. Low light-grown cells only showed responses when the starting intensity was low and therefore the final intensity was extremely low, below the intensity required to saturate photosynthesis in these cells (Grishanin *et al.*, 1997). *R. sphaeroides* is therefore, sensing a change in photosynthetic electron transport, and this only occurs when the reduction in light means that the final intensity falls below that needed to saturate photosynthetic electron transport. As had been seen by Clayton, the response depends not only on the size of the change in light intensity but also on the starting intensity, and it must therefore, depend directly on photosynthetic activity rather than any specific receptors.

(*a*)

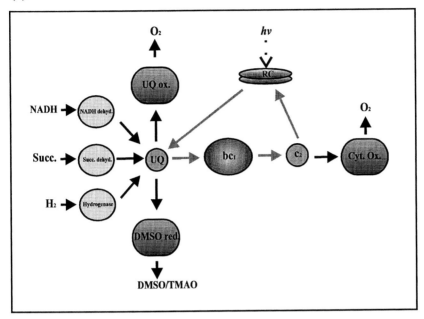

Fig. 3. Interaction of electron transport-dependent stimuli and their effect on responses (*a*) Simplified schematic of the electron transport pathway of *R. sphaeroides*. Succ = succinate, UQ ox = ubiquinone oxidase, DMSO red = dimethyl sulphoxide reductase, RC = reaction centre, bc_1, c_2 and cyt.ox = cytochromes.

Whether photoresponses themselves have a role in natural environments where many photosynthetic species grow in very dim light requires more research into the natural intensities and the gradients experiences in sediments and mats. It may be, however, that the response to light is part of a co-ordinated response to electron transfer activities as a whole.

INTEGRATION WITH OTHER ELECTRON TRANSPORT PATHWAYS

Photosynthetic electron transport shares electron transport components with both aerobic and anaerobic respiratory pathways (Fig. 3(*a*)). How do different stimuli causing behavioural responses but dependent on the shared electron transport chain interact with each other? Early behavioural studies had shown that photosynthetic bacteria moved away from air bubbles when illuminated, suggesting some sort of integration between respiration and photosynthesis. Recent research has gone further and suggests that bacteria are probably not sensing specific electron transport-dependent stimuli at all, but only sense the change in the rate of electron transport, moving to an

(*b*)

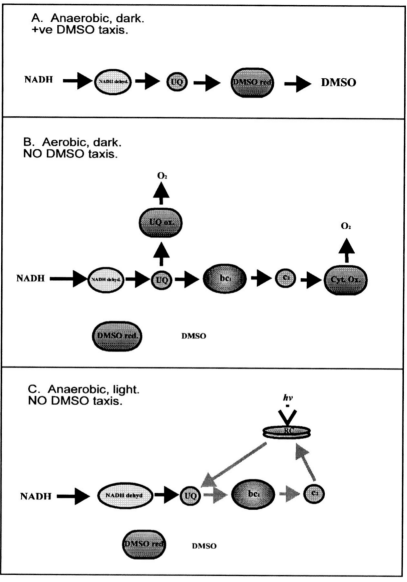

Fig. 3(*b*). Electron transport activity and chemotactic responses to DMSO showing how electron transport dependent responses integrate. A. Electron transport to DMSO anaerobically in the dark; cells show positive responses to DMSO gradients. Under (B) aerobic or (C) anaerobic light conditions both electron transport to DMSO and taxis to DMSO are lost (from Armitage, 1997, with permission).

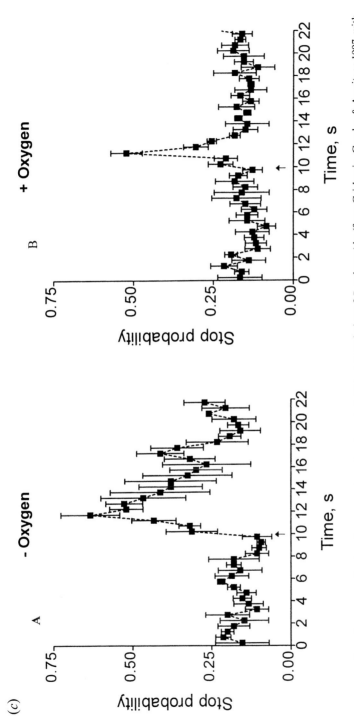

Fig. 3(c). Effect of oxygen on the size of a response to a step-down in light by a population of *R. sphaeroides* (from Grishanin, Gauden & Armitage 1997, with permission).

environment which maintains a maximum rate. Once there they only respond if that rate changes.

R. *sphaeroides* shows a positive response to the anaerobic electron acceptor dimethyl sulphoxide, DMSO, either in swarm agar plates or when tethered. Swarm plates are of two types either (i) soft agar plates in which motile bacteria are suspended, hard agar blocks containing attractants are inserted into the plates and the accumulation of bacteria around the plugs measured after a few hours or (ii) cells are inoculated into soft agar containing the attractant and swarm out as they metabolize the attractant and create a gradient. Alternatively for quantitative measurements, cells can be tethered to microscope slides via anti-flagella antibody and changes in the rotational behaviour of the cell bodies measured in response to changes in effector concentration or a reduction in light intensity.

DMSO is a terminal electron acceptor, using the DMSO reductase enzyme and branching from the ubiquinone. R. *sphaeroides* is only attracted to DMSO if incubated in the dark, anaerobically. The response depends on the enzyme DMSO reductase as DMSO reductase mutants lose both electron transport to DMSO and any behavioural response to changes in its concentration. The bacteria no longer accumulate around DMSO-containing agar plugs if the plates are placed either in the light, or in an oxygenated environment, although the cells remain motile. Although the enzyme is present and functional, if oxygen or light are present, electron transport to the DMSO reductase is lost (Gauden & Armitage, 1995) (Fig. 3(*b*)). This suggests that the response is a result of electron transport increase in response to the DMSO gradient; or, to be more precise, the decrease in the rate of electron transfer if the cells move down the gradient. If electron transfer to DMSO is no longer occurring, because there is either respiratory or photosynthetic electron transfer happening, both of which have higher kinetic rates, then the response to that acceptor is lost and the cells respond to the new driving force for electron transfer.

As suggested in the very early experiments, light and oxygen interact in a similar way. R. *sphaeroides* shows a stop response to a step-down in light. However, if oxygen is present, the response is greatly reduced, suggesting that electron transfer to the terminal oxidases reduces the photosynthetic electron transfer rate (Fig. 3(*c*)). The effect of reducing the light intensity on the electron transfer rate is less than if oxygen is present. Similarly, the addition of oxygen to light incubated cells will cause a brief response, suggesting that the oxygen has briefly reduced the rate of photosynthetic electron transfer and this has resulted in a response. This interaction of electron transfer chains would mean that bacteria will only be attracted to environments supporting the maximum rate of electron transfer, whatever that environment may be. Physiologically it could mean that bacteria would not move up a light gradient into an aerobic (and therefore toxic) environment as the oxygen would reduce the rate of electron transfer and result in a reversal or stop, and reorient the cells back to a light, but anoxic, environment.

WHAT IS THE SIGNAL?

What signal is sent from the electron transport chain to the bacterial flagellar motor? Over the years there has been speculation as to whether the signal for these types of responses might be a change in the Δp, a protometer sensing changes in the proton motive force directly (Taylor, 1983; Glagolev, 1984), or whether the rate of electron transfer, i.e. a redox sensor is involved. Δp as a signal has been implicated in responses of *Azospirillum* to oxygen (Zhulin *et al.*, 1996). There is some controversy over whether the ArcA/ArcB system in *E. coli*, responsible for controlling expression of redox pathways, can sense Δp or is attached to some, unidentified, redox centre (Iuchi & Lin, 1992, 1995; Bogachev, Murtazina & Skulachev, 1993). Many of the operons coding for photosynthesis proteins appear to be controlled by redox sensing proteins, from cytoplasmic proteins like FNR (Zeilstra-Ryalls & Kaplan, 1995) to membrane-spanning sensors which probably sense electron transport dependent redox changes such as Prr and MgpS (Eraso & Kaplan, 1995; Sabaty & Kaplan, 1996). Recently a gene has been identified in *E. coli* which codes for a protein, Aer, looking very similar at the C-terminal end to the signalling domain of an *E. coli* membrane-spanning chemosensory receptor, an MCP (Bibikov *et al.*, 1997; Rebbapragada *et al.*, 1997). However, the N-terminal end does not look like an MCP, but shows homology to redox sensing flavoproteins and haem binding proteins such as NifL and FixL (Agron & Helinski, 1995; Hill *et al.*, 1996). Overexpression of the gene produced yellow *E. coli* cells with an absorption spectrum suggesting an FAD moiety is linked to the Aer protein. Deletion of the gene resulted in *E. coli* cells unable to accumulate around oxygen bubbles. It seems likely therefore, that aerotaxis in *E. coli* relies, at least partially, on an FAD-containing redox protein signalling to the flagellar motor.

No homologue to Aer has yet been identified as being involved in redox sensing in any photosynthetic species. However, experiments on *R. sphaeroides* do suggest that it is the rate of electron transfer, i.e. the redox state of some protein which is part of, or connected to, the electron transfer chain, that signals to the flagellar motor. Evidence therefore, suggests that the responses are linked to changes in the rate of electron transfer, and the discovery of an aerotaxis signalling protein in *E. coli* that probably senses changes in the rate of respiratory electron flow through a redox centre supports the possibility that photoresponses in phototrophic bacteria may also depend on redox sensing; particularly as the aerotactic responses have been shown to interact with photoresponses.

There is some direct evidence that *R. sphaeroides* does respond to a redox signal rather than changes in Δp (Grishanin *et al.*, 1997). As described above,

R. sphaeroides only responds to a reduction in a stimulus, not to an increase in stimulus. The addition of the proton ionophore FCCP causes a reduction in the Δp, by moving protons directly through the membrane. This results not only in a decrease in Δp, but an increase in the rate of electron transfer to try and compensate for the reduction in Δp. The removal of FCCP has the reverse effect, the Δp returns to normal and therefore increases, and the electron transport rate decreases as protons are no longer being lost directly through the membrane. The carotenoid bandshift can be used to titrate the amount of FCCP added, such that there is a fall in Δp on the addition of a certain concentration, but the Δp still remains above the level required for normal swimming. If *R. sphaeroides* responds to a change in Δp, the reduction caused by the addition of FCCP should cause a stop response, however, it does not. *R. sphaeroides* only responds to the removal of FCCP i.e. a reduction in electron transfer rate, and they respond to its removal by increased stopping and adaptation. A reduction in electron transport rate therefore causes a response equivalent to that seen on a step-down in light. These data are confirmed by examining behaviour of cells subjected to FCCP to cause a reduction in Δp, in parallel with cells in which both the Δp and the electron transfer rate were reduced by reducing the light intensity. *R. sphaeroides* responded when the light intensity was reduced, but not when FCCP was added to cause an identical fall in Δp. The only difference in the two conditions was that in one the electron transfer rate also fell, and in the other it rose.

It therefore seems that photosynthetic bacteria respond to a change in the rate of electron transfer via a redox sensing system which signals to the flagellar motor (Fig. 4).

HOW DOES THE SIGNAL REACH THE FLAGELLAR MOTOR?

The chemosensory pathway in *E. coli* is probably the best understood intracellular signalling pathway in biology (Amsler & Matsumura, 1995; Blair, 1995). Recent work has shown that bacteria belonging to the α-subgroup, which includes many of the motile, photosynthetic species, have a similar chemosensory system, but somewhat more complex (Armitage & Schmitt, 1997). In *E. coli*, membrane-spanning chemoreceptor proteins, methyl accepting chemotaxis proteins (MCPs), sense whether the binding of specific chemoeffectors to the periplasmic domain has changed. If the binding is reduced, the conformation of the MCP dimer changes and this change is transmitted across the membrane and via a pair of linking, CheW, proteins to a dimer of the histidine protein kinase, CheA. CheA autophosphorylates at a conserved histidine. The phosphate can then be transferred to one of two response regulators. CheY is a small 14 kD protein which, when phosphory-lated, can bind to the motor and cause switching in the direction of flagellar rotation. CheB is a specific methyl esterase, responsible with a transferase,

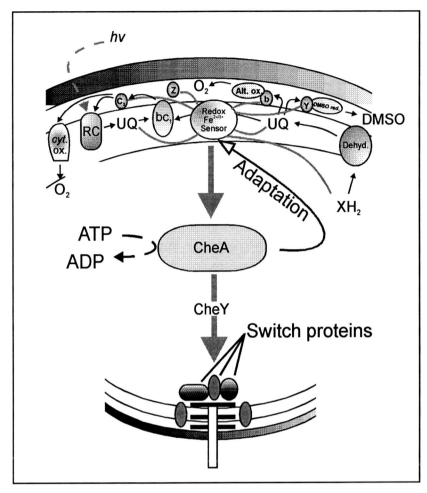

Fig. 4. Possible sensory mechanism involved in electron transport dependent responses. Model shows a putative redox sensor responding to a change in electron transport rate signalling through a phosphorelay system to the flagellar motor. Abbreviations as in Fig. 2 and in text.

CheR, for resetting the MCP by controlling the level of posttranslational methylation of specific glutamate residues on the cytoplasmic side of the MCPs. The rate of CheY-P dephosphorylation is increased by a protein CheZ, which competes with the motor protein FliM, and this terminates the signal (Fig. 5(a)). This is a relatively straightforward phosphorelay system controlling the switching frequency of the motor in response to chemosensory gradients.

In the past two or so years genes coding for homologous proteins have been identified in *R. sphaeroides* and *R. centenum* (Ward *et al.*, 1995a; Hamblin *et al.* (in press); Jiang & Bauer, 1997). In both cases the system

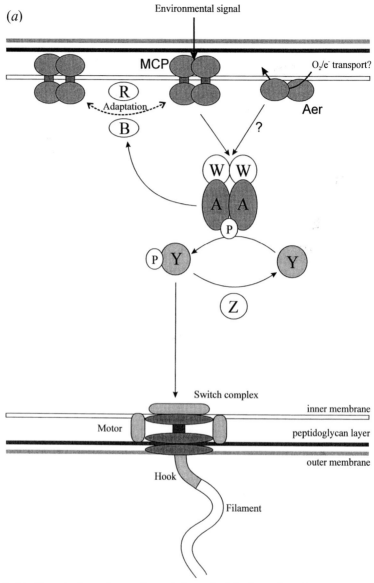

Fig. 5. (*a*) Chemotactic pathway of enteric bacteria. Membrane-bound receptors, MCPs, transmit information to a phosphorelay system which controls switching of the flagellar motor. The signal is terminated by CheZ and adaptation is brought about by a methyl esterase, CheB and a methyl transferase, CheR. Oxygen is sensed through an MCP homologue, Aer, which controls CheA, but it has no adaptation sites.

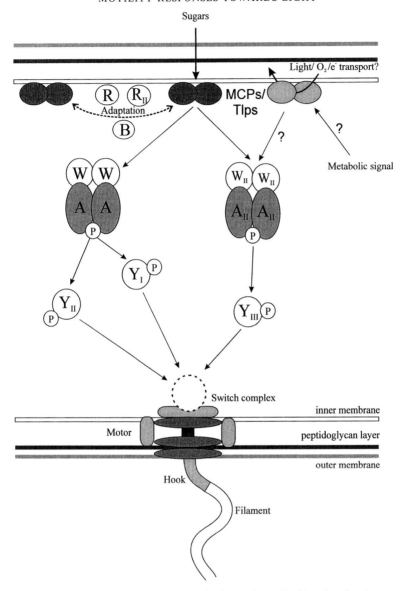

Fig. 5(b). Possible scheme for sensory transduction in *R. sphaeroides*. Two phosphorelay systems control the single motor. Multiple CheY homologues control flagellar rotation, possibly controlling both stopping and speed in response to signals. Photo and oxygen responses are fed through one phosphorelay system to integrate at the motor. Abbreviations are described in the text.

appears somewhat more complex. In neither species has a CheZ homologue been identified, suggesting that signal termination is different (no CheZ homologue has yet been identified in any member of the α-subgroup of bacteria). *R. centenum* has homologues of all the genes in the *E. coli* pathway, but CheA is fused to a second copy of CheY. A similar fusion has been identified in the gliding bacterium *Myxococcus xanthus* (Zusman *et al.*, 1990; Acuña *et al.*, 1995). Two copies of the *cheY* gene have been found in every member of the α-subgroup examined (Greck *et al.*, 1995; Ward *et al.*, 1995*a*, and D. Alley, personal communication). The second copy of CheY and no CheZ suggests that signal termination, required for gradient sensing, may occur by competition between the two CheY proteins for the phosphodonor. In *R. centenum*, mutation in the *che* genes resulted not only in the loss of chemotactic behaviour, but also the positive photoresponse of the colony suggesting that the signals are integrated through the chemosensory pathway, as the signal from Aer is in the aerotactic signalling pathway of *E. coli* (Jiang *et al.*, 1997).

R. sphaeroides is even more complex. A gene cluster was identified encoding homologues of CheA, CheW, CheR and two CheY proteins as well as two MCP-like proteins (now renamed 'transducer-like proteins', Tlps). These coded for proteins involved in chemotaxis as they either complemented *E. coli* mutants or interfered with chemotaxis when expressed in *E. coli* (Hamblin *et al.*, 1997). However, in-frame deletion of the individual genes, or the whole operon, had no effect on either chemotaxis or photoresponses. Mutations in TlpA (formerly McpA) did, however, result in a reduction in chemotactic swarming under aerobic conditions (Ward *et al.*, 1995*a*). Although this gene shows homology to Aer there is no evidence of FAD binding.

Transposon mutagenesis on the mutant in which the complete *che* operon had been deleted did, however, result in mutants in both chemotaxis and photoresponses. When the interrupted genes were cloned they were found to code for a complete second set of *che* proteins; CheAII, CheWII, CheWIII, CheRII, CheB, CheYIII and another Tlp. Deletion of these genes and the transfer of these mutations back into the wild-type strain results in altered chemotactic and phototactic responses (Hamblin *et al.*, 1997; Fig. 5(*b*)). Unlike *E. coli* behavioural mutants which are either exclusively smooth swimming or tumbly, all *R. sphaeroides* mutants produced so far swim normally, stopping and starting in the same way as wild-type cells.

The discovery that unlike *R. centenum*, *R. sphaeroides* has two, and possibly three, functional chemosensory phosphorelay systems explains why the genes were not found by simple transposon mutagenesis; the mechanism used to identify the genes in *E. coli*, as in *R. sphaeroides* the pathways can, to some extent, compensate for each other. Interestingly, when individual cells were examined by motion analysis, the chemosensory response of the mutant in *cheAII* was found to be inverted. Although the cells no longer swarm on swarm plates (in which a chemosensory gradient is set up as the central colony metabolizes the effector), the cells do respond to changes in concentration. When examined microscopically, a step-up in chemoeffector concentration resulted in a stop followed by adaptation; the reverse of the wild-type response. The responses to changes in light intensity were also found to be reversed, the cells stopping in response to an increase in light and swimming in response to a decrease (S. Romagnoli, P. A. Hamblin and J. P. Armitage, unpublished observations). This suggests that the second pathway may be involved in both chemosensory and photosensory responses.

The mutant cells also swim faster, suggesting that they may be locked in chemokinesis. Several members of the α-subgroup have been shown to exhibit Δp-independent changes in swimming speed under certain conditions. A range of chemical effectors can cause a long term increase in the unstimulated swimming speed of R. sphaeroides, and this could serve to spread a population in times of plenty (Brown et al., 1993; Packer & Armitage, 1994). The change in swimming speed when chemosensory pathway components are deleted suggests that speed may also be involved in behavioural responses, with signals from the sensory pathway slowing flagellar rotation. This suggests that taxis in these species may, in part, be the result of not only a change in stopping frequency, but also a change in flagellar rotation rate. This is similar to a mechanism suggested for behaviour in the related, non-photosynthetic, Rhizobium meliloti (Sourjik & Schmitt, 1996).

The data therefore, show that chemotaxis and photoresponses must both be integrated through the same Che pathway, but why a mutation in this pathway results in an inverted response is unknown, unless the response depends on a balance between a number of pathways. Recent work has suggested that there may be yet a third CheA, whose interruption results in the complete loss of responses to chemoattractant and light, but how this relates is still being investigated (these mutants cells again still swim normally, unlike equivalent E. coli mutants). Why R. sphaeroides, with only one flagellum, has at least two and probably three chemosensory phosphorelay systems remains to be identified. It is possible that the number of pathways relates to different growth conditions as appears to be the case for the chemotaxis receptor proteins.

Recently E. coli-like MCPs have been identified by immunoelectron microscopy, fluorescence microscopy and Western blot analysis in R. sphaeroides. It was found that aerobic grown cells had far more MCPs than phototrophically grown cells. Aerobic cells had clusters of MCPs at the poles of the cells, apparently targeted to the septation sites, with a few clustered around an electron dense body in the cytoplasm. Anaerobically grown cells had about 20 × fewer MCPs and these were primarily around the cytoplasmic structure. It may be that there are also different phosphorelay systems for different growth conditions (D. M. Harrison, J. P. Armitage & J. R. Maddock, unpublished observations). This is under investigation.

CONCLUSIONS

What is clear from this work is that all the signals from changes in electron transfer, be they light, oxygen or another terminal acceptor, are integrated, probably through a redox-sensing protein. This signal is then integrated through chemosensory phosphorelay systems and these control the rotation of the flagellum, making sure that, overall, the motility of the cell is biased towards an optimum environment for growth. There are still many unanswered questions. If there is a redox-sensing protein, it is still to be identified and the mechanisms involved in transmitting any conformational change to the phosphorelay system characterized. The interaction with the flagellar motor of the CheY proteins to control motor rotation and why there are so many CheYs in these bacteria when enteric species have only one, remains to be understood. From a physiological point of view, the role of the different gradient sensing mechanisms in the natural environment is fascinating, how are photoresponses and chemoresponses balanced and integrated, and what are the adaptation mechanisms involved in photoresponses? Hopefully over the next few years the combination of molecular genetics and physiological studies will provide answers to some of these questions.

ACKNOWLEDGEMENTS

I would like to thank Carl Bauer and Howard Gest, Sandy Parkinson and Barry Taylor and Dickon Alley for unpublished data. I would also like to thank the Microbial Diversity Course, Woods Hole 1996, for the facilities to test light gradient sensing. The BBSRC and Wellcome Trust supported much of the work described from my lab. on *R. sphaeroides* over the years. I would also like to thank Paul Hamblin for help with the figures.

REFERENCES

Acuña, G., Shi, W., Trudeau, K. & Zusman, D. R. (1995). The 'CheA' and 'CheY' domains of *Myxococcus xanthus* FrzE function independently *in vitro* as an autokinase and a phosphate acceptor, respectively. *FEBS Letters*, **358**, 31–3.

Agron, P. G. & Helinski, D. R. (1995). Symbiotic expression of *Rhizobium meliloti* nitrogen fixation genes is regulated by oxygen. In *Two-component Signal Transduction*, ed. J. A. Hoch & T. J. Silhavy, pp. 275–87. Washington, DC: ASM.

Amsler, C. D. & Matsumura, P. (1995). Chemotactic signal transduction in *Escherichia coli* and *Salmonella typhimurium*. In *Two-component Signal Transduction*, ed. J. A. Hoch & T. J. Silhavy, pp. 89–103. Washington, DC: ASM.

Armitage, J. P. (1997). Behavioural responses of bacteria to light and oxygen. *Archives of Microbiology*, **168**, 249–61.

Armitage, J. P. & Evans, M. C. W. (1981). The reaction centre in the phototactic and chemotactic responses of *Rhodopseudomonas sphaeroides*. *FEMS Microbiology Letters*, **11**, 89–92.

Armitage, J. P. & Macnab, R. M. (1987). Unidirectional intermittent rotation of the flagellum of *Rhodobacter sphaeroides*. *Journal of Bacteriology*, **169**, 514–18.

Armitage, J. P. & Schmitt, R. (1997). Bacterial chemotaxis: *Rhodobacter sphaeroides* and *Sinorhizobium meliloti*-variations on a theme. *Microbiology*, **143**, 3671–82.

Bibikov, S. I., Biran, R., Rudd, K. E. & Parkinson, J. S. (1997). A signal transducer for aerotaxis in *Escherichia coli*. *Journal of Bacteriology*, **179**, 4075–9.

Blair, D. F. (1995). How bacteria sense and swim. *Annual Review of Microbiology*, **49**, 489–522.

Bogachev, A. V., Murtazina, R. A. & Skulachev, V. P. (1993). Cytochrome *d* induction in *Escherichia coli* growing under unfavorable conditions. *FEBS Letters*, **336**, 75–8.

Brown, S., Poole, P. S., Jeziorska, W. & Armitage, J. P. (1993). Chemokinesis in *Rhodobacter sphaeroides* is the result of a long term increase in the rate of flagellar rotation. *Biochimica et Biophysica Acta*, **1141**, 309–12.

Clark, A. J. & Jackson, J. B. (1981). The measurement of membrane potential during photosynthesis and during respiration in intact cells of *Rhodopseudomonas capsulata* by both electrochromism and by permanent ion distribution. *Biochemical Journal*, **200**, 389–97.

Clayton, R. K. (1953*a*). Studies in the phototaxis of *Rhodospirillum rubrum*. I. Action spectrum, growth in green light and Weber-law adherence. *Archiv für Mikrobiologie*, **19**, 107–24.

Clayton, R. K. (1953*b*). Studies in the phototaxis of *Rhodospirillum rubrum*. II. The relation between phototaxis and photosynthesis. *Archiv für Mikrobiologie*, **19**, 125–40.

Clayton, R. K. (1953*c*). Studies in the phototaxis of *Rhodospirillum rubrum*. III. Quantitative relationship between stimulus and response. *Archiv für Mikrobiologie*, **19**, 141–65.

Clayton, R. K. (1958). On the interplay of environmental factors affecting taxis and mobility in *Rhodospirillum rubrum*. *Archiv für Mikrobiologie*, **29**, 189–212.

Ehrenberg, G. S. (1883). *Die Infusionstrierchen als Vollkommene Organismen*, ed. T. W. Engelmann, p. 15. Leipzig.

Engelmann, T. W. (1881). Bacterium photometricum: An article on the comparative physiology of the sense for light and colour. *Archiv Gesamte Physiologie Bonn*, **30**, 95–124.

Engelmann, T. W. (1883). *Bakterium photometricum*. Ein Beitrag zur vergleichenden Physiologie des Licht-und Farbensinnes. *Pfuegers Archiv Gesamte Physiologie Menschen Tiere*, **42**, 183–6.

Eraso, J. M. & Kaplan, S. (1995). Oxygen-insensitive synthesis of the photosynthetic membranes of *Rhodobacter sphaeroides*: a mutant histidine kinase. *Journal of Bacteriology*, **177**, 2695–706.

Gauden, D. E. & Armitage, J. P. (1995). Electron transport-dependent taxis in *Rhodobacter sphaeroides*. *Journal of Bacteriology*, **177**, 5853–9.

Glagolev, A. N. (1984). Bacterial H$^+$-sensing. *Trends in Biochemical Sciences*, **9**, 397–400.

Greck, M., Platzer, J., Sourjik, V. & Schmitt, R. (1995). Analysis of a chemotaxis operon in *Rhizobium meliloti*. *Molecular Microbiology*, **15**, 989–1000.

Grishanin, R. N., Gauden, D. E. & Armitage, J. P. (1997). Photoresponses in *Rhodobacter sphaeroides*: Role of photosynthetic electron transport. *Journal of Bacteriology*, **179**, 24–30.

Häder, D. P. (1987). Photosensory behavior in prokaryotes. *Microbiological Reviews*, **51**, 1–21.

Hamblin, P. A., Bourne, N. A. & Armitage, J. P. (1997). Characterisation of the chemotaxis protein CheW from *Rhodobacter sphaeroides* and its effect on the behaviour of *Escherichia coli*. *Molecular Microbiology*, **24**, 41–51.

Hamblin, P. A., Maguire, B. A., Grishanin, R. N. & Armitage, J. P. (1998). Evidence for two chemosensory pathways in *Rhodobacter sphaeroides*. *Molecular Microbiology*, **27** (in press).

Harayama, S. (1977). Phototaxis and membrane potential in the photosynthetic bacterium *Rhodospirillum rubrum*. *Journal of Bacteriology*, **131**, 34–41.

Hill, S., Austin, S., Eydmann, T., Jones, T. & Dixon, R. (1996). *Azotobacter vinelandii* NIFL is a flavoprotein that modulates transcriptional activation of nitrogen fixation genes via a redox-sensitive switch. *Proceedings of the National Academy of Sciences, USA*, **93**, 2143–8.

Iuchi, S. & Lin, E. C. C. (1992). Mutational analysis of signal transduction by ArcB, a membrane sensor protein responsible for anaerobic repression of operons involved in the central aerobic pathways in *Escherichia coli*. *Journal of Bacteriology*, **174**, 3972–80.

Iuchi, S. & Lin, E. C. C. (1995). Signal transduction in the Arc system for control of operons encoding aerobic respiratory enzymes. In *Two-component Signal Transduction*, ed. J. A. Hoch & T. J. Silhavy, pp. 223–32. Washington DC: ASM.

Jiang, Z. Y. & Bauer, C. E. (1997). Chemosensory and photosensory perception in purple photosynthetic bacteria that utilize common signal transduction components. *Journal of Bacteriology*, **179**, 5720–7.

Jones, C. J. & Aizawa, S. (1991). The bacterial flagellum and flagellar motor: structure, assembly and function. *Advances in Microbial Physiology*, **32**, 109–72.

Mitchell, J. G., Martinez-Alonso, M., Lalucat, J., Esteve, I. & Brown, S. (1991). Velocity changes, long runs and reversals in *Chromatium minus* swimming responses. *Journal of Bacteriology*, **173**, 997–1003.

Nultsch, W. & Häder, D. (1988). Photomovements in motile microorganisms-II. *Photochemistry and Photobiology*, **47**, 837–69.

Packer, H. L. & Armitage, J. P. (1994). The chemokinetic and chemotactic behavior of *Rhodobacter sphaeroides*: two independent responses. *Journal of Bacteriology*, **176**, 206–12.

Packer, H. L., Gauden, D. E. & Armitage, J. P. (1996). The behavioral response of anaerobic *Rhodobacter sphaeroides* to temporal stimuli. *Microbiology*, **142**, 593–9.

Pfennig, N. (1968). *Chromatium akenii* (*Thiorhodaceae*). Institut für den Wissenschaftlichen Film, Gottingen, pp. 3–9.

Ragatz, L., Jiang, Z-Y., Bauer, C. & Gest, H. (1994). Phototactic purple bacteria. *Nature (London)*, **370**, 104.

Ragatz, L., Jiang, Z-Y., Bauer, C. E. & Gest, H. (1995). Macroscopic phototactic behavior of the purple photosynthetic bacterium *Rhodospirillum centenum*. *Archives for Microbiology*, **163**, 1–6.

Rebbapregada, A., Johnson, M. S., Harding, G. P., Zuccarelli, A. J., Fletcher, H. M., Zhulin, I. B. & Taylor, B. L. (1997). The Aer protein and the serine chemoreceptor Tsr independently sense the intracellular energy levels and transduce oxygen, redox, and energy signals for *Escherichia coli* behavior. *Proceedings of the National Academy of Sciences, USA*, **94**, 10541–6.

Romagnoli, S., Hochkoeppler, A., Damgaard, L. & Zannoni, D. (1997). The effects of respiration on the phototactic behaviour of the purple non-sulfur bacterium *Rhodospirillum centenum*. *Archives for Microbiology*, **167**, 99–105.

Sabaty, M. & Kaplan S. (1996). MgpS, a complex regulatory locus involved in the transcriptional control of the *puc* and *puf* operons in *Rhodobacter sphaeroides* 2.4.1. *Journal of Bacteriology*, **178**, 35–45.

Sackett, M. J., Armitage, J. P., Sherwood, E. E. & Pitta, T. P. (1997). Photoresponses of the purple non-sulfur bacteria *Rhodospirillum centenum* and *Rhodobacter sphaeroides*. *Journal of Bacteriology*, **179**, 6764–8.

Sourjik, V. & Schmitt, R. (1996). Different roles of CheY1 and CheY2 in the chemotaxis of *Rhizobium meliloti*. *Molecular Microbiology*, **22**, 427–36.

Stock, J. B., Surette, M. G., Levit, M. & Park, P. (1995). Two-component signal transduction systems: structure–function relationships and mechanisms of catalysis. In *Two-component Signal Transduction*, ed. J. A. Hoch & T. J. Silhavy, pp. 25–51. Washington, DC: ASM.

Taylor, B. L. (1983). Role of proton motive force in sensory transduction in bacteria. *Annual Review of Microbiology*, **37**, 551–73.

Ward, M. J., Bell, A. W., Hamblin, P. A., Packer, H. L. & Armitage, J. P. (1995*a*). Identification of a chemotaxis operon with two *cheY* genes in *Rhodobacter sphaeroides*. *Molecular Microbiology*, **17**, 357–66.

Ward, M. J., Harrison, D. M., Ebner, M. J. & Armitage, J. P. (1995*b*). Identification of a methyl-accepting chemotaxis protein in *Rhodobacter sphaeroides*. *Molecular Microbiology*, **18**, 115–21.

Zeilstra-Ryalls, J. H. & Kaplan, S. (1995). Aerobic and anaerobic regulation in *Rhodobacter sphaeroides* 2.4.1: the role of the *fnr*L gene. *Journal of Bacteriology*, **177**, 6422–31.

Zhulin, I. B., Bespalov, V. A., Johnson, M. S. & Taylor, B. L. (1996). Oxygen taxis and proton motive force in *Azospirillum brasilense*. *Journal of Bacteriology*, **178**, 5199–204.

Zusman, D. R., McBride, M. J., McCleary, W. R. & O'Connor, K. A. (1990). Control of directed motility in *Myxococcus xanthus*. In *Biology of the Chemotactic Response*, ed. J. P. Armitage & J. M. Lackie, pp. 199–219. Cambridge, UK: Cambridge University Press.

ARCHAEAL RHODOPSINS USE THE SAME LIGHT-TRIGGERED MOLECULAR SWITCH FOR ION TRANSPORT AND PHOTOTAXIS SIGNALLING

JOHN L. SPUDICH

Department of Microbiology and Molecular Genetics, The University of Texas-Houston Health Science Center, Medical School, Houston, Texas 77030, USA

INTRODUCTION

At first glance, the two major roles of light in our biosphere appear to be distinct and separate. Photosynthetic bacteria, algae, and higher plants capture light as the elemental fuel for life's processes, whereas we and other animals use light rather for the information it carries as it illuminates the environment. A variety of photoactive proteins are dedicated to either energy transduction, for example, the photoreaction centre complex of *Rhodobacter* (Cogdell, this volume), or sensory transduction, for example, human rod rhodopsin and phytochrome in microbes and higher plants.

Although differing in the consequences of their photoactivity, photoenergy and photosensory transducing pigments have fundamental tasks in common. They must tune their chromophores for optimal absorption in the spectral range abundant on the earth, convert the energy of the photon into a chemical form, and use the chemical potential thereby generated at the chromophoric site to activate other sites on the protein, either for catalysis of energy-conserving reactions or for signalling.

In principle, then, the same chromophore with perhaps only minor modifications in the surrounding protein, could be used for both photo-energy and photosensory transduction. Such appears to be the case in the salt-loving halobacteria, archaeal organisms that inhabit the intensely sunlit Dead Sea, solar evaporation ponds, and other waters with nearly saturated salt content. The best studied, *Halobacterium salinarum*, uses a family of four structurally similar membrane proteins, the archaeal rhodopsins, to (i) capture light energy by photon-driven electrogenic ion transport and (ii) use light for information to control its motility (phototaxis). The proteins are structurally (and as discussed below, also mechanically) similar to animal visual pigments, consisting of seven transmembrane α-helices forming an internal pocket for the chromophore retinal.

Bacteriorhodopsin (BR, Krebs & Khorana, 1993; Lanyi, 1995) and halorhodopsin (HR; Lanyi, 1990; Oesterhelt, 1995) are ion pumps for

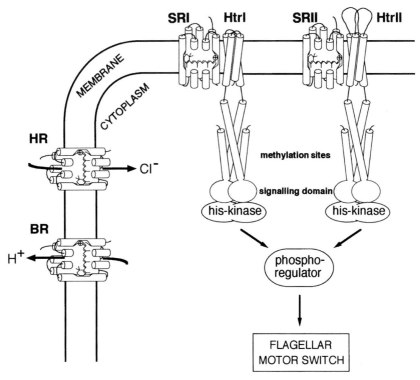

Fig. 1. The four archaeal rhodopsins in *H. salinarum*. The transport rhodopsins BR (a proton pump) and HR (a chloride pump) are shown in addition to the sensory rhodopsins SRI and SRII with components in their signal transduction chains. Each rhodopsin consists of seven transmembrane α-helices enclosing a retinal chromophore linked through a protonated Schiff base to a lysine residue in Helix G. The sensory rhodopsins are complexed to their corresponding transducer proteins HtrI and HtrII, which have conserved methylation and histidine kinase-binding domains (Yao & Spudich, 1992; Zhang *et al.*, 1996) that modulate kinase activity which in turn controls flagellar motor switching through a cytoplasmic phosphoregulator (Rudolph *et al.*, 1995). The structures drawn for the Htr transducers are only approximate, since crystal structures are not available. The transducers are represented as dimers based on the dimeric structure of the homologous *Escherichia coli* aspartate chemotaxis receptor Tar (Scott *et al.*, 1993) and on quantitative disulphide crosslinking into dimers observed following oxidation of HtrI containing engineered cysteine residues (Zhang & Spudich, unpublished observations). The oligomeric state of the SRs, assumed to be monomeric in the drawing, is not known. The four-helix bundle structure of the transmembrane and methylation domains is based on the structures of the corresponding domains for Tar, for which extensive evidence exists (for review see Stock & Mowbray, 1995; Stock & Surette, 1996). The depicted structure of the Htrs between the membrane and methylation domains (including the stimulus relay domain (Hoff *et al.*, 1997)) is that predicted from the Chou–Fasman algorithm. The relative positions of Htr and SR helices are arbitrary and chosen for illustration only.

protons and chloride ions, respectively. Sensory rhodopsins I and II (SRI and SRII; Hoff, Jung & Spudich, 1997) are phototaxis receptors found tightly bound to transducer proteins (HtrI and HtrII) that control a phosphorylation cascade modulating the cell's flagellar motors (Fig. 1).

Recent studies indicate that the sensory rhodopsins use for signalling to their transducers some of the same chemistry used by transport rhodopsins for ion pumping. This idea was strongly suggested by the observation that removal of HtrI from SRI by genetic deletion converted SRI into a BR-like electrogenic proton pump (Bogomolni *et al.*, 1994). SRI complexed with HtrI does not carry out electrogenic ion transport (evidence summarized in Spudich, 1994). Therefore, essential features of the BR mechanism must be conserved in SRI, but blocked by HtrI interaction. Recently, HtrII-free SRII from another halobacterial species, *Natronobacterium pharaonis*, has also been found to pump protons (M. Engelhard, personal communication). As discussed below, these and other observations can be explained by a model for signalling in which the Htr transducers are coupled to a light-controlled, electrostatically constrained, cytoplasmic channel in the SRs that is the same as exists in BR as a crucial component of its pumping mechanism.

HALOBACTERIAL PHOTOPHYSIOLOGY

The two sensory rhodopsins provide the cells with the ability to respond to light as an attractant or repellent depending on is colour and the stage of growth of the cells (Hoff *et al.*, 1997). *H. salinarum* grows at its maximum rate chemoheterotrophically in aerobic conditions. When oxygen and respiratory substrates are plentiful, the cells exhibit avoidance motility responses to sunlight, presumably to escape potential photooxidative damage (there are various terminologies for motility behaviour in use; here these responses will be called 'repellent phototaxis'). To accomplish this, they synthesize as their only rhodopsin the repellent receptor SRII (also known as phoborhodopsin). SRII (λ_{max} 487 nm) absorbs blue-green light in the energy peak of the solar spectrum at the Earth's surface. Hence, its wavelength sensitivity is tuned strategically to be maximally effective for seeking the dark.

A drop in oxygen tension suppresses SRII production and induces synthesis of BR (λ_{max} 568 nm) and HR (λ_{max} 576 nm), enabling orange light absorbed by these pumps to be used as an energy source. Like the respiratory chain, BR ejects protons out of the cell, generating inwardly directed proton motive force ($\Delta\mu_H +$) needed for ATP synthesis, active transport, and motility. HR is an inwardly directed pump, electrogenically transporting chloride into the cell. Like cation ejection, anion uptake hyperpolarizes the membrane positive-outside. Therefore, the inward transport of chloride by HR contributes to the membrane potential component of $\Delta\mu_H +$ without loss of cytoplasmic protons. This transport augments the $\Delta\mu_H +$ and also helps maintain pH homeostasis by avoiding cytoplasmic alkalization.

Production of SRI (λ_{max} 587 nm) is induced under the same semi-anaerobic conditions that induce BR and HR synthesis. SRI mediates attractant motility responses to orange light ('attractant phototaxis'), facilitating migration into illuminated regions where ion pumping by BR and HR will be maximally activated. SRI exhibits a second signalling activity to ensure it will not guide the cells perilously into higher energy light. A long-lived photointermediate of SRI, a species called S_{373} (subscript denotes λ_{max} in nm), absorbs near-UV photons and mediates a strong repellent response. The colour-sensitive signals from SRI therefore, attract the cells into a region containing orange light only if this region is relatively free of near-UV photons. When back in a rich aerobic environment, the *H. salinarum* cells switch off BR, HR, and SRI synthesis and switch on SRII production.

Although the sensory rhodopsins are responsible for phototaxis under most conditions, some earlier work, especially action spectroscopy, suggested that BR could mediate attractant responses. More recent studies have confirmed attractant responses to orange light due to light-driven photon pumping by BR (Yan *et al.*, 1992; Bibikov *et al.*, 1993). The BR-mediated responses occur at high light intensities and are most evident in partially de-energized cells. Aerotaxis, which occurs in *H. salinarum*, has been attributed to $\Delta\mu_H +$ or membrane potential $(\Delta\Psi)$ changes, and hence BR may contribute via these parameters.

TOWARD A UNIFIED MOLECULAR MECHANISM

Recent work shows that transport and sensory rhodopsins use modifications of the same phototransduction mechanism to carry out their different functions. One common element appears to be an *interhelical salt-bridge-locked conformational switch* that is released by photoisomerization of retinal. The lock is between an anionic aspartyl residue on Helix C and the protonated Schiff base attachment site of the retinal on Helix G. In BR, disruption of the lock opens a cytoplasmic half-channel that ensures uptake of the transported proton from the cytoplasmic side of the membrane at a critical time in the pumping cycle. In SRs, evidence points to the same mechanism operating to modulate interaction with a cytoplasmic domain of the Htr proteins when the receptor signalling states are formed.

HtrI blocks a BR-like cytoplasmic channel in SRI

Orange light converts SRI in <1 ms from its dark-adapted form SR_{587} to a near-UV-absorbing species S_{373}. S_{373} thermally returns to SR_{587} with a halftime of ~ 1 s, completing the SRI photocycle. The S_{373} species is the attractant signalling state that, during its transient existence, transmits through HtrI signals that inhibit swimming reversals. The most direct evidence for this is that retinal analogues that increase the lifetime of S_{373} increase the sensitivity of the cells (that is, the signalling efficiency of the receptor) in proportion to the amount of S_{373} generated by an orange light stimulus (Yan & Spudich, 1991).

Formation of S_{373} is accompanied by release of the proton from the Schiff base nitrogen in the attachment site of the retinal within the photoactive site of SRI (Haupts *et al.*, 1994). A proton returns to the Schiff base nitrogen in the second half of the photocycle during the thermal S_{373}-to-SR_{587} conversion. In *H. salinarum* membranes containing HtrI-free SRI, S_{373} formation results in a stoichiometric release of one proton/S_{373} into the aqueous phase (Olson & Spudich, 1993). A proton returns from the aqueous phase when S_{373} returns to the SR_{587} state (Spudich & Spudich, 1993). The proton release is either to the cytoplasmic side or the extracellular side, depending on pH and the ionization state of Asp76, a residue near the Schiff base (evidence summarized by Spudich, 1994). The return is from the cytoplasmic side of the SRI protein. Therefore, under conditions in which the proton is released to the outside, BR-like vectorial proton transport from the cytoplasm to the extracellular medium occurs (Bogomolni *et al.*, 1994).

HtrI has major effects on the proton transfers during the SRI photocycle described in detail in Spudich (1994) and Hoff *et al.* (1997). In particular, measurements of proton movements (Olson & Spudich, 1993) and their pH dependence (Spudich & Spudich, 1993) established that HtrI blocks the cytoplasmic proton-conducting channel of SRI. This is a key piece of evidence for the channel-coupling model elaborated below. The N-terminal half of HtrI containing the two transmembrane segments (TM1 and TM2) and the first half of the cytoplasmic domain (Fig. 1) is sufficient to block the channel (Perazzona, Spudich & Spudich, 1996), and mutations near the cytoplasmic end of TM2 influence SRI photocycling kinetics (Jung & Spudich, 1996), suggesting this region of HtrI interacts with SRI.

The channel in SRI that is closed by HtrI interaction is functionally similar to the cytoplasmic channel of BR, which has been characterized in more structural detail as discussed in the next section.

The cytoplasmic channel of BR is constrained by a salt bridge between helices C and G

The proton pump BR opens a cytoplasmic channel in the latter half of its photocycle. This process is thought to play an important role in the transport process by facilitating proton uptake from the cytoplasmic side after proton release to the outside in the first half of the photocycle. The opening of the channel by light has been inferred from cryoelectron microscopy images of the protein (Subramaniam *et al.*, 1993). In the mutant D96N, in which the M_{412} intermediate (corresponding to the S_{373} intermediate of SRI) is stabilized and can be freeze-trapped after illumination, Helix F is observed to be tilted and displaced toward the periphery of the protein, opening the structure on the cytoplasmic side.

A similar opening seen by X-ray diffraction of the D85N-D96N double mutant in the dark and D85N at alkaline pH argues that the salt-bridge between Asp85 on Helix C and the protonated Schiff base on Helix G is largely responsible for constraining the channel in the closed conformation (Lanyi, 1995; Spudich & Lanyi, 1996). In wild-type BR, this salt-bridge is disrupted by the transfer of the Schiff base proton to Asp85 during the formation of M_{412} in the photocycle. Therefore, in a recent model, the opening of the channel and switch in accessibility of the Schiff base from the outside to the cytoplasmic side were proposed to occur as a consequence of the salt-bridge disruption and possibly other structural changes due to photoisomerization (Spudich & Lanyi, 1996). The cause/effect relationship in this sequence of events has been questioned and an alternative model proposed in which the deprotonation of the Schiff base and the accessibility switch are assumed to be kinetically independent processes (Haupts et al., 1997). Recently, an X-ray diffraction and FTIR study of tertiary structural changes in the BR M state further supports the model in which the large structural change in BR is the result of the charge redistribution initiated by proton movement from the Schiff base to Asp85 (Sass et al., 1997). Also, as discussed in the next section, disruption of the homologous salt bridge by mutagenesis of SRII constitutively activates the protein in the dark, arguing that breaking the salt-bridge causes a significant conformational change in this protein.

The conformation of SRII is also constrained by the salt bridge between helices C and G

Asp73 on Helix C in SRII (Zhang et al., 1996) is the residue homologous to Asp85, the salt bridge anion and proton acceptor in BR. Asp73 has been concluded to similarly form a salt bridge with the protonated Schiff base in SRII based on spectroscopic effects of Asp73 mutagenic replacements and pH titrations (Zhu et al., 1997; Spudich et al., 1997). Moreover, flash photolysis (Spudich et al., 1997) and Fourier transform infrared spectroscopy (of *N. pharaonis* SRII, Engelhard, Scharf & Siebert, 1996) strongly indicate that Asp73, like BR Asp85, is the proton acceptor from the Schiff base during the SRII photocycle.

The substitution of Asp73 with Asn in SRII was engineered to test whether disruption of the salt bridge affected signalling. Coexpression of the genes encoding D73N and HtrII in *H. salinarum* cells resulted in a three-fold higher swimming reversal frequency in the dark than that from coexpression of native SRII and HtrII, and the mutation causes demethylation of HtrII in the dark, showing that D73N produces repellent signals in its unphotostimulated state (Spudich et al., 1997). Therefore this mutation and, by inference, disruption of the salt bridge of SRII constitutively activates the receptor.

The activating role of Asp73 is analogous to that of Glu113 in human rod rhodopsin (Rao & Oprian, 1996). Glu113, on Helix C, forms a salt bridge with the Schiff base on Helix G, and serves as the proton acceptor during photoconversion to the G protein-activating state Metarhodopsin-II$_{380}$. Disruption of the Glu113-protonated Schiff base salt bridge by mutagenic replacement of Glu113 with Gln constitutively activates the rhodopsin apoprotein. This observation supports the notion that in rhodopsin the counterion-Schiff base salt bridge constrains the protein in an inactive conformation and the constraint is released by its light-induced disruption. The similar observation in SRII argues for generality of this mechanism in retinylidene receptors from archaea to man.

Consistent with dark activation by the substitution of Asn for Asp73 in SRII, cells carrying this mutation exhibit a strongly reduced, but still detectable, taxis response. Therefore, salt bridge disruption appears to be sufficient for shifting the conformational equilibrium toward the signalling conformation, but another consequence of photoisomerization of retinal must also contribute to this shift. This result strengthens the analogy to visual pigments, in which proton transfer from the Schiff base to Glu113 is an important factor in stabilizing the G protein-activating state (Rao & Oprian, 1996) and to which other determinants also contribute significantly (Hofmann, Jager & Ernst, 1995; Fahmy, Siebert & Sakmar, 1995).

A unified molecular mechanism

The findings cited above suggest a mechanism for signalling by SRII (Fig. 2) that shares its main feature with the proton transport mechanism of BR, as well as with human rhodopsin: namely a photolabile, interhelical, salt bridge-controlled conformational switch. In BR, SRII, and rhodopsin, photoisomerization of retinal disrupts the salt bridge by inducing the transfer of the proton from the Schiff base on Helix G to the Helix C aspartate. The consequence in BR as well as in the proton-pumping Htr-free SRs is a conformational change that opens a proton-conducting cytoplasmic channel. The assumption in the channel-coupling model is that, in the SR–Htr complexes, this process (together with the other consequence of photo-isomerization detected in the D73N mutant of SRII) alter the structure of the receptors at the SR–Htr interface, thereby communicating the signal to the Htr proteins.

In SRI the corresponding salt bridge, formed by Asp76 and the Schiff base, is present in the proton-pumping form of HtrI-free SRI (Rath et al., 1996), but in the SRI–HtrI complex Asp76 is already protonated in the dark (that is, in SR$_{587}$; Rath et al., 1994). Therefore, Asp76 is not the proton acceptor from the Schiff base upon S$_{373}$ formation in the complex, but it is the acceptor in HtrI-free SRI at neutral and higher pH where proton pumping occurs (Rath

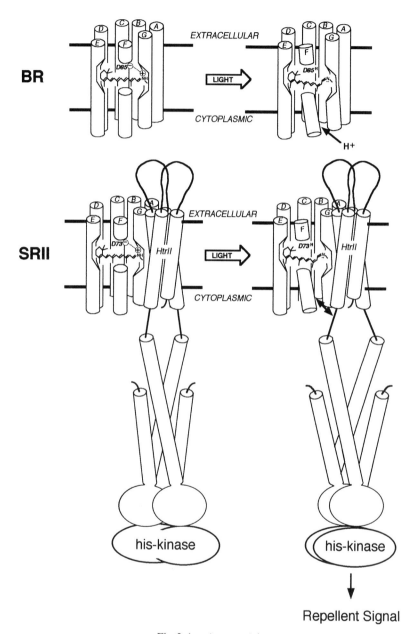

Fig. 2. (*caption opposite*).

et al., 1996). His166, a residue on the cytoplasmic side in SRI and not conserved in BR or SRII, is implicated in the Schiff base deprotonation path upon S_{373} formation in the SRI–HtrI complex (Zhang & Spudich, 1997).

In other words, SRI starts out with the salt bridge already disrupted in the dark. An explanation of this intriguing difference of SRI from BR, SRII, and rhodopsin may be found in the unique colour-sensitive dual signalling by SRI. In any two-conformation model of SRI signalling, the receptor must undergo a shift of the conformational equilibrium toward one conformer upon one-photon activation (that is, upon formation of S_{373}, the attractant signalling state, from SR_{587}) and toward the other conformer upon two-photon activation (that is, upon formation of S^b_{510}, the repellent signalling state, from SR_{587}). Therefore, the SR_{587} form must be poised in an intermediate position of the equilibrium in the dark in order to shift toward one or other extreme of the equilibrium. In terms of the channel-coupling model, the channel must be partially open in the dark (or equivalently in a metastable equilibrium between the open and closed channel forms). A simple mechanism for accomplishing this would be to prevent formation of the Asp76/Schiff base salt bridge in the dark form of the pigment. This is the case, since interaction with HtrI raises the pK of Asp76 to 8.5 in SR_{587}, thereby ensuring the Asp residue will not be ionized in the usual near neutral pH of the cell and its environment.

The residual signalling by SRII D73N shows that in addition to salt bridge disruption, a second unknown consequence of photoisomerization also contributes to receptor activation. This second factor is interpreted in the model as contributing to the stabilization of the open channel conformation, and is proposed to explain the attractant signalling by SRI. To accommodate the repellent signal, the model predicts two-photon activation of SR_{587} will close the channel, either by deprotonating Asp76 and forming the salt bridge with the protonated Schiff base in S^b_{510} or by an alternative process modulating the conformer equilibrium.

Fig. 2. Model for the role of light-induced disruption of the Asp/Schiff base salt bridge in BR and in SRII. The main feature of the model is that the same conformational change operates in proton transport and signaling: the opening of a cytoplasmic channel constrained by a Helix C/Helix G salt bridge, which is disrupted by photoisomerization-induced proton transfer from the protonated Schiff base to Asp85 (on BR) or Asp73 (on SRII), allows cytoplasmic proton uptake by BR and exposes new binding sites on SRII to the cytoplasmic region of HtrII. *BR:* The drawings are only to illustrate and emphasize the outward movement of Helix F on the cytoplasmic side and are not intended to be an accurate depiction of the near atomic resolution structures determined by cryoelectron crystallography (see text). *SRII:* The drawing is to illustrate a possible coupling mechanism, and the structure of HtrII is a first approximation based on the information described in the legend to Fig. 1. The predicted region near the cytoplasmic end of the second transmembrane segment of HtrI is implicated in SRI interaction by site-directed mutagenesis (Jung & Spudich, 1996) and suppressor mutant analysis (Jung & Spudich, unpublished observations). On this basis the corresponding region of HtrII is depicted as physically interacting with SRII in the light-activated form.

A specific prediction of the two-conformation shuttling aspect of the SRI model (put forth by Spudich & Lanyi, 1996) has recently been confirmed by a suppressor mutant study. The existence of mutations that shift the dark equilibrium toward either the open or closed channel conformation would be expected. A shift to the open conformation would eliminate the one-photon attractant signalling but leave the two-photon repellent signalling intact. Three mutants of SRI with this phenotype are D201N (Olson, Zhang & Spudich, 1995), H166A and H166S (Zhang & Spudich, 1997). Conversely, a mutation strongly shifting the dark equilibrium toward the closed channel form would restore the one-photon attractant signalling (and therefore be a second-site suppressor of D201N, H166A and H166S), and, accordingly to the model, would eliminate the two-photon repellent signalling. Mutants with the predicted phenotype have been found: for example, a second-site suppressor mutation in SRI, R215W, restores attractant responses to D201N and to H166A and H166S cells, and also eliminates the two-photon repellent response to the mutants and in a wild-type background (Jung & Spudich, unpublished observations).

The model presented here serves to emphasize that energy and sensory transducing proteins can evolve from a common progenitor and share much of their detailed mechanism despite their different functions. Further testing and refinement of the model will likely come from genetic and biochemical studies of isolated SR–Htr complexes exposed to various photostimuli, as well as from crystallographic analysis.

ACKNOWLEDGEMENT

Work cited here that was carried out in the author's laboratory was supported by National Institutes of Health grant R01-GM27750.

REFERENCES

Bibikov, S. I., Grishanin, R. N., Kaulen, A. D., Marwan, W., Oesterhelt, D. & Skulachev, V. P. (1993). Bacteriorhodopsin is involved in halobacterial photo-reception. *Proceedings of the National Academy of Sciences, USA*, **90**, 9446–50.

Bogomolni, R. A., Stoeckenius, W., Szundi, I., Perozo, E., Olson, K. D. & Spudich, J. L. (1994). Removal of transducer HtrI allows electrogenic proton translocation by sensory rhodopsin I. *Proceedings of the National Academy of Sciences, USA*, **91**, 10188–92.

Engelhard, M., Scharf, B. & Siebert, F. (1996). Protonation changes during the photocyle of sensory rhodopsin II from *Natronobacterium pharaonis*. *FEBS Letters*, **395**, 195–8.

Fahmy, K., Siebert, F. & Sakmar, T. P. (1995). Photoactivated state of rhodopsin and how it can form. *Biophysical Chemistry*, **56**, 171–81.

Haupts, U., Eisfeld, W., Stockburger, M. & Oesterhelt, D. (1994). Sensory rhodopsin I photocycle intermediate SRI_{380} contains 13-*cis* retinal bound via unprotonated Schiff base. *FEBS Letters*, **356**, 25–9.

Haupts, U., Tittor, J., Bamberg, E. & Oesterhelt, D. (1997). General concept for ion translocation by halobacterial retinal proteins: the isomerization/switch/transfer (IST) model. *Biochemistry*, **36**, 2–7.

Hoff, W. D., Jung, K-H. & Spudich, J. L. (1997). Molecular mechanism of photosignaling by archaeal sensory rhodopsins. *Annual Reviews of Biophysics and Biomolecular Structure*, **26**, 223–58.

Hofmann, K. P., Jäger, S. & Ernst, O. P. (1995). Structure and function of activated rhodopsin. *Israel Journal of Chemistry*, **35**, 339–55.

Jung, K-H. & Spudich, J. L. (1996). Protonatable residues of the cytoplasmic end of transmembrane helix-2 in the signal transducer HtrI control photochemistry and function of sensory rhodopsin I. *Proceedings of the National Academy of Sciences, USA*, **93**, 6557–61.

Krebs, M. P. & Khorana, H. G. (1993). Mechanism of light-dependent proton translocation by bacteriorhodopsin. *Journal of Bacteriology*, **175**, 1555–60.

Lanyi, J. K. (1990). Halorhodopsin: a light-driven electrogenic chloride transport system. *Physiology Reviews*, **70**, 319–30.

Lanyi, J. K. (1995). Bacteriorhodopsin as a model for proton pumps. *Nature*, **375**, 461–3.

Oesterhelt, D. (1995). Structure and function of halorhodopsin. *Israel Journal of Chemistry*, **35**, 475–94.

Olson, K. D. & Spudich, J. L. (1993). Removal of the transducer protein from sensory rhodopsin I exposes sites of proton release and uptake during the receptor photocycle. *Biophysical Journal*, **65**, 2578–85.

Olson, K. D., Zhang, X-N. & Spudich, J. L. (1995). Residue replacements of buried aspartyl and related residues in sensory rhodopsin I: D201N produces inverted phototaxis signals. *Proceedings of the National Academy of Sciences, USA*, **92**, 3185–9.

Perazzona, B., Spudich, E. N. & Spudich, J. L. (1996). Deletion mapping of the sites on the HtrI transducer for sensory rhodopsin I interaction. *Journal of Bacteriology*, **178**, 6475–8.

Rao, R. & Oprian, D. D. (1996). Activating mutations of rhodopsin and other G protein-coupled receptors. *Annual Reviews of Biophysics and Biomolecular Structure*, **25**, 287–314.

Rath, P., Olson, K. D., Spudich, J. L. & Rothschild, K. J. (1994). The Schiff base counterion of bacteriorhodopsin is protonated in sensory rhodopsin I: Spectroscopic and functional characterization of the mutated proteins D76N and D76A. *Biochemistry*, **33**, 5600–6.

Rath, P., Spudich, E. N., Neal, D. D., Spudich, J. L. & Rothschild, K. J. (1996). Asp76 is the Schiff base counterion and proton acceptor in the proton translocating form of sensory rhodopsin I. *Biochemistry*, **35**, 6690–6.

Rudolph, J., Tolliday, N., Schmitt, C., Schuster, S. C. & Oesterhelt, D. (1995). Phosphorylation in halobacterial signal transduction. *EMBO Journal*, **14**, 4249–57.

Sass, H. J., Schachowa, I. W., Rapp, G., Koch, M. H. J., Oesterhelt, D., Dencher, N. A. & Büldt, G. (1997). The tertiary structural changes in bacteriorhodopsin occur between M states – X-ray diffraction and Fourier transform infrared spectroscopy. *EMBO Journal*, **16**, 1484–91.

Scott, W. G., Milligan, D. L., Milburn, M. V., Prive, G. G., Yeh, J., Koshland, D. E., Jr. & Kim, S-H. (1993). Refined structures of the ligand-binding domain of the aspartate receptor from *Salmonella typhimurium*. *Journal of Molecular Biology*, **232**, 555–73.

Spudich, J. L. (1994). Protein–protein interaction converts a proton pump into a sensory receptor. *Cell*, **79**, 747–50.

Spudich, J. L. & Lanyi, J. K. (1996). Shuttling between protein conformations: The common mechanism for sensory transduction and ion transport. *Current Opinion in Cell Biology*, **8**, 452–7.

Spudich, E. N. & Spudich, J. L. (1993). The photochemical reactions of sensory rhodopsin I are altered by its transducer. *Journal of Biological Chemistry*, **268**, 16095–7.

Spudich, E. N., Zhang, W., Alam, M. & Spudich, J. L. (1997). Constitutive signaling of the phototaxis receptor sensory rhodopsin II from disruption of its protonated Schiff base-Asp73 salt bridge. *Proceedings of the National Academy of Sciences, USA*, **94**, 4960–5.

Stock, A. M. & Mowbray, S. L. (1995). Bacterial chemotaxis: a field in motion. *Current Opinion in Structural Biology*, **5**, 744–51.

Stock, J. B. & Surette, M. G. (1996). Chemotaxis. In *Escherichia coli and Salmonella typhimurium*, ed. F. C. Neidhardt, pp. 1103–29. Washington, DC: ASM Press.

Subramaniam, S., Gerstein, M., Oesterhelt, D. & Henderson, R. (1993). Electron diffraction analysis of structural changes in the photocycle of bacteriorhodopsin. *EMBO Journal*, **12**, 1–8.

Yan, B. & Spudich, J. L. (1991). Evidence that the repellent receptor form of sensory rhodopsin I is an attractant signaling state. *Photochemistry and Photobiology*, **54**, 1023–6.

Yan, B., Cline, S. W., Doolittle, W. F. & Spudich, J. L. (1992). Transformation of a BOP$^-$HOP$^-$ SOP-I$^-$SOP-II$^-$ *Halobacteriuim halobium* mutant to BOP$^+$: effects of bacteriorhodopsin photoactivation on cellular proton fluxes and swimming behavior. *Photochemistry and Photobiology*, **56**, 553–61.

Yao, V. J. & Spudich, J. L. (1992). Primary structure of an archaebacterial transducer, a methyl-accepting protein associated with sensory rhodopsin I. *Proceedings of the National Academy of Sciences, USA*, **89**, 11915–19.

Zhang, X-N. & Spudich, J. L. (1997). His-166 is critical for active site proton transfer and phototaxis signaling by sensory rhodopsin I. *Biophysical Journal*, **73**, 1516–23.

Zhang, W., Brooun, A., Müller, M. M. & Alam, M. (1996). The primary structures of the Archaeon *Halobacterium salinarium* blue light receptor sensory rhodopsin II and its transducer, a methyl-accepting protein. *Proceedings of the National Academy of Sciences, USA*, **93**, 8230–5.

Zhu, J., Spudich, E. N., Alam, M. & Spudich, J. L. (1997). Effects of substitutions D73E, D73N, D103N, and V106M on signaling and pH titration of sensory rhodopsin II. *Photochemistry and Photobiology*, **66**, 788–91.

GAS VESICLES AND BUOYANCY IN CYANOBACTERIA: INTERRELATIONS WITH LIGHT

A. E. WALSBY

Department of Botany, School of Biological Sciences, University of Bristol, Bristol BS8 IUG, UK

INTRODUCTION

As sunlight passes through water it is attenuated by absorption and scattering. Photosynthetic organisms that depend on light for their sustenance are restricted to a zone near the water surface where the irradiance is high enough to support a photosynthetic rate that exceeds respiratory losses. This *euphotic zone* is generally assumed to extend to a depth where the irradiance is about 1% of that at the water surface (Kirk, 1994). The actual value varies with daylength, because the photosynthate accumulated during the day must sustain the organism through the ensuing night, and it depends on the efficiency of the organism in harvesting light energy and incorporating it into organic matter. Nevertheless, the photosynthetic mode of existence, if not supplemented by other modes of nutrition, is everywhere restricted to a relatively shallow layer of the Earth's lakes and seas.

The majority of phytoplankton indulge in a lottery to make their living: they are slowly sinking organisms that take their chance in being returned by water currents to the euphotic zone with sufficient frequency to win enough photosynthetic gains to pay off respiratory losses. There are some types of phytoplankton, however, that insure themselves against the chance of loss to the darkness of the *aphotic* zone by investing in flotation devices that enable them to rise back towards the water surface when conditions are calm enough. The most efficient of these devices is the gas vesicle, a structure produced by many aquatic bacteria and even some archaea (Walsby, 1994).

This chapter deals with gas vesicles in cyanobacteria. It first describes the structure and properties of the gas vesicle, essential to the provision of buoyancy; next, it considers mixed seas and stratified lakes in which the buoyancy provided by gas vesicles affects the distribution of cyanobacteria in relation to the light field and determines the primary productivity of the population; and finally it considers buoyancy regulation and gas vesicle production in response to light.

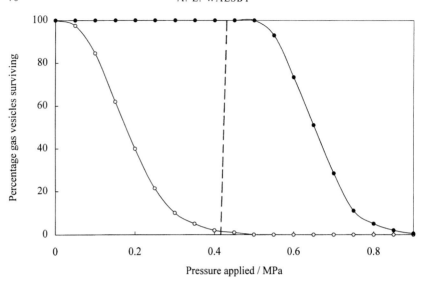

Fig. 1. The distribution of critical collapse pressures of gas vesicles in *Anabaena flos-aquae*: (●) the true critical pressure, p_c, with cells suspended in hypertonic sucrose solution to remove turgor pressure; (○) the apparent critical pressure, p_a, in turgid cells suspended in culture medium. The difference between the two curves (broken line) indicates the turgor pressure, p_t, which falls slightly as gas vesicles are collapsed. Data from Walsby (1971).

The structure and properties of gas vesicles

The gas vesicle is a hollow, cylindrical structure closed at each end with a conical cap. The space inside the structure is occupied by gases that diffuse into it across the gas-permeable wall. The gas is therefore usually air at atmospheric pressure, in equilibrium with the air dissolved in the cytoplasm of the cell (Walsby, 1994). Although permeable to gases, the gas vesicle wall is impermeable to liquid water. Waterproofing is provided by the hydrophobic nature of the inner, gas-facing surface, which prevents water droplets entering pores by surface tension. In these respects the gas vesicle wall resembles the man-made membrane Goretex, used in breathable rainproof garments.

Another important property of the wall is its rigidity, which minimises the elastic compression of the gas vesicle. Gas vesicles from the cyanobacterium *Microcystis* sp. decrease in volume by only 1 part in a thousand when subjected to a pressure increase of 1 bar. There is a negligible loss of buoyancy therefore, in gas-vesicle containing cells that are mixed down to considerable depths, as long as their gas vesicles remain intact.

The gas vesicle is a brittle structure, however, and at a certain critical collapse pressure the wall breaks and the structure collapses irreversibly (Fig. 1). The critical pressure depends mainly on the cylinder diameter of the gas vesicle and, as discussed below, varies considerably between different cyanobacteria. If a gas vesicle is to remain intact the critical pressure (p_c) must exceed the sum of the turgor pressure of the cell (p_t) and the hydrostatic head of the water column (p_h).

The two proteins, GvpA and GvpC, that form gas vesicles

The properties of the gas vesicle wall are provided by proteins, which are the sole constituents of the structure. The principal component of the wall is the gas vesicle protein GvpA, a small molecule of only 70 amino acids (Tandeau de Marsac et al., 1985; Hayes, Walsby & Walker, 1986). The amino acid composition suggests that this is one of the most hydrophobic proteins known. Electron diffraction studies indicate that the hydrophobic residues, which have a lower electron density, are more abundant at the inner, gas-facing surface (Blaurock & Walsby, 1976). Neutron diffraction studies of layers of collapsed gas vesicles confirms that liquid water is excluded from the inner surface (Worcester, 1975).

GvpA assembles to form ribs that are oriented at right angles to the principal axis of the intact cylinder. The GvpA molecules can be thought of as parallelogram-shaped tiles stacked in rows that curve around to form the cylinder (Blaurock & Walsby, 1976); the width of each tile is the 1.15 nm distance between the adjacent pairs of β-chains into which the GvpA polypeptide chain is folded, and the angle at which the slanting sides are tilted is determined by the simple fact that the adjacent chains differ in length by a single dipeptide unit. In this arrangement the hydrogen bonds that tie together adjacent polypeptide chains are orientated at an angle close to the unique 'magic' angle of 55° at which the longitudinal and transverse stresses in a cylinder are equal; the significance of this is that the structure is not distorted when pressure is applied to it (Walsby, 1991).

Each rib formed by GvpA is thought to be tied to the ribs on either side by a second type of gas vesicle protein, GvpC, which has a very different structure. In *Anabaena flos-aquae* the amino acid sequence of GvpC reveals a chain of 33 amino acid residues that is repeated five times (Hayes et al., 1988). Each 33-residue repeat (33RR) is thought to form an α-helix of 5 nm length, long enough to cross each 4.6 nm wide rib at such an angle that it would touch five of the slanting GvpA molecules. The five 33RRs within a GvpC molecule would contact 25 GvpA molecules in the five adjacent ribs; 25:1 is in fact the ratio of GvpA:GvpC found in isolated gas vesicles (Buchholz, Hayes & Walsby, 1993). The removal of GvpC with detergents or 6M urea greatly weakens the gas vesicles (Walsby & Hayes, 1988); this weakening can be reversed by reattaching the protein (Hayes, Buchholz & Walsby, 1992).

Together, GvpA and GvpC form most or all of the biomass of the gas vesicle wall and their synthesis therefore constitutes the principal cost in material and energy of providing buoyancy. There must, however, be some additional costs in carrying the genetic information for the gas vesicle and in assembling the structure.

In cyanobacteria the genes that encode these two proteins, *gvpA* and *gvpC*, form part of an operon in which there may be two or more copies of *gvpA* in tandem repeat (Damerval *et al.*, 1987; Hayes & Powell, 1995). Studies with halophilic archaea of the genus *Halobacterium* show homologues of the same two genes; a cluster of 14 genes is implicated in gas vesicle formation in these organisms, of which at least eight (GvpA, GvpC and six others) are indispensable to the process (Horne *et al.*, 1991; Jones, Young & DasSarma, 1991). Sequences homologous to those in the six additional genes are found in four genes downstream of the *gvpAC* operon in *Anabaena flos-aquae* (Kinsman & Hayes, 1997): by inference they are also involved in the assembly or regulation of gas vesicle production in cyanobacteria. Each of these genes imposes a genetic burden, the cost of replicating this information; if there are 3000 genes in the average cyanobacterium then eight gas vesicle genes represent about 0.3% of the genetic load. The protein burden is much higher, however; the two structural proteins GvpA and GvpC alone may account for as much as 10% of the total cell protein (Hayes & Walsby, 1984), a major item in the housekeeping budget of planktonic cyanobacteria. Buoyancy does not come cheaply. The widths of gas vesicles in other cyanobacteria varies, from only 45 nm in *Trichodesmium* sp. from the deep oceans (Gantt, Okhi & Fujita, 1984) to 110 nm in *Dactylococcopsis salina* from a shallow desert brine pool. The critical collapse pressure of gas vesicles varies inversely as the cylinder width (Walsby, 1971, 1991; Hayes & Walsby, 1986; Walsby & Bleything, 1988) and this has driven natural selection for gas vesicles of decreasing width and hence increasing critical pressure in waters of increasing depth (Walsby, 1994). The strongest gas vesicles, with P_c exceeding 3 MPa (30 bar), occur in oceanic *Trichodesmium* species (Walsby, 1978).

Interactions between gas vesicles, gas vacuoles and light

Isolated gas vesicles form colloidal suspensions that show light scattering, which increases strongly with the frequency of radiation. Gas vesicles in cells of cyanobacteria are clustered together in arrays called gas vacuoles, which are visible by light microscopy; gas vacuoles show frequency-independent light refraction (see Walsby, 1994). Gas vacuoles are usually distributed throughout the cell but in *Pseudanabaena* and in some species of *Oscillatoria* they occur adjacent to the cross walls (Meffert, 1971). In *Anabaena flos-aquae* the gas vacuoles occur throughout the cells when grown at low irradiance but in cells grown at high irradiance they are mainly located at the cell periphery

(Shear & Walsby, 1975). This rearrangement suggested that gas vacuoles might provide light shielding. Some support for this idea is provided by the changes in absorption spectra that occur when gas vesicles are collapsed by pressure (Waaland, Waaland & Branton, 1972; Ogawa, Sekine & Aiba, 1979), but gas vacuole collapse results in little change in photosynthetic rate at limiting irradiances, which indicates that the degree of shielding is not of great importance (Shear & Walsby, 1975; Walsby, 1994). Gas vacuoles in cyanobacterial suspensions (Porter & Jost, 1976; Van Liere & Walsby, 1982) and natural populations (Ganf, Oliver & Walsby, 1988) do contribute to the attenuation of light, however, and in this respect they play a part in the competition for light between gas-vacuolate cyanobacteria, which float up in calm conditions, and other phytoplankton, which remain dispersed throughout the epilimnion (see below).

THE BENEFITS OF FLOATING IN RELATION TO PHOTOSYNTHESIS IN THE VERTICAL GRADIENTS OF LIGHT

The buoyancy provided by gas vesicles affects the vertical distribution of planktonic cyanobacteria in different ways: in transparent stratified lakes it enables filaments to position themselves near the thermocline region; in other lakes, seas and oceans it enables cyanobacteria to remain in the surface mixed layer. The understanding of this requires some explanation of light penetration and mixing processes in natural waters.

Filaments stratifying in the metalimnion

In lakes that become thermally stratified during the summer months planktonic cyanobacteria may take up residence near the thermocline, the region of steepest temperature change between the warm surface layer (epilimnion) and the cold lower layer (hypolimnion) of the lake (Thomas & Märki, 1949; Baker & Brook, 1969; Klemer, 1976; Konopka, 1982). In freshwater lakes the thermocline is the region of steepest density change in the lake and is therefore the most stable layer; it takes more energy to lift the cold dense water of the hypolimnion vertically into the warm less dense water of the epilimnion.

If sufficient light penetrates to the thermocline to support the growth of cyanobacteria, populations of the organisms will gradually develop there. The most commonly encountered are species of *Oscillatoria* (*Planktothrix*), and particularly red-coloured forms such as *O. agardhii* and *O. rubescens* (see below). In addition to cyanophycin, the blue-coloured photosynthetic pigment present in almost all cyanobacteria, they possess phycoerythrin, a red pigment which absorbs strongly in the green region of the spectrum. As sunlight penetrates more deeply into water, the proportion of green light increases because chlorophyll-containing phytoplankton in the surface layers remove more of the blue and red wavelengths. The *Oscillatoria* filaments are opportunists that scavenge these remnants of the solar irradiation, growing at irradiances of green light that may be insufficient to support net photosynthesis by green algae (see below). Nevertheless, they probably grow quite slowly, perhaps with a generation time of about 3 days or more. The population can increase in these unmixed layers only if the average residence time of the filaments exceeds the generation time. The majority of other phytoplankton organisms, which are denser than water, would sink from this stable metalimnion into the hypolimnion and then have no means of returning.

The buoyancy provided by gas vesicles enables *Oscillatoria* filaments to hover in these unmixed layers and fulfil the residence requirement. Their buoyant density is adjusted approximately to that of the water in which they are suspended, although this requires continual fine tuning. During the day the density of the filaments increases as they photosynthesise and accumulate carbohydrate. At night, carbohydrate is respired or converted into less dense protein, and buoyancy increases. The net result of these buoyancy changes is that filaments tend to converge at a particular depth determined by the irradiance averaged over the day. A filament that rises above this depth will receive more light during the day, accumulate more carbohydrate, lose buoyancy and then sink further down. Conversely, a filament that sinks below this layer will receive less light, make less carbohydrate, become more buoyant and then float up. The system is self-regulating. The mechanisms of buoyancy regulation by light are discussed in more detail below.

A requirement for stratification is that the organisms must be small: only single cells or separate filaments can do it. If the cells or filaments bundle together in aggregates or grow to produce colonies, they sink down or float up so quickly, as they adjust their density, that they continually overshoot the equilibrium depth and they either spend much unproductive time in the darkened hypolimnion or become caught up in the water circulating within the unstable epilimnion. The explanation for the faster movement with increasing size is found in the Stokes equation: the sinking velocity of a sphere increases as the square of its radius. This powerful effect of size also holds for objects of other shapes, though most sink rather more slowly than spheres of the same volume and density.

Another requirement for stratification is that sufficient light must penetrate to the thermocline to drive these buoyancy-regulating responses. If there is insufficient light, the cyanobacteria will float further up into the epilimnion where the thermal gradients are too weak to prevent sporadic mixing of the cyanobacteria within the surface water layer (Walsby *et al.*, 1989). This situation arises not only in very large lakes and seas, where wind mixing deepens the epilimnion, but also in eutrophic waters, where self-shading by the concentrated growths of phytoplankton themselves so attenuate the light that the cyanobacteria would have to float into the surface mixed layer before they encounter irradiances high enough to cause buoyancy losses. In such lakes different types of planktonic cyanobacteria are found.

Cyanobacteria floating in the surface mixed layer

Buoyancy regulation is a fascinating phenomenon but in the majority of lakes the simple provision of buoyancy by gas vesicles in planktonic cyanobacteria is of more importance. In brief, in the majority of waterbodies gas vesicles serve to maintain cyanobacteria within the epilimnion. When the epilimnion is mixed all phytoplankton cells are circulated by the water movements, which usually exceed the fastest speeds of floating or swimming by an order of magnitude or more. Under such conditions there is little advantage in either of these activities. During the summer months, however, when the temperature gradients are strongest, the water column may frequently stabilize; when this happens buoyant cyanobacteria immediately begin to float up whereas other non-buoyant organisms begin to sink down. Because the irradiance decreases exponentially with depth, the consequence is that floating results in increased photosynthesis by gas-vacuolate cyano-bacteria while sinking results in a decrease in many other organisms. In the extreme case the sinking organisms may settle through the thermocline and when this happens they have no way of returning.

Before assessing the benefits of floating in more detail it is useful to summarize the mixing processes in lakes in this summer period, because they may counteract the floating movements of the cyanobacteria. During the day-time the sun shines on the water, heating the surface layers. About half of the energy is in infra-red wavelengths, much of which are absorbed within the top 2 m. The temperature of this layer may rise by as much as 5 °C and a diurnal thermocline develops. The temperature increase stabilises the layer and cyanobacteria that float into it may rise to the surface forming a waterbloom that is clearly visible. The heat input ceases at dusk and the surface layer of water is then cooled by evaporation and by convection to the cooler air. The surface water therefore becomes more dense than the water layer below and it mixes down. By the end of the night, convective cooling may have caused uniform mixing to a depth of 2 metres. Such daily cycles of heating and cooling are repeated in even the most sheltered lakes (Spigel & Imberger, 1987; Reynolds, 1997). Organisms that float up again each day will continue to benefit from higher irradiances near the surface.

The other principal agent of mixing is wind. When wind blows over water a small proportion of the kinetic energy is transferred by frictional drag and the water column begins to mix. The depth of mixing depends on the wind speed, the fetch (the length of the lake in the wind direction) and the stability of the water column, which is determined by the temperature gradient. In summary, large and weakly stratified waterbodies mix to a greater depth and with greater velocity than small, strongly stratified ones; small lakes in wooded surroundings are also more sheltered from wind (Spigel & Imberger, 1987). Wind is sporadic. When it abates, the residual movements in the water gradually cease and stability returns with solar heating. Cyanobacteria can immediately start to rise up again while non-buoyant organisms begin to sink or at best remain where they were left when mixing stopped.

The time taken by an organism to return to the euphotic zone when stability returns is crucial to its success. The deeper that mixing occurs, the faster an organism must float up to regain access to light. In less stable waterbodies there is therefore selection for cyanobacteria that form aggregates or colonies, which float (or sink) much more rapidly than separate filaments or cells.

With these general principles established, it is possible to explain the light-regulated movements of cyanobacteria in different types of lakes, and to consider the benefits that buoyancy brings. This is best done by reference to specific examples.

INTERACTIONS OF PLANKTONIC CYANOBACTERIA WITH LIGHT IN DIFFERENT WATER BODIES

Lake Gjersjøen, Norway

In Gjersjøen, a lake of 70 m depth, a red-pigmented variety of *Oscillatoria agardhii* occurred throughout the year. It stratified at a depth of 6 m near the thermocline during the summer, was circulated to the lake bottom during the early winter, floated up when ice formed, was mixed down during the second period of mixing at ice melt and finally returned to the metalimnion in June to form the inoculum for the next season's growth. During the summer, the filaments in the metalimnion population regulated their buoyancy according to the irradiance. The buoyancy showed a general increase with depth and it increased during the night at all depths. In culture, the organism gained buoyancy when kept at photon irradiances below a threshold value of 5 μmol m^{-2} s^{-1} and lost buoyancy when left at irradiances above this value (Walsby, Utkilen & Johnsen, 1983). This threshold corresponded to the mean irradiance at the metalimnion.

Based on measurements of the rates of change in buoyant density of *Oscillatoria agardhii* at different irradiances, some detailed calculations have been made of the expected changes in depth at which filaments occurred on the vertical light gradient in Lake Gjersjøen (Walsby, 1988). They indicated that the position of a filament in the metalimnion would oscillate with an amplitude of only a few centimetres over a day as it regulated its buoyancy in response to the ambient light field. These computer calculations produced a realistic model of the movements of filaments observed by Walsby *et al.* (1983) in the lake (Fig. 2(*a*)).

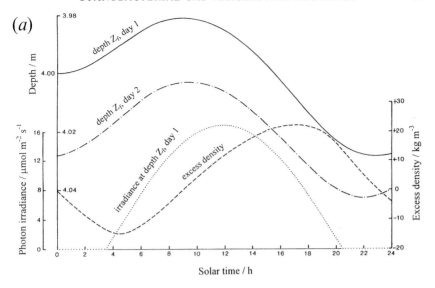

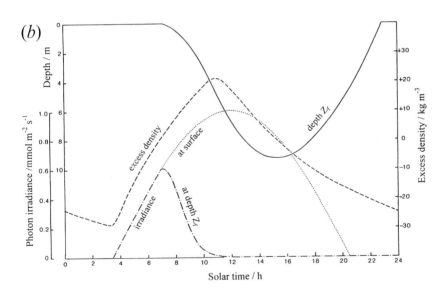

Fig. 2. A computer simulation of the changes in depth, Z_f, at which *Oscillatoria agardhii* occurs in Lake Gjersjøen, Norway, predicted by a computer model from rates of buoyant density change and hydrodynamics of the organism, as it responds to the irradiance at depth Z_f: (*a*) separate filaments, which slowly float up or sink down a few centimetres each day as they oscillate about depth of 4 m; (*b*) aggregates of the same type of filaments, which, because of their larger size and resulting faster movement, migrate between the surface and a depth of about 9 metres. Note the different depth and photon irradiance scales in each figure. From Walsby (1988).

The same calculations also produced a quantitative explanation for a rather extraordinary phenomenon at the lake: at dawn on consecutive mornings the lake surface was covered by pink fluffy specks, which were aggregates of the *Oscillatoria agardhii* filaments that had evidently formed in the metalimnion and then floated up; by mid morning they disappeared, sinking back down in the lake (Walsby *et al.*, 1983). The computer model showed that the filaments had changed their density in response to irradiance in the same way as the separate filaments, but owing to their larger size the aggregates oscillated over a much greater depth range (Fig. 2(*b*)).

More refined models of buoyancy change have now been produced: Kromkamp & Walsby (1990) based their model on equations describing actual measurements of the rate of density change by *Oscillatoria agardhii* at different irradiances, and they used it to investigate the effects of colony size and nutrient availability on the vertical migration of the organism. Similar procedures based on models of carbohydrate metabolism have recently been developed by Howard, Irish & Reynolds (1996) and applied to movements of other cyanobacteria.

Lake Zürich, Switzerland

During the summer a similar phenomenon occurs in Lake Zürich (Thomas and Märki, 1949): the red-coloured *Oscillatoria rubescens* forms a population that also stratifies in the metalimnion but at greater depths, sometimes exceeding 15 m. The mechanism is similar to that in Lake Gjersjøen (Fig. 2(*a*)) but operates at a greater depth owing to the greater transparency of the water column. It might appear remarkable that there is sufficient light for photoautotrophic growth at such depths, but detailed calculations of the daily integral of photosynthesis of the population have shown that it has a high productivity.

The daily integral of photosynthesis is calculated in six steps. (i) The relationship between photosynthesis (P) and photon irradiance (I) is determined by measurements of the biomass-specific or chlorophyll-specific molar oxygen production rate (equivalent to the molar carbon fixation rate), at a temperature Θ'. (ii) The cyanobacterial biomass or chlorophyll concentration (N) is measured in samples taken at intervals of depth. (iii) The vertical profile of temperature, Θ, is also measured. (iv) The light attenuation coefficient (K_d) is determined from vertical profiles of photon irradiance. (v) The surface irradiance (and wind speed) is measured by continuously recording instruments, and from these data is calculated the irradiance immediately under the water surface (I_0), allowing for losses by reflection from the surface, adjusted for roughening by wind. (vi) The final step is to combine all of these measurements to calculate the photosynthesis at each depth throughout the water column and each time interval during the day; by adding the results together the total daily integral of photosynthesis is determined (Walsby, 1997).

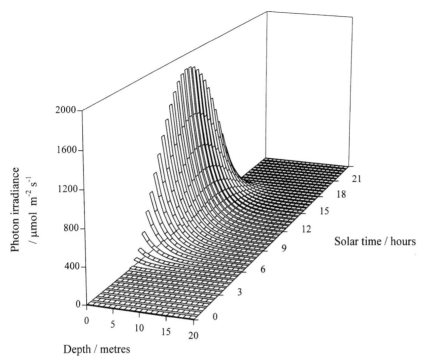

Fig. 3. Photon irradiance at different depths and times in Lake Zürich, on 3 August 1995, calculated from the vertical profile of light attenuation and continuous measurements of the irradiance at the water surface. Data from S. Micheletti, F. Schanz & A. E. Walsby (in preparation).

Two stages of the calculation are illustrated in Figs. 3 and 4. Figure 3 shows a 3-dimensional plot of the changes in photon irradiance with time throughout the day and depth down the water column. The irradiance at the surface on this nearly cloudless day follows the sine curve of the elevation of the sun; at each time, the irradiance decreases exponentially with depth. By combining these measurements of irradiance with measurements of the abundance of the cyanobacteria through the water column, and measurements of the P/I curve, the photosynthetic rate at each depth and time is calculated (Fig. 4). These calculations show the high productivity of *Oscillatoria rubescens* in the metalimnion (S. Micheletti, F. Schanz & A. E. Walsby, in preparation).

Lake Rotongaio, New Zealand

Rotongaio is a small lake (0.34 km²), with a mean depth of 11 m, adjacent to Lake Taupo in New Zealand. Formed in the crater of an extinct volcano, it is sheltered from wind and becomes strongly stratified, with a seasonal thermocline at a depth of about 6 m. The water is rich in dissolved reactive

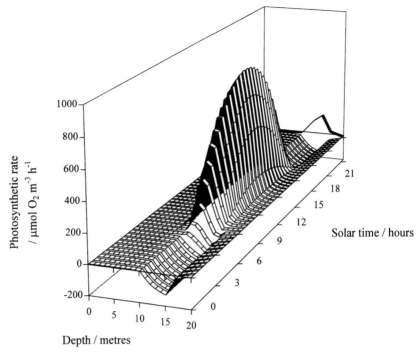

Fig. 4. Rates of photosynthetic oxygen production by filaments of *Oscillatoria rubescens* at different depths and times in Lake Zürich, on 3 August 1995. The last profile (black ribbon) shows the relative filament concentration. Data from S. Micheletti, F. Schanz & A. E. Walsby (in preparation).

phosphorus, of volcanic origin, but relatively impoverished in combined nitrogen, with a very low N:P ratio of 0.5 (Vincent, 1989). This explains the dominance of a nitrogen-fixing cyanobacterium, *Anabaena minutissima*, which produces small separate filaments homogeneously suspended in the water. The high phosphorus concentration supports the development of a high concentration of cyanobacteria, which results in a high light attenuation coefficient, $K_d = 3.4 \, \mathrm{m}^{-1}$; this, in turn, causes heating of the surface layer so that an additional strong thermal gradient forms rapidly during calm periods at depths of 1 to 2 m (Vincent, 1989).

Owing to the steep light attenuation, even those filaments suspended in the top 2 m experience too low an irradiance to lose buoyancy and they remain entrained in the surface mixed layer; for much of the time more than 80% of the filaments were present within the top 2 m. Occasionally, a storm causes mixing to depths of 6 m or more; the filaments then float slowly back to the surface layer, at a velocity of about 0.1 m d^{-1}. Colonies, with a faster rate of floating, would have returned more quickly on such occasions. For much of the time, however, colonies would have floated up into the surface film from where they would have experienced greater losses by photooxidation and by skimming from the surface into reed beds and down-wind into an outflowing stream.

Lake Okaro, New Zealand

Only 65 km from Rotongaio is another lake, Lake Okaro, of similar size (0.28 km^2) and mean depth (12 m), also stratified with a seasonal thermocline at about 6 m (Vincent, 1989). A major difference between the two lakes was the lower phosphate concentration of Lake Okaro, and this caused the following chain of events: a smaller standing crop of phytoplankton was produced; there was less attenuation of light near the surface ($K_d = 1.0\,\text{m}^{-1}$); there was less heating of the surface layer; and no thermal gradient developed in the surface layers. Mixing induced by wind therefore penetrated more readily to the 6 m deep seasonal thermocline. The dominant cyanobacterium in the lake at this time was a species of *Microcystis* that formed large spherical colonies, which were able to float back to the lake surface after deep mixing events. The mean floating velocities varied with time of day, from 10 to 58 m d^{-1}, two orders of magnitude faster than the *Anabaena minutissima* in Rotongaio (Walsby & McAllister, 1987). These *Microcystis* colonies showed the capacity to lose buoyancy in high irradiance. Under calm conditions the colonies were abundant at the lake surface in early morning but disappeared during the afternoon. In this way they avoided prolonged exposure to damaging high irradiance, and by sinking they may have gained access to nutrients below the thermocline.

This brief comparison of the two lakes illustrates that responses to light cannot be considered in isolation of other factors. There is an interesting rider to this. The *Microcystis* population in Lake Okaro was later replaced by a population of *Anabaena spiroides*, which produced a larger biomass than *Microcystis* and, as a consequence, steeper light attenuation and a more restricted epilimnion (Viner, 1989). The larger filaments, which floated faster (1.5 m d^{-1}) than those of *A. minutissima*, can be seen as adapted to this intermediate situation. Clearly, the conditions in a lake are subject to change, as are the inhabiting organisms.

The Baltic Sea

The Baltic Sea is the largest area of brackish water on Earth. Its salinity varies from about 0.1% in the Gulf of Bothnia, to about 1.3% approaching the Skagerak, where it joins the North Sea (salinity 3.3%). In summer, much of the southern basin of the Baltic Sea develops waterblooms of gas-vacuolate cyanobacteria, which are present in such high concentrations that they are visible in satellite photographs. The principal cyanobacteria are species of *Nodularia*, *Aphanizomenon* and *Anabaena* (Walsby, Hayes & Boje, 1995). These organisms form colonies or tangled aggregates that float rapidly (at 20–35 m d^{-1}) to the water surface when calm conditions allow. Because of

(a)

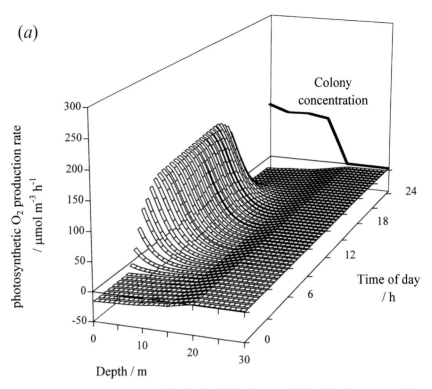

Fig. 5. Changes in photosynthesis with depth and time by a population of *Aphanizomenon flos-aquae* in the Baltic Sea and the relative filament concentration (last profile, black ribbon) in early August 1993: (a) on 1 August, the *Aphanizomenon* colonies have been mixed down and the productivity of the population is low, $3.4\,\mathrm{mmol\ m^{-2}\,d^{-1}}$. From data in Walsby *et al.* (1997).

the large size of the water mass, even moderate winds cause mixing to depths of several metres and even relatively short-lived gales can mix the epilimnion to the thermocline at 20–25 m, well below the limit of the euphotic zone at about 13 m ($K_d = 0.35\ \mathrm{m^{-1}}$). The cyanobacteria, like other phytoplankton, are dispersed throughout the epilimnion during wind-mixing, but they rapidly float up to exploit light in the upper euphotic zone during calm periods. The extent of the benefits brought by buoyancy during these episodes has been determined by calculating the daily integral of photosynthesis by the population (as described above) from continuous measurements made over a 9-day period in which a population was tracked with a drogue (Walsby *et al.*, 1997).

The benefits are graphically illustrated in Fig. 5, which shows plots of photosynthesis versus depth and time for an *Aphanizomenon flos-aquae* population on two days. In the first, the organism had been mixed down to a depth of 18 m by a storm; in the second, a few days later, the population

(*b*)

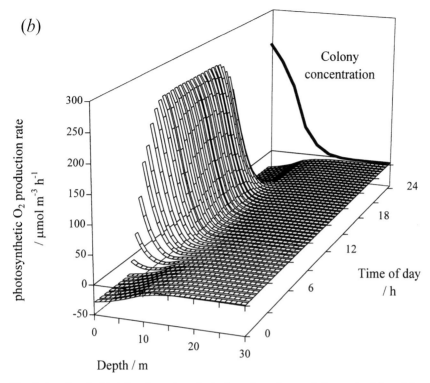

Fig. 5. (*continued*). (*b*) 4 days later the colonies have floated up and the productivity of the population has increased to 9.5 mmol m⁻² d⁻¹. The two data sets are standardized for the same daily insolation, light attenuation and colony concentration. From data in Walsby *et al.* (1997).

had become redistributed towards the surface and the mean depth of the population had been halved. The result of this buoyancy-dependent movement was a 2.8-fold increase in photosynthesis. Over the whole 9-day period of measurement it is estimated that the upward movements of the population provided by gas vesicles resulted in a two-fold increase in primary production. These benefits can be contrasted with the costs of making gas vesicles, less than 10% of the cost of making other cell components (Walsby, 1994).

Tropical oceans

The tropical oceans contain some of the clearest water on Earth. The reason for this is that, owing to the warmth and high solar irradiance, they are permanently stratified and they lose much of their particulate matter by sedimentation. Among the material lost are many phytoplankton cells, which carry with them nutrients they have absorbed from the surface layers; the surface water is therefore oligotrophic and impoverished in phytoplankton that contribute to light absorption.

Transparent though the water is, the depth of the thermocline, at about 110 m, exceeds the euphotic depth, at 50–75 m, and there are still gains to be made in floating back towards the water surface when calm conditions allow. Several species of *Trichodesmium* do this. They produce some of the largest and fastest floating colonies of any planktonic cyanobacteria, rising by as much as 200 m d^{-1} (Walsby, 1978). These velocities enable the colonies to float back up after profound mixing by tropical storms. As indicated above, their gas vesicles are specially adapted to withstand the pressures at great depths.

Absence from temperate oceans

The range of adaptability indicated by the presence of *Trichodesmium* at the extremes of depth in the tropical oceans suggests that gas-vacuolate cyanobacteria can evolve in most types of water body. Apart from remnants of *Trichodesmium* populations that sometimes stray up from lower latitudes, however, gas-vacuolate cyanobacteria are absent from temperate seas. The surface layers of these seas are so often mixed throughout the euphotic zone that there is little opportunity for buoyancy to benefit the organisms. In winter, surface cooling results in deep mixing, which redistributes nutrients throughout the water column. The greater nutrient loading supports much higher crops of phytoplankton in the summer and this attenuates the euphotic zone.

MECHANISMS OF BUOYANCY REGULATION IN RESPONSE TO LIGHT

Much of the foregoing relates to the consequences of gas vesicle production and buoyancy regulation for light absorption. Here we consider the mechanisms of buoyancy regulation in more detail and speculate on how these responses developed. A more detailed review of this subject is given by Oliver (1994).

Three methods of light-mediated buoyancy regulation have been described: the accumulation of carbohydrate; the collapse of gas vesicles by turgor pressure; and the regulation of gas vesicle production. These three responses have broadly the same effect of decreasing buoyancy at high irradiance but their kinetics may differ (Fig. 6).

Carbohydrate accumulation

A method of buoyancy regulation that occurs in all the planktonic cyanobacteria involves the accumulation of dense carbohydrate, which provides ballast that offsets the buoyancy provided by gas vesicles (Oliver & Walsby, 1984; Utkilen, Oliver & Walsby, 1985; Thomas & Walsby, 1985; Kromkamp, Konopka & Mur, 1986; Konopka, Kromkamp & Mur, 1987*a*). In high

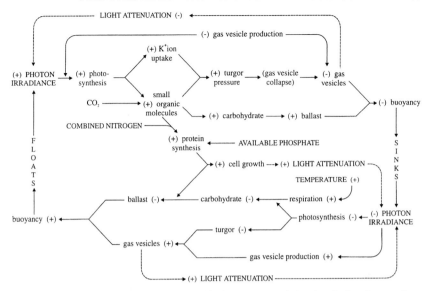

Fig. 6. A summary of the processes involved in buoyancy regulation by planktonic cyanobacteria. The signs indicate whether the following factor shows an increase (+) or decrease (−) in response to the change (+ or −) in the previous factor. Redrawn from Walsby (1988).

irradiance excess photosynthate is stored as a branched polymer of glucose, which resembles glycogen; the carbohydrate content of the cells increases from about 30% to as much as 60% of the dry mass. Polyglucoses have a specific gravity (relative to water) of about 1.6 and without a matching accumulation of gas vesicles this production causes the cells to sink. A return to lower irradiances results in a gradual loss of this carbohydrate reserve as the material is either respired or converted into the carbon skeletons of amino acids that form protein, which has a lower specific gravity of about 1.3. In the natural diel light–dark cycle, carbohydrates typically exhibit this diurnal increase and nocturnal decrease in all cyanobacteria; in planktonic cyanobacteria that possess gas vesicles these diel changes may give rise to an oscillation in sinking and floating movements.

Gas vesicle collapse by turgor pressure

In most bacteria and plants the difference in water potential between the outside and inside of the cell gives rise to a hydrostatic pressure, known as turgor pressure, inside cells. This pressure acts outwards on the cell wall, which is stretched, and inwards on the gas vesicles, compressing them (Walsby, 1971). In cyanobacteria from freshwater habitats the turgor pressure ranges from 0.1–0.6 MPa (1–6 bar) and is usually within the range of 0.2–0.4 MPa (2–4 bar); in some species a rise of turgor pressure may be sufficient to cause collapse of some of the gas vesicles. Because the p_c

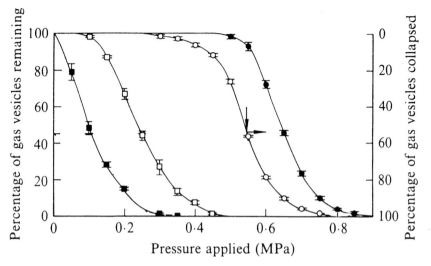

Pressure applied (MPa)

Fig. 7. Critical pressure distributions of gas vesicles in *Anabaena flos-aquae*: (■, □) apparent critical pressure, p_a, of cells suspended in water; (●, ○) true critical pressure, p_c, of cells suspended in hypertonic sucrose solution to remove turgor pressure. Open symbols, in cells grown at low irradiance; filled symbols, after 16 h exposure to a photon irradiance of 135 μmol $m^{-2} s^{-1}$. The change in p_c indicates the loss of the weaker gas vesicles, 56% (horizontal arrow) of which would have been collapsed by the turgor pressure of 0.55 MPa (vertical arrow) at 16 h. From Kinsman, Ibelings & Walsby (1991).

distribution of gas vesicles shows considerable variation (Fig. 7) the effect of turgor rise is progressive and rarely results in collapse of more than half of the gas vesicles (see, for example, Kinsman, Ibelings & Walsby, 1991).

Turgor pressure may rise when the irradiance increases, for one of two reasons. First, at limiting irradiances, the rate of photosynthesis increases and this may sustain a higher concentration of photosynthetic products such as sugars and organic acids. Second, the rate of potassium ion uptake may increase. In *Anabaena flos-aquae* these two processes contribute to a turgor pressure rise of about 0.1 MPa (1 bar) within an hour. In this organism turgor pressure rise can cause collapse of over 50% of the gas vesicles and result in loss of buoyancy (Oliver & Walsby, 1984; Kinsman, Ibelings & Walsby, 1991).

The regulation of gas vesicle collapse is possible only in those cyanobacteria that have relatively weak gas vesicles. In cyanobacteria like the species of *Oscillatoria* from deep lakes, the highest turgor pressure observed, about 0.5 MPa (5 bar), would be insufficient by itself to cause collapse of even the weakest gas vesicles.

Regulation of gas vesicle production

There are two ways in which the rate of gas vesicle production might be affected by irradiance, by the energy requirement of the process or by an evolved behavioural response. Photoautotrophy provides the principal or only source of energy for protein synthesis in cyanobacteria and the continued production of gas vesicles is therefore dependent on light. At low, limiting irradiances the rate of gas vesicle production is dependent on the irradiance (Kromkamp, van den Heuvel & Mur, 1989; Deacon & Walsby, 1990). At higher irradiances, however, the rate of gas vesicle production falls below the cell growth rate in *Aphanizomenon flos-aquae* (Kromkamp, Konopka & Mur, 1986, 1988; Konopka, Kromkamp & Mur, 1987a), although in various strains of *Microcystis* the relative gas vesicle content showed little change (Thomas & Walsby, 1985; Konopka, Kromkamp & Mur, 1987b; Kromkamp, Konopka & Mur, 1988).

Extrapolating from knowledge of the regulation of many other processes in microbiology, it might be expected that the regulation of gas vesicle gene expression would play a central role in the regulation of gas vesicle production, and thence in buoyancy regulation, by light. At present, however, there is little direct evidence on its role.

In *Microcystis* sp. the loss of buoyancy in high irradiance was effected entirely by changes in carbohydrate, and there was no change in the gas vesicle content of the cells (Thomas & Walsby, 1985). In *Anabaena flos-aquae* a rise in carbohydrate and a loss of gas vesicles by turgor rise both contribute to buoyancy loss (Oliver & Walsby, 1984; Kinsman *et al.*, 1991); it was not possible to determine if additional changes in gas vesicle content were caused by changes in gas vesicle production rate. In *Oscillatoria agardhii*, however, where gas vesicles were too strong to be collapsed by turgor, it was possible to demonstrate a cessation in gas vesicle formation at higher irradiances in a manner that partly contributes to light-mediated regulation of buoyancy (Utkilen, Oliver & Walsby, 1985); it seems likely that these changes could be explained by the control of gene expression in response to light, but this needs investigating.

To date, the only demonstration of the regulation of gas vesicle gene expression by light is in the filamentous cyanobacterium *Pseudanabaena* sp. It produces gas vesicles only under low irradiance, and then only in small numbers at the ends of the cells. It is not known if sufficient gas vesicles are produced to provide this organism with positive buoyancy and the consequences of this control for buoyancy regulation are unknown. Damerval *et al.* (1991) showed by Northern hybridization that, when *Pseudanabaena* sp. was transferred from low irradiance (5 μmol m^{-2} s^{-1}) to higher irradiance (50 μmol m^{-2} s^{-1}), the production of the single 0.4 kb transcript of *gvpA* gene ceased. Expression was resumed on returning the filaments to low irradiance. Damerval *et al.* (1991) found evidence of only a single *gvpA* in this organism; the failure of the *gvpC* gene from *Calothrix* to hybridize with genomic DNA of *Pseudanabaena* was interpreted as indicating that the gas vesicles of this organism contained only GvpA. It is possible, however, that *Pseudanabaena* has a *gvpC* but, like the *gvpCs* of *Oscillatoria* spp. (Griffiths *et al.*, 1992), it possesses insufficient homology to allow hybridization with the *Calothrix* gene. The genetic control of gas vesicle expression requires investigation of all of the six *gvp* genes now identified in cyanobacteria (Kinsman & Hayes, 1997).

Determining the controlling mechanism

Each of the above mechanisms may make a contribution to the loss of buoyancy that occurs at high irradiance. Both the increase in carbohydrate content and turgor pressure rise are inevitable consequences of increased photosynthetic activity and have been observed in a number of cyanobacteria. Oliver & Walsby (1984) devised a procedure, the ballast balance sheet, for determining the mechanism that caused the buoyancy loss. They measured the mass of the principal cell components and calculated the mass of water that each displaced. The excess mass of each component contributes to the ballast of the cell: carbohydrate, protein and lipid contribute (in descending order) to the positive ballast; only the gas vesicles, which have a buoyant density of between 10 and 20% of that of water (Walsby & Bleything, 1988), provide a negative ballast. By measuring the changes in ballast contributed by each of the components the cause of the overall change in buoyant density of the cells can be determined. This procedure has been used to determine the principal component responsible for buoyancy loss in species of *Anabaena* (Oliver & Walsby, 1984), *Aphanizomenon* (Kromkamp *et al.*, 1986), *Microcystis* (Thomas & Walsby, 1985) and *Oscillatoria* (Utkilen *et al.*, 1985).

In most cases the ballast balance sheet has been used to determine the ballast changes immediately before and at some time after the increase in irradiance. In order to resolve a controversy over whether turgor rise or carbohydrate increase caused buoyancy loss in *Anabaena flos-aquae*, Kinsman *et al.* (1991) made a time course of such measurements. They showed that, after 16 h at high irradiance, the rise in turgor pressure had caused collapse of 56% of the gas vesicles, much more than required to cause buoyancy loss. Nevertheless, 80% of the filaments had already lost buoyancy within 4.5 h of the increase in irradiance; all of this early buoyancy loss was explained by the rise in carbohydrate (Fig. 8).

Limitations of regulating buoyancy loss by gas vesicle expression

It is premature to speculate on what studies of *gvp* expression will reveal but I would like to conclude by considering the limitations of controlling buoyancy by regulating gas vesicle gene expression.

In general, the regulation of a metabolic activity requires turnover of the participating proteins. The concentration of a particular protein is then

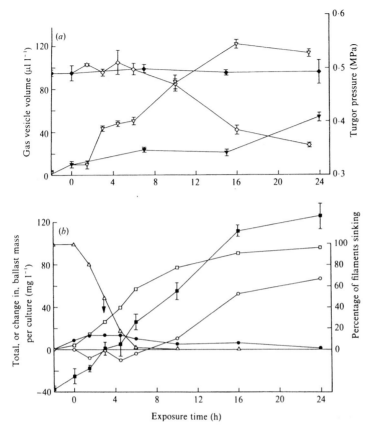

Fig. 8. The time course of changes in cells of *Anabaena flos-aquae* after an increase in photon irradiance from 5 to 135 μmol m⁻² s⁻¹. (a) Gas vesicles (◇) are collapsed after 6 h as the turgor pressure (▽) exceeds 0.4 MPa; after 16 hours 56% of the gas vesicles are collapsed. In low irradiance, gas vesicle volume (◆) and turgor pressure (▼) show less change. (b) Almost complete loss of buoyancy (△ occurs within 6 h, before gas vesicles are lost. At 6 hour the net ballast (■) of the cells changes from negative to positive: the early increase is due to carbohydrate ballast (□); protein ballast (●) shows little change; the increase in ballast due to gas vesicle loss (○) does not occur until after 6 h. From Kinsman, Ibelings & Walsby (1991).

determined by the relative rates of its synthesis and degradation; the rate of change will depend on the rapidity of each of these processes. For a protein that accounts for only 0.03% of the total protein (the average amount expressed by each of 3000 genes) the losses involved in the turnover process can be sustained.

For gas vesicle proteins, accounting for a large proportion of the cell protein, however, the cost of rapid turnover would be unsustainable. Consider the consequences of modulating *gvp* gene expression in the absence of gas vesicle collapse. At steady state growth conditions, under which gas vesicles account for 10% of the cell protein (as in *Anabaena flos-aquae*, Hayes & Walsby, 1984), the maintenance of gas vesicle content consumes 10% of the investment in protein, and perhaps 7% of the total organic investment (see values given by Oliver & Walsby, 1984). Without gas vesicle collapse, it will take a minimum of one cell generation time for the relative gas vesicle content of the cells to fall to half the original level. For an organism attempting to respond to the changing light regime in a lake this response is too slow: the minimum doubling time is 20 h in *Anabaena flos-aquae* (Shear & Walsby, 1975) and the value observed in lakes is commonly 3 days or more (Reynolds, 1997). The regulation of *gvp* expression cannot, therefore, by itself account for the buoyancy losses that have been observed to occur within 1–4 hours in several cyanobacteria. Isotopic labelling experiments in *Anabaena flos-aquae* indicate that there is negligible turnover of the gas vesicle protein in gas vesicles as long as the structures remain intact (Hayes & Walsby, 1984).

Although inhibition of gas vesicle gene expression at high irradiances would not provide a fast enough change in buoyancy, increased expression at low irradiances may be of value. The recovery of buoyancy of a cyanobacterium that has sunk below the euphotic zone may be of paramount importance to its survival. For this reason the cyanobacterium should, perhaps, devote more of its dwindling resources to the production of more gas vesicles.

Without such a response, the role of the gas vesicle becomes a static one, providing 'background' buoyancy which, in ballooning jargon, can be trimmed to allow floating or sinking by the disposal or accumulation of ballast. Further investigations of the regulation of gas vesicle production by planktonic cyanobacteria are now needed and, with the discovery of the new gas vesicle genes, are now more feasible.

ACKNOWLEDGEMENTS

I am very grateful to Dr Paul Hayes for many discussions of this subject. These studies are supported by grants from the Natural Environment Research Council and the Environment Programme of the European Union.

REFERENCES

Baker, A. L. & Brook, A. J. (1969). Optical density profiles as an aid to the study of microstratified phytoplankton populations in lakes. *Archiv für Hydrobiologie*, **69**, 214–33.

Blaurock, A. E. & Walsby, A. E. (1976). Crystalline structure of the gas vesicle wall from *Anabaena flos-aquae*. *Journal of Molecular Biology*, **105**, 183–99.

Buchholz, B. E. E., Hayes, P. K. & Walsby, A. E. (1993). The distribution of the outer gas vesicle protein, GvpC, on the *Anabaena* gas vesicle, and its ratio to GvpA. *Journal of General Microbiology*, **139**, 2352–63.

Damerval, T., Houmard, J., Guglielmi, G., Csiszàr, K. & Tandeau De Marsac, N.

(1987). A developmentally regulated *gvpABC* operon is involved in the formation of gas vesicles in the cyanobacterium *Calothrix* 7601. *Gene*, **54**, 83–97.

Damerval, T., Castets, A-M., Houmard, J. & Tandeau de Marsac, N. (1991). Gas vesicle synthesis in the cyanobacterium *Pseudanabaena* sp.: occurrence of a single photoregulated gene. *Molecular Microbiology*, **5**, 657–64.

Deacon, C. & Walsby, A. E. (1990). Gas vesicle formation in the dark, and in light of different irradiances, by the cyanobacterium *Microcystis* sp. *British Phycological Journal*, **25**, 133–9.

Ganf, G. G., Oliver, R. L. & Walsby, A. E. (1989). Optical properties of gas-vacuolate cells and colonies of *Microcystis* in relation to light attenuation in a turbid, stratified reservoir (Mount Bold Reservoir, South Australia). *Australian Journal of Marine and Freshwater Research*, **40**, 595–611.

Gantt, E., Ohki, K. & Fujita, Y. (1984). *Trichodesmium thiebautii*; structure of a nitrogen-fixing marine blue-green alga (*Cyanophyta*). *Protoplasma*, **119**, 188–96.

Griffiths, A. E., Walsby, A. E. & Hayes, P. K. (1992). The homologies of gas vesicle proteins. *Journal of General Microbiology*, **138**, 1243–50.

Hayes, P. K. & Powell, R. S. (1995). The *gvpA/C* cluster of *Anabaena flos-aquae* has multiple copies of a gene encoding GvpA. *Archives of Microbiology*, **164**, 50–57.

Hayes, P. K. & Walsby, A. E. (1984). An investigation into the recycling of gas vesicle protein derived from collapsed gas vesicles. *Journal of General Microbiology*, **130**, 1591–6.

Hayes, P. K. & Walsby, A. E. (1986). The inverse correlation between width and strength of gas vesicles in cyanobacteria. *British Phycological Journal*, **21**, 191–7.

Hayes, P. K., Walsby, A. E. & Walker, J. E. (1986). Complete amino acid sequence of cyanobacterial gas-vesicle protein indicates a 70-residue molecule that corresponds in size to the crystallographic unit cell. *Biochemical Journal*, **236**, 31–6.

Hayes, P. K., Lazarus, C. M., Bees, A., Walker, J. E. & Walsby, A. E. (1988). The protein encoded by *gvpC* is a minor component of gas vesicles isolated from the cyanobacteria *Anabaena flos-aquae* and *Microcystis* sp. *Molecular Microbiology*, **2**, 545–52.

Hayes, P. K., Buchholz, B. & Walsby, A. E. (1992). Gas vesicles are strengthened by the outer-surface protein, GvpC. *Archives of Microbiology*, **157**, 229–34.

Horne, M., Englert, C., Wimmer, C. & Pfeifer, F. (1991). A DNA region of 9 kbp contains all genes necessary for gas vesicle synthesis in halophilic archaebacteria. *Molecular Microbiology*, **5**, 1159–74.

Howard, A., Irish, A. & Reynolds, C. S. (1996). A new simulation of cyanobacterial underwater movement (SCUM'96). *Journal of Plankton Research*, **18**, 1375–85.

Jones, J. G., Young, D. C. & DasSarma, S. (1991). Structure and organization of the gas vesicle gene cluster on the *Halobacterium halobium* plasmid pNCR100. *Gene*, **102**, 1017–22.

Kinsman, R. & Hayes, P. K. (1997). Genes encoding proteins homologous to halobacterial Gvps N, J, K, F, & L are located downstream of *gvpC* in the cyanobacterium *Anabaena flos-aquae*. *DNA Sequence – The Journal of Sequencing and Mapping*, **7**, 97–106.

Kinsman, R., Ibelings, B. W. & Walsby, A. E. (1991). Gas vesicle collapse by turgor pressure and its role in buoyancy regulation by *Anabaena flos-aquae*. *Journal of General Microbiology*, **137**, 1171–8.

Kirk, J. T. O. (1994). *Light and Photosynthesis in Aquatic Ecosystems*. Cambridge: Cambridge University Press.

Klemer, A. R. (1976). The vertical distribution of *Oscillatoria agardhii* var. *isothrix*. *Archives für Hydrobiologie*, **78**, 343–62.

92 A. E. WALSBY

Konopka, A. (1982). Buoyancy regulation and vertical migration by *Oscillatoria rubescens* in Crooked Lake, Indiana. *British Phycological Journal*, **17**, 427–42.

Konopka, A., Kromkamp, J. & Mur, L. R. (1987*a*). Regulation of gas vesicle content and buoyancy in light- or phosphate-limited cultures of *Aphanizomenon flos-aquae* (Cyanophyta). *Journal of Phycology*, **23**, 70–8.

Konopka, A., Kromkamp, J. C. & Mur, L. R. (1987*b*). Buoyancy regulation in phosphate-limited cultures of *Microcystis aeruginosa*. *FEMS Microbiology Ecology*, **45**, 135–42.

Kromkamp, J. & Walsby, A. E. (1990). A computer model of buoyancy and vertical migration in cyanobacteria. *Journal of Plankton Research*, **12**, 161–83.

Kromkamp, J., Konopka, A. & Mur, L. R. (1986). Buoyancy regulation in a strain of *Aphanizomenon flos-aquae* (Cyanophyceae): the importance of carbohydrate accumulation and gas vesicle collapse. *Journal of General Microbiology*, **132**, 2113–21.

Kromkamp, J., Konopka, A. & Mur, L. R. (1988). Buoyancy regulation in light-limited continuous cultures of *Microcystis aeruginosa*. *Journal of Plankton Research*, **10**, 171–83.

Kromkamp, J., van den Heuvel, A. & Mur, L. R. (1989). Formation of gas vesicles in phosphorus-limited cultures of *Microcystis aeruginosa*. *Journal of General Microbiology*, **135**, 1933–9.

Meffert, M-E. (1971). Cultivation and growth of two planktonic *Oscillatoria* species. *Mitt. Internat. Verein Limnol.* **19**, 189–205.

Ogawa, T., Sekine, T. & Aiba, S. (1979). Reappraisal of the so-called light shielding of gas vacuoles in *Microcystis aeruginosa*. *Archives of Microbiology*, **122**, 57–60.

Oliver, R. L. (1994). Floating and sinking in gas-vacuolate cyanobacteria. *Journal of Phycology*, **30**, 161–73.

Oliver, R. L. & Walsby, A. E. (1984). Direct evidence for the role of light-mediated gas vesicle collapse in the buoyancy regulation of *Anabaena flos-aquae* (cyanobacteria). *Limnology and Oceanography*, **29**, 879–86.

Porter, J. & Jost, M. (1976). Physiological effects of the presence and absence of gas vacuoles in the blue-green alga, *Microcystis aeruginosa* Kuetz. emend. Elenkin. *Archives of Microbiology*, **110**, 225–31.

Reynolds, C. S. (1984). *The Ecology of Freshwater Phytoplankton*. Cambridge: Cambridge University Press.

Reynolds, C. S. (1987). Cyanobacterial water-blooms. *Advances in Botanical Research*, **13**, 67–143.

Reynolds, C. S. (1997). *Vegetation Processes in the Pelagic: A Model for Ecosystem Theory*, 371 pp. Oldendorf: Ecology Institute.

Shear, H. & Walsby, A. E. (1975). An investigation into the possible light-shielding role of gas vacuoles in a planktonic blue-green alga. *British Phycological Journal*, **10**, 241–51.

Spigel, R. H. & Imberger, J. (1987). Mixing processes relevant to phytoplankton dynamics in lakes. *New Zealand Journal of Marine and Freshwater Research*, **21**, 361–77.

Tandeau De Marsac, N., Mazel, D., Bryant, D. A., Houmard, J. (1985). Molecular cloning and nucleotide sequence of a developmentally regulated gene from the cyanobacterium *Calothrix* PCC 7601, a gas vesicle protein gene. *Nucleic Acids Research*, **13**, 7223–36.

Thomas, E. A. & Märki, E. (1949). Der heutige Zustand des Zürichsees. Mitteilungen der internationalen Vereinigung für theoretisch und angewandte Limnologie, **10**, 476–88.

Thomas, R. H. & Walsby, A. E. (1985). Buoyancy regulation in a strain of *Microcystis*. *Journal of General Microbiology*, **131**, 799–809.

Utkilen, H. C., Oliver, R. L. & Walsby, A. E. (1985). Buoyancy regulation in a red *Oscillatoria* unable to collapse gas vacuoles by turgor pressure. *Archiv für Hydrobiologie*, **102**, 319–29.

Van Liere, L. & Walsby, A. E. (1982). Interactions of cyanobacteria with light, pp. 9–45. In *The Biology of Cyanobacteria*, eds N. G. Carr & B. A. Whitton. Oxford: Blackwell Scientific Publications.

Vincent, W. F. (1989). Cyanobacterial growth and dominance in two eutrophic lakes: review and synthesis. *Archiv für Hydrobiologie, Ergebnisse der Limnologie*, **32**, 239–54.

Viner, A. B. (1989). Buoyancy and vertical distribution of *Anabaena spiroides* in Lake Okaro (New Zealand). *Archiv für Hydrobiologie, Ergebnisse der Limnologie*, **32**, 221–38.

Waaland, J. R., Waaland, S. D. & Branton, D. (1971). Gas vacuoles. Light shielding in a blue-green alga. *Journal of Cell Biology*, **48**, 212–15.

Walsby, A. E. (1971). The pressure relationships of gas vacuoles. *Proceedings of the Royal Society of London*, B, **178**, 301–26.

Walsby, A. E. (1978). The properties and buoyancy-providing role of gas vacuoles in *Trichodesmium* Ehrenberg. *British Phycological Journal*, **13**, 103–16.

Walsby, A. E. (1988). Homeostasis in buoyancy regulation by planktonic cyanobacteria. In *Homeostatic Mechanisms in Micro-organisms*, ed. R. Whittenbury, G. W. Gould, J. G. Banks & R. G. Board. *FEMS Symposium No.* **44**, pp. 99–116. Bath: Bath University Press.

Walsby, A. E. (1991). The mechanical properties of the *Microcystis* gas vesicle. *Journal of General Microbiology*, **137**, 2401–8.

Walsby, A. E. (1994). Gas vesicles. *Microbiological Reviews*, **58**, 94–114.

Walsby, A. E. (1997). Numerical integration of phytoplankton photosynthesis through time and depth in a water column. *New Phytologist*, **136**, 189–209.

Walsby, A. E. & Bleything, A. (1988). The dimensions of cyanobacterial gas vesicles in relation to their efficiency in providing buoyancy and withstanding pressure. *Journal of General Microbiology*, **134**, 2635–45.

Walsby, A. E. & Hayes, P. K. (1988). The minor cyanobacterial gas vesicle protein, GVPc, is attached to the outer surface of the gas vesicle. *Journal of General Microbiology*, **134**, 2647–57.

Walsby, A. E. & McAllister, G. K. (1987). Buoyancy regulation by *Microcystis* in Lake Okaro. *New Zealand Journal of Marine and Freshwater Research*, **21**, 521–4.

Walsby, A. E., Hayes, P. K. & Boje, R. (1995). The gas vesicles, buoyancy and vertical distribution of cyanobacteria in the Baltic Sea. *European Journal of Phycology*, **30**, 87–94.

Walsby, A. E., Utkilen, H. C. & Johnsen, I. J. (1983). Buoyancy changes of a red coloured *Oscillatoria agardhii* in Lake Gjersjøen, Norway. *Archiv für Hydrobiologie*, **97**, 18–38.

Walsby, A. E., Reynolds, C. S., Oliver, R. L. & Kromkamp, J. (1989). The role of gas vacuoles and carbohydrate content in the buoyancy and vertical distribution of *Anabaena minutissima* in Lake Rotongaio, New Zealand. *Archiv für Hydrobiologie, Ergebnisse der Limnologie*, **32**, 1–25.

Walsby, A. E., Hayes, P. K., Boje, R. & Stal, L. J. (1997). The selective advantage of buoyancy provided by gas vesicles for planktonic cyanobacteria in the Baltic Sea. *New Phytologist*, **136**, 407–17.

Worcester, D. L. (1975). Neutron diffraction studies of biological membrane components. In *Neutron Scattering for the Analysis of Biological Structures. Brookhaven Symposia in Biology*, **27**, III, pp. 37–57.

HOW MICROALGAE SEE THE LIGHT

PETER HEGEMANN[1] AND HARTMANN HARZ[2]

[1]*Institut für Biochemie 1, Universitätsstr. 31, 93040 Regensburg, Germany*
[2]*Botanisches Institut der Ludwig-Maximilians, Universität München, Menzinger Straße 67, 80638 München, Germany*

INTRODUCTION

Eyes are important tools for orientation for most higher animals. Unicellular flagellate algae, such as *Chlamydomonas*, also possess eyes, which allow these plant species to access environments which are optimal for photosynthetic growth. These eyes are only 1 μm in diameter. Their mirror optic (Fig. 1) modulates the colour, the amplitude and the polarization of the light according to the angle of light incidence. The eyes enable the algae not only to detect the intensity of the ambient light but also the direction of its origin. Since the cells are more or less transparent, with a total optical density of only 0.2 OD at 500 nm (Beckmann & Hegemann, unpublished observations), they developed this optical system to enhance light coming from one side of the cell striking the eyespot directly and to attenuate the light coming from the other side. The intact eye is composed of the optical system together with the rhodopsin photoreceptor (Deininger *et al.*, 1995) and other biochemical components which are responsible for signal transduction and adaptation. The eyes of chlorophycean algae, for example, *Chlamydomonas*, appear as orange spots when observed using bright field microscopy. The eyespot reflects the light and changes its polarization plane. Thus, they appear as bright spots on dark background under epi-illumination or in cross-polarized transmitted light.

When studying any organism which possesses eyes, it becomes immediately clear that the intensity detected by the eye has something to do with the orientation of the organism respective to the light source. The strength of the perceived light signal depends on the angle of light incidence, the efficiency of the optical system and its spectral properties. This is also true for flagellate algae. Since green flagellates are not able to place many photoreceptor cells in an array – especially if they are unicellular themselves – the construction of algal eyes must be different from eyes found in higher animals or in humans. The function of these eyes is based on certain physical principles such as reflection, interference, and polarization. The optical principles of these eyes were first brought to a wider audience by Foster and Smyth (1980). Algal eyes, which operate on the basis of reflection and interference, are not as unusual as one might assume. There are also some remarkable examples of

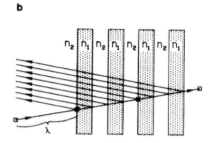

Fig. 1 a. Thin section through the eyespot of *Chlamydomonas reinhardtii* showing four layers of pigmented globules. The layers show a slightly negative curvature (taken from Foster & Smyth, 1980). b. Multilayer reflection as realized in the eyes of unicellular flagellates. Constructive interference occurs when the optical path of both high and low refractive index layers (n_1 and n_2) are close to $\lambda/4$.

animal eyes which can be explained using the principles of mirror optics. Among them are the simple eyes of scallops and the compound eyes of shrimps, crayfish and lobsters (Land, 1972).

The strength of the perceived light signal and its variation during body rotation are responsible for the orientation of the flagellate. In this chapter aspects of light modulation during cell rotation are discussed which are, in principle, well known but have never been surveyed in such a context. First, the angle of light incidence was calculated for a rotating cell swimming at different angles relative to the light source. Next, the light reflectance was determined during rotation by treating the optical apparatus as a quarter wave plate. Finally, both results were combined. A simplified plot was generated, which illustrates how the amount of reflected light at the site of the rhodopsin photoreceptor changes during cell rotation. However, it should always be kept in mind that a flagellate eye never constitutes an even nearly ideal reflector due to its small size, which is in the range of the reflected light itself, and, in addition, biological membranes and organelles of sizes below 1 μm are never ideally shaped in a physical sense.

THE ANGLE OF LIGHT INCIDENCE

When the tracking direction of a unicellular alga is identical with the axis of rotation and perpendicular to the direction of light incidence, the change of

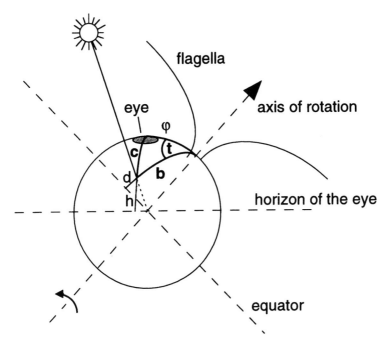

Fig. 2. The spheric triangle (b,c,φ) of a navigating cell connects the system of algal co-ordinates with those of the horizon and the light source. φ is the scan angle, i.e. the angle between tracking direction and the position of the eye. b is the tracking angle, the angle between tracking direction and direction of the light source; d is the declination of the light source and defines the angular distance of the light source from the equator. h is the height of the light source and defines the angle of the light incidence onto the eye. c is its complementary angle to 90°. t is the angle between the projection of the light source, the pole, and the eye. t varies continuously with the rotation of the cell. Note that the eye advances the flagellar beating plane by 30° to 40°.

the angle of light incidence on to the eyespot during rotation can be described by a simple sinusoidal function. Yet, when the tracking direction deviates from the direction of light incidence and is not perpendicular, the modulation of the light incidence is less obvious. To illustrate this, a general situation will be considered in which a cell with a non-equatorial eyespot swims at any given angle between the tracking direction and the light incidence. The situation may be compared to a human being standing at a given place on earth, observing the height of the sun. The relation between the axis of rotation (tracking direction), the position of the eye and the direction of light incidence can be described by spherical angles defined at the centre of the cell (see Fig. 2). Of major interest is the angle of light incidence, h, between the horizon of the eye and a distant light source. It can be described by means of spherical geometry:

$$\cos c = \sin h = \cos \varphi \times \cos b + \sin \varphi \times \sin b \times \cos \omega t \qquad (1)$$

φ is the scan angle, the angle between tracking direction and the position of the eye. b is the tracking angle, the angle between tracking direction and direction of the light source, t is the angle between the projection of the light source, the pole, and the eye. t varies continuously with the rotation of the cell. Equation 1 now describes the variation of the light incidence during rotation. The angle h originally defined at the centre of the cell is the same for the height of the light source at the eyespot surface as long as the diameter of the cell is small compared to the distance between cell and light source. The equation is only appropriate for non-helical movement. Helical swimming is minimal at low light and low Ca^{2+} concentrations (Hegemann & Bruck, 1989). Three special conditions will now be considered.

(i) $\varphi = 90°, 0° < b < 90°$, for algae with equatorial eyes swimming towards the light. Equation 1 can now be simplified to: $\sin h = \sin b \times \cos \omega t$.

(ii) $\varphi = 90°$, b = 90°, for algae with equatorial eyes swimming perpendicular to the light. The equation can be further simplified to: $\sin h = \cos \omega t$.

(iii) $\varphi = 90°$, b = 0°, for algae with equatorial eyes swimming towards the light. Here, $\sin h$ equals zero. The angle of light incidence does not change during rotation and the light source is always seen at the horizon: h = 0.

It is true for all algae with eyes located between the equator and the flagellar base that the closer the cells swim to the light the smaller is the modulation amplitude. In algae with eyes located close to the flagella, the maximal modulation amplitude is lower than in cells with equatorial eyes, but they collect more light when they swim directly towards the light source. In contrast, species with equatorial eyes receive a signal with a higher modulation amplitude, especially when they swim perpendicular to the light. Thus, they are good at hunting the light at low light levels, whereas species with eyespots close to the flagellar base are good trackers when they are already swimming towards the light source.

OPTICAL DESIGN OF THE EYE

The ideal situation described above is far from the real situation in a living alga because the modulation of the light incidence alone does not render the modulation of the perceived light signals. Although the eyespot is only 1 μm in size, it is large compared with the whole organism. Thus, the angle of light incidence is not constant at all places of the eye surface. The function of the optical system is to increase the amount of light falling onto the photo-receptor from the side where the eyespot is facing and to reduce absorption of light coming from the other side. However, the front-to-back contrast of the cell is not infinite. Light coming from the back side of the cell is also absorbed to some extent by the photoreceptor. Below, elementary optical

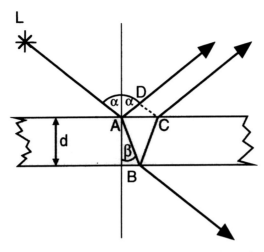

Fig. 3. Rise of interference in an 'ideal' thin film system. The scheme shows a plane parallel plate of the refractive index n and the thickness d. A light wave originating from the light source L hits the plate under an angle that differs from 90° by α. This wave is partially reflected at the point A, whereas the other part is refracted into the plate under the angle β. Part of the resulting light wave is reflected at point B, hits the surface at point C and leaves the plate at point C. Since both reflected waves result from the same original wave, they are coherent and may interfere if they cross each other somewhere. The geometrical path difference of the light waves is AB + BC − AD.

properties of algal eyes are considered for normal and non-normal light incidence in order to describe the modulation of the light signal at the site of the photoreceptor.

The eyespot structure consists of one or several alternating layers of high and low refractive index material, whose thickness is comparable to the wavelength of the visible light. Carotenoids packed into high density vesicles are responsible for the high refractive index. The most common form of eye is found in freshwater algae such as *Chlamydomonas moewusii* (formerly *eugametos*) and *Haematococcus*, or in gametes of marine algae such as *Acetabularia* and *Ulva*. Their eyes contain a single layer of pigmented granules which, in an idealized form, can be treated as a plane parallel plate (Fig. 3). In contrast, *Chlamydomonas reinhardtii* and *Hafniomonas reticulata* have eyespots containing up to seven globule layers (Foster & Smyth, 1980; Kreimer & Melkonian, 1990), which operate as a quarter wave stack (Fig. 1).

In a plane parallel plate (Fig. 3), the geometrical path difference between two light waves reflected at neighbouring low-to-high and high-to-low interfaces is AB + BC − AD. The optical path difference Δ is n(AB + BC) − AD or n(AB + BC) = 2dn/cosβ. The difference can be defined more generally:

$$\Delta = 2d\sqrt{n^2 - \sin^2\alpha} + \lambda/2 \qquad (2)$$

where n is the refractive index of the plate and $\lambda/2$ accounts for the phase shift that occurs when the light wave is reflected at the surface of a dense medium. Although light coming from a conventional light source is incoherent, the reflected waves interfere if the interfering light waves originate from a single-emission event. Crossing of the reflected light waves is further facilitated by the negative curvature of the vesicular layers found in most algal species (Kreimer & Melkonian, 1990). Enhanced reflection has been measured experimentally using confocal microscopy. Constructive interference occurs when the difference of the optical path of two reflected waves is n $\times$ λ (a multiple of λ) (Kreimer & Melkonian, 1990).

REFLECTION AND INTERFERENCE AT NORMAL LIGHT INCIDENCE

When considering the simplest case, in which the light meets a single layered reflector at normal light incidence, constructive interference occurs when the length of the optical path through one layer is $\lambda/4$ (quarter wave plate). A path difference of λ results when part of the light wave penetrates the layer and returns, and another part is reflected at the surface undergoing a phaseshift of $\lambda/2$. In a multi-layered stack, the optical path in all layers has to be $\lambda/4$ in order to make reflected waves from the different layers interfere. Consequently, the non-pigmented layers having low optical density need to be thicker than the pigmented ones. The maximum interference is observed under the light microscope at a distance of $\lambda/4$ away from the locus of the outermost reflecting surface when double-cross polarized light is used. Other maxima can be seen outside the cell. These maxima, however, are of lower intensity and sharpness due to the non-ideal conditions (Harz & Hegemann, unpublished observations). Foster and Smyth (1980) were already aware of these properties and thus were able to predict that the part of the plasma membrane which overlays the eyespot must be the locus of the photorecep-tor. Meanwhile, in *Chlamydomonas* the rhodopsin was, in fact, localized within the eyespot overlying membranes by immunofluorescence (Deininger *et al.*, 1995). However, it is still unclear whether the rhodopsin is located in the outer chloroplast membrane or in the plasma membrane.

There are two reasons for the increase of the total reflectance on a thin film system. First, the increase of the number of layers and secondly, the difference in refractive indices of adjacent layers, which are the carotenoid layers and the spaces in-between (Land, 1972). Originally, Foster and Smyth (1980) assumed values of 1.55 for the carotenoid vesicles and 1.30 for the spaces (close to water). Later, Foster and Saranak (1989) presented values of 1.9 and 1.35. With the indices of 1.55 and 1.33, five layers reflect around 50% of the light, whereas with indices of 1.9 and 1.35, 80% reflection can be achieved by the same stack. The total front-to-back contrast of a single cell at the site of the photoreceptor has been determined experimentally by comparing photo-currents in response to flashes applied directly onto the eye (h = 90°) with those in response to flashes hitting the eye from behind. The stimulus–response curves were shifted by a factor of three in *Haematococcus* and by a factor or eight in *Chlamydomonas* (Sineshchekov, 1991; Harz, Nonnengäßer & Hegemann, 1992). The high contrast in *Chlamydomonas* is consistent with the refractive indices of 1.9 and 1.35.

The spectral band width of the principal peak becomes smaller as the number of layers increase. The band enlarges, however, when the difference in the refractive indices increases. For a *Chlamydomonas* eye with five layers and refractive indices of 1.9 and 1.35, the spectral reflectance curve has a half-width of about 200 nm, if the eye is assumed to be an ideal reflector (Land, 1972). The reflector is not ideal when n × d differs in both layers. Then the band width continues to decrease (Kreimer & Melkonian, 1990) and a reflection maximum of second order appears at the half wavelength of the principal peak. For an algal reflector with a reflection maximum of approximately 500 nm, the second-order maximum would appear at a shorter wavelength close to the protein absorption band. Foster, Saranak and Zarrilli (1988) reported a second peak near 280 nm within the phototaxis action spectrum of cells with unpigmented eyes. To what extent the reflection maximum of second order contributes to the second peak of the action spectrum is, however, unclear. None the less, one should keep in mind that algal eyes are by no means ideal reflectors, so that all these considerations are at best approximations of real optics.

AMPLITUDE AND COLOUR MODULATION DURING CELL ROTATION

As previously mentioned, a rotating algal cell only rarely detects perpendicular light incidence, whereas oblique light incidence is the common situation. Therefore, the reflectance and interference should be calculated for every angle of light incidence. The reflectance was calculated again for an idealized single layered eye. First, the path difference between both reflected waves was calculated for different light incidences according to Equation 2 anticipating a Δn of 0.2, d = 100 nm and a reflectivity of 0.25 at both layer transitions. Secondly, the total intensity of the interfering waves was calculated for all angles of light incidence (Eicher, 1987). Using these values and knowing the change of light incidence during rotation, the change of total reflection was calculated for algae with different scan angles and for several tracking angles. Four conditions were considered in detail.

(i) The only situation where the angle of light incidence is directly proportional to the angle of rotation is when cells with an equatorial eye ($\varphi = 90°$) swim perpendicular to the light beam (b = 90°) (Fig. 4A). The reflectance is plotted vs. the angle of rotation for the wavelengths 400 nm, 500 nm and 600 nm. At perpendicular light incidence, green light of 500 nm (the wavelength close to $2dn + \lambda/2$) is reflected the best.

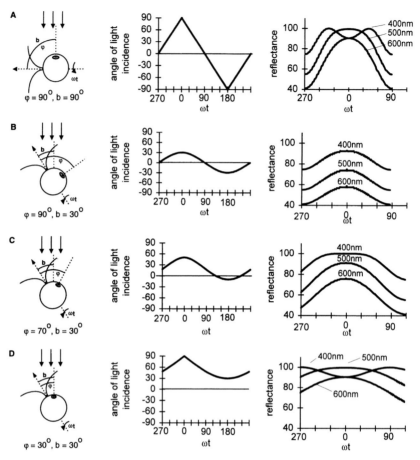

Fig. 4. For four different combinations of scan angle φ and tracking angle b, the angle of light incidence h and the total reflectance was calculated as a function of the angle of rotation ωt. The refractive indices were set to $n_a = 1.0$ and $n_b = 1.2$. The thickness of the layer was set to 100 nm.

For 600 nm red light, reflectance is reduced at all light incidences and rotational angles. In contrast, 400 nm blue light is not well reflected at perpendicular light incidence but significantly better at more tilted angles. Two maxima appear. For eyes with thinner layers, the trough is less pronounced, whereas it is amplified for thicker layers.

(ii) When cells with an equatorial eyespot swim neither perpendicular nor parallel to the light beam ($0 < b < 90°$) (Fig. 4B), the angle of light incidence is not linearly proportional to the angle of rotation. Instead, it changes in a sinusoidal fashion. The reflectance curves resemble the curve for the light incidence qualitatively with some colour dependent features. Since perpendicular light incidence never occurs unless the cell changes the swimming direction, there is a clear preference for blue light compared to green and red light. For equatorial eyes, light and dark periods are still equal in length. Only when cells swim directly to or away from light, light and dark periods are not defined.

(iii) In cells with non-equatorial eyes ($\varphi < 90°$), the light antennae gain new properties. When these cells swim towards the light ($0° < b < 90°$) the light and dark periods are not of equal length any more (Fig. 4C). The light phases becomes longer than the dark phases, whereas in cells swimming away from light, the light period appears shorter than the dark phase.

(iv) When species with non-equatorial eyes track the light at an angle b that equalizes the scan angle φ (Fig. 4D), h is always positive. The eye sees a modulated light period without dark intervals. The light incidence is almost directly proportional to the rotational angle between 270° and 90°, whereas it is not between 90° and 270°. The light incidence reaches 90° at 0° ωt, and the rhodopsin receives the maximal light signal in the green at least during part of its rotation. During the rest of the rotation cycle blue light reflection is preferred.

As a summary, at *all* flat angles, the reflectance increases for shorter wavelengths and decreases for the longer ones. This results not only in a colour and intensity modulation of the detected light signal during rotation, but in addition, the whole signal shifts towards the blue when the algae approach the direction of the light source. These eyespot properties have significant consequences for action spectroscopy. The ideal wavelength for perpendicular light incidence seems to match the maximum of the rhodopsin absorbance. At low light levels, the cells only occasionally catch photons, preferably when the eye is facing the light. Consequently, the action spectrum for phototaxis at low light is close to the wavelength dependence of the reflector at normal light incidence and close to the rhodopsin absorption spectrum. In contrast, action spectra constructed from measurements at higher light levels, where cells swim more or less directly to the light source, should be blue-shifted or express side bands in the blue. The shift should be most pronounced in spectra from algae with equatorial eyes. When these swim directly to the light source as they do at high light levels, the light always hits the eye at very flat angles which makes blue light more efficient than green light. High intensity action spectra for *Chlamydomonas* photo-taxis show a pronounced shoulder at the blue side of the spectrum (Nultsch, Throm & von Rimscha, 1971) and a predicted sharp cut off at the red side of the spectrum.

Precise calculations for the reflectance at oblique light incidence are difficult to generate because the system becomes less and less ideal. The path through both the high and the low refractive index media no longer match each other and reflection coefficients theoretically change with the angle of incidence. At flat angles, reflection depends largely on the polarization of the light because only light waves which oscillate in the same plane as the reflector layers can be reflected. In other words, when non-polarized light is used for stimulation, the reflected and transmitted light beams are polarized linearly. Due to the thickness of the layers, polarization also shows a strong wavelength dependence. Denton, Nichol and Nichol (1965) measured the reflection of the silvery gill cover of the bleak *Alburnus alburnus*, which is also based on thin film reflection. To their surprise, they found that the reflected light becomes increasingly polarized when the angle of light incidence increases, at a range where the reflectance changes only very little. Buder (1917) had already pointed out that algal eyes have a lower angular resolution than lens eyes. This is especially true for single-layered eyes and means that these eyes are designed for tracking diffuse light rather than point sources. Foster and Smyth (1980) estimated that the half beam width of the antenna is in the range of 60°. Once again, this beam width is not constant over the whole spectrum. The angle is wider in the blue and more narrow in the red.

At higher refractive index n, the relative intensity of the reflected light increases over the whole angular range and produces a clear break at 180°. However, at higher refractive indices, the intensity and colour modulation of the reflected light decreases. The cell only sees bright light at one side and dim light at the other with a sharp break at horizontal light incidence. Such properties would lead to an all-or-none response, but would not allow the cell to find its path smoothly.

Yoshimura (1994) studied photo-currents in *Chlamydomonas* upon stimulation with light applied under different angles of incidence. He confirmed earlier results about the contrast. In addition, he found that the retinal chromophore of the rhodopsin is arranged in a parallel fashion to the membrane and thus enhances the contrast and the directivity of the eye. This is especially significant in mutants with non-pigmented eyes. These strains are still phototactic with an unchanged absolute light sensitivity but their phototactic rate is strongly reduced. Therefore, there must be some front-to-back contrast also in white mutants. This contrast is considerably enhanced by the neatly designed optical system in the wild-type.

REFERENCES

Buder, J. (1917). Zur Kenntnis der phototaktischen Richtungsbewegung. *Jahrbuch Wissenschaftliche Botanik*, **58**, 105–220.

Denton, E. J., Nichol, F. R. S. & Nichol, J. A. C. (1965). Polarization of light reflected from the silvery exterior of the bleak *Alburnus alburnus*. *Journal of the Marine Biology Association, UK*, **45**, 705–9.

Deininger, W., Kröger, P., Hegemann, U., Lottspeich, F. & Hegemann, P. (1995). Chlamyrhodopsin represents a new type of sensory photoreceptor. *EMBO Journal*, **14**, 5849–58.

Eichler, H. J. (1987). Interferenz und Beugung. In *Lehrbuch der Experimentalphysik, Band III: Optik 8. Auflage*, ed. H. Gobrecht, pp. 325–480. Berlin, New York: de Gryter.

Foster, K. W. & Smyth, R. D. (1980). Light antennas in phototactic algae. *Microbiological Review*, **44**, 572–630.

Foster, K. W. & Saranak, J. (1989). The *Chlamydomonas* (chlorophyceae) eye as a model of cellular structure, intracellular signalling and rhodopsin activation. In *Algae as Experimental Systems*, ed. A. W. Coleman, L. J. Goff & J. R. Stein-Taylor, pp. 215–30. New York: Alan R. Liss, Inc.

Foster, K. W., Saranak, J. & Zarrilli, G. (1988). Autoregulation of rhodopsin synthesis in *Chlamydomonas reinhardtii*. *Proceedings of the National Academy of Sciences, USA*, **85**, 6379–83.

Harz, H., Nonnengäßer, C. & Hegemann, P. (1992). The photoreceptor current of the green alga *Chlamydomonas*. *Philosophical Transactions of the Royal Society London B*, **338**, 39–52.

Hegemann, P. & Bruck, B. (1989). Light-induced stop responses in *Chlamydomonas reinhardtii*: occurrence and adaptation phenomena. *Cell Motility and Cytoskeleton*, **14**, 501–15.

Kreimer, G. & Melkonian, M. (1990). Reflection confocal laser microscopy of eyespots in flagellated green algae. *European Journal of Cell Biology*, **53**, 101–11.

Land, M. (1972). The physics and biology of animal reflectors. *Progress in Biophysics and Molecular Biology*, **24**, 75–106.

Lawson, M. A., Zacks, D. N., Derguini, F., Nakanishi, K. & Spudich, J. L. (1991). Retinal analog restoration of phobic responses in a blind *Chlamydomonas reinhardtii* mutant. *Biophysical Journal*, **60**, 1490–8.

Nultsch, W., Throm, G. & von Rimscha, I. (1971). Phototaktische Untersuchungen an *Chlamydomonas reinhardtii* Dangeard in homokontinuierlicher Kultur. *Archives of Microbiology*, **80**, 351–69.

Sineshchekov, O. A. (1991). Electrophysiology of photomovements in flagellated algae. In *Biophysics of Photoreceptors and Photomovements in Microorganisms*, ed. F. Lenci, F. Ghetti, G. Colombetti, D-P. Häder & P. S. Song, pp. 191–202. New York: Plenum Press.

Yoshimura, K. (1994). Chromophore orientation in the photoreceptor of *Chlamydomonas* as probed by stimulation with polarized light. *Photochemistry and Photobiology*, **60**, 594–7.

NEGATIVE PHOTOTAXIS IN PHOTOSYNTHETIC BACTERIA

K. J. HELLINGWERF, R. KORT AND W. CRIELAARD

E. C. Slater Institute Plantage, Muidergracht 12, 1018 TV Amsterdam, The Netherlands

INTRODUCTION

In members of the Archaea, positive and negative phototactic responses, mediated according to a mechanism that shows many similarities with the mechanism of enterobacterial chemotaxis, have been documented thoroughly. In both responses, at least one retinal-containing sensory rhodopsin functions as a photoreceptor. In the Domain of the Bacteria the situation is less well resolved. Reports of tactic responses in photosynthetic bacteria, which lead to accumulation of those organisms in photosynthetically active irradiation, can be considered as 'classics' among the reports on tactic responses of Prokaryotes. These responses are still under intense investigation. Only recently, however, it has been reported that, in this family of organisms (in the purple- or proteobacteria), a second type of phototactic response occurs. Motion analysis of individual cells has revealed that blue light of physiological intensities elicits a repellent response, with adaptation characteristics typical of enterobacterial chemotaxis. The photoreceptor that, presumably, mediates this response is the Photoactive Yellow Protein, a member of the xanthopsins. This family of photoreceptors consists of 4-hydroxycinnamate containing water-soluble proteins, for which detail of structure and function is available. Here the photoresponses mediated by these xanthopsins are reviewed, together with their structure and function, and the molecular genetic studies aimed at the characterization of the underlying signal transduction chain.

Light plays two roles in the life of microorganisms. First, and by far the most important, light supplies the organisms that have learned to live the phototrophic mode of life with free energy for maintenance and growth. Secondly, light functions as a source of information. Light, however, is not only useful, it is dangerous too. Upon absorption of a visible photon, for instance, in a photoactive protein the immediate surroundings of a light-absorbing chromophore heats up some 200°C within a nanosecond. Because of this temperature shock, photoactive proteins are prone to damage. However, additional types of damage can occur. The singlet- and/or triplet-state of many chromophores can react with oxygen, giving rise to the highly reactive superoxides, which may covalently modify proteins in a cell at random. Furthermore, radiation from the (near) UV-part of the visible spectrum may cause damaging photochemical reactions in nucleic acids. Significantly, these latter two modes of light-induced damage are counteracted by specific light-dependent protection- and repair-mechanisms, respectively (for a review, see Hellingwerf, Hoff & Crielaard, 1996).

The beneficial and detrimental effects of electromagnetic radiation urge many organisms to respond specifically to their ambient light climate, which has three aspects: intensity, daily periodicity and spectral composition. The responses can take two forms: cells may migrate to or from a particular environment, or they may induce or repress specific sets of genes and/or modulate the activity of specific enzymes. Below, aspects of the first type of response are discussed.

Prokaryotes (members of the Domains of the Archaea or the Bacteria; Woese, 1987) has developed several ways to move towards a more optimal environment. These are: (i) twitching motility with use of fimbriae (Henrichsen, 1983), (ii) gliding motility (see, for example, Häder, 1987), (iii) flotation regulated by the buoyancy of the cell (Imhoff, 1992) and (iv) flagella-based swimming (see below). Current knowledge about the ways in which light is used as a source of information to guide cells or cell filaments through space, is unevenly distributed over these four types of migration. Tactic responses, induced by either chemical, or physical stimuli have not been reported for twitching motility, although *che*-like genes have been identified near the region encoding the genes required for twitching motility in *Pseudomonas aeruginosa* (Mattick, Whitchurch & Alm, 1996). In buoyancy regulation the entire photosynthetic machinery plus carbohydrate metabolism of the cell is involved. This process is discussed further in Walsby, this volume. Gliding motility is a common form of migration in cyanobacteria. In this class of organisms many light responses have been described, towards, as well as away from, the light source, mostly for filamentous species. Two basically different types of phototactic migration can be distinguished (Häder, 1987). A trial and error-like pattern in *Phormidium*, and related organisms, and a direct movement in response to the direction of the incoming light in organisms like *Anabaena*. From the wavelength dependence of these responses it has been inferred which of the bulk pigments of the cells could be involved in these phototactic responses. However, the recent characterization of a number of specific low-abundance photoreceptors in cyanobacteria has established that it is less likely that these bulk pigments of the organism are involved. Furthermore, since little is known of the underlying molecular mechanism of both gliding motility itself and the signal transduction pathway that modulates it, these aspects will not be discussed further here.

Both among the Bacteria and the Archaea there are many representatives that swim through rotation of their flagellum/flagella, in order to migrate to a more optimal environment. The archetype of this response is the chemotactic response in *Escherichia coli* (for recent reviews see Armitage, 1992; Parkinson, 1993; Stock & Mowbray, 1995). Net migration of organisms using this type of mechanism is achieved by a random walk of short 'runs', spaced by 'tumbles' in which the flagella either rotate in opposite directions or pause. In this pattern the length of the runs is biased by the chemical and physical stimuli from the environment of the cell through specific receptors, called methyl-accepting chemotaxis proteins. Tactic migration in which light signals are processed, in a mechanism that is very similar to the mechanism of chemotaxis in *E. coli*, occurs in representatives of the halophilic branch of the Archaea (for a recent review see Hoff, Jung & Spudich, 1997). Despite earlier discussions about terminology (see below), this process is now generally referred to as phototaxis; in many of these organisms both positive- and negative phototactic responses can be discriminated. Both responses compete and are integrated at the level of the intracellular signal transduction pathway (Spudich, this volume; Armitage, this volume).

The description of light-induced tactic migration of photosynthetic bacteria has a long history. In 1883 Engelmann reported that motile, photosynthetic purple bacteria (most likely *Chromatium* cells) accumulate at specific wavelengths in a dispersed spectrum, including the infrared. The accumulation patterns appeared to follow the absorbance spectra of the photosynthetic pigments of these bacteria and could be explained as the result of photokinesis, that is the effect of light on the velocity of movement. Another important observation to explain these accumulation patterns was that the *Chromatium* cells reversed their swimming direction when they entered a region of reduced light intensity (Engelmann, 1883). This has become known as the photophobic, or more correctly, the scotophobic response (Ragatz *et al.*, 1994). So, photokinesis is responsible for movement towards an optimal light environment and the scotophobic response allows the bacterium to stay in that environment.

The observations made by Engelmann strongly suggested that the light-induced tactic response in *Chromatium* cells occurs as a result of changes in photosynthetic electron transport, rather than excitation of specific photoreceptors. The linkage between photosensory responses and the photosynthetic pigments in purple bacteria was confirmed by the work of Clayton (1953), who showed that the scotophobic response of *Rhodospirillum rubrum* only occurred with light that is photosynthetically active. In addition, no light-induced tactic responses could be identified in photosynthetic reaction centre mutants of *Rhodobacter sphaeroides, Rhodobacter capsulatus* and *Rsp. rubrum*, even though these mutants showed normal motility and chemotaxis (Armitage & Evans, 1981). Such mutants still absorb light via their photosynthetic antenna pigments, but cannot use the associated free energy for electron transport. Since bacteria use the free energy of the electrochemical proton gradient for flagellar rotation (Glagolev & Skulachev, 1978), photokinesis is probably a direct result of a change in the rate of electron transfer, whereas the scotophobic response occurs as a result of sensory signalling, through alteration of the flagellar switching frequency. An intriguing problem, which still has to be solved, is the exact nature of the sensing that is required for this tactic process. Recent inhibitor studies on the tactic response of tethered *Rb. sphaeroides* cells suggest that changes in the rate of electron transport mediate this regulation (Gauden & Armitage, 1995). Several *che*-like genes were recently identified *Rb. sphaeroides*. However, their role in phototaxis has not yet been elucidated (this positive phototactic response is discussed in detail in Armitage, this volume).

Light can be detected by its intensity, colour and direction. The latter aspect particularly, has led to the introduction of rather complicated terminology regarding the characterization of tactic responses of microorganisms. Purple bacteria are only capable of sensing light intensity and colour. Recently, however, it was reported that colonies of *Rhodospirillum centenum* are capable of sensing the *direction* of light; this may provide the first report of true phototaxis in prokaryotes (Ragatz *et al*, 1994, 1995; Gest, 1995). However, to further substantiate this possibility, motion analysis of single cells is required to exclude that colony migration towards the light is a result of sensing differences in light intensity caused by shading within a colony, rather than sensing of the direction of illumination by individual cells. Both a positive and a negative phototactic response have been observed in *Rsp. centenum*, depending on the light intensity used (Ragatz *et al.*, 1995). At low light intensity the positive phototactic response was observed, with a wavelength dependence that suggested that this response is mediated through the photosynthesis pigments. The tactic response of *Rsp. centenum* at high light intensities, which caused the cells to migrate away from the light source, was elicited mainly by light in the wavelength region between 550 and 600 nm (see further below). In the following, the term 'phototaxis' will be used quite loosely, to refer to processes in which individual bacteria show a net migration in response to changes in their ambient light climate.

NEGATIVE PHOTOTAXIS IN ANOXYPHOTOBACTERIA

In 1993 a new type of light-induced repellent response was reported in the halophilic purple sulphur bacterium *Ectothiorhodospira halophila* (Sprenger *et al.*, 1993). The initial observation that led to these studies was that, in a light spot of red- or infrared light (in other words light that can be absorbed by the photosynthesis machinery), cells of this species accumulate, whereas a different response is observed in a blue light spot. With light of the latter colour, cells accumulate at the edge of the spot, indicating that, besides an attractant response, selectively elicited by red light, these cells additionally display a repellent response towards blue light. Subsequent motion analyses of *E. halophila* cells showed a relative increase in the number of reversals of swimming direction, upon a step-up in the intensity of incoming blue light, in the physiological range of light intensities. This response to blue light showed

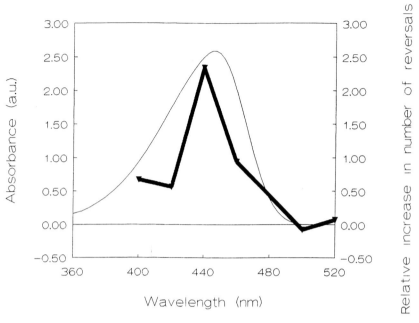

Fig. 1. Action spectrum of the blue-light induced repellent response in *E. halophila*. The triangles and thick line represent the measured reversal frequencies at the indicated wavelengths. For comparison the visible absorption spectrum is shown (thin line). From Sprenger *et al.* (1993), with permission.

adaptation, with kinetics similar to the kinetics of adaptation in chemotaxis of enterobacteria. The phenomenon of adaptation in tactic responses of bacteria strongly expands the dynamic range of attractant and repellent concentrations to which such organisms can respond, but is also essential in the temporal response to a changing concentration of a particular modulator. Its mechanism is based upon the reversible methylation of specific glutamate side chains in the cytoplasmic domain of chemo- and phototactic receptor proteins, which counteracts the conformational changes induced by attractant and repellent binding. CheR functions as the methylating enzyme in this response and CheB as the regulated (by phosphorylation; see below) methyl-esterase. The adaptation process in enterobacterial chemotaxis typically takes place in the seconds time-scale.

Because it was known at that time that a low-abundant, highly absorbing photoactive protein was present in *E. halophila*, which is called Photoactive Yellow Protein (PYP), the wavelength dependence of this response was subsequently investigated. Light of a wavelength longer than 500 nm did not elicit any increase in the probability of directional switching of the cells, whereas light of 450 nm elicited a maximal effect (Fig. 1). Thus, this new repellent response is not mediated by the photosynthesis machinery as the primary photoreceptor, like the attractant response in this and other purple bacteria (Hustede, Liebergesell & Schlegel, 1989). In contrast, its wavelength dependence matches the absorption spectrum of PYP, which makes this latter protein the designated candidate for the photoreceptor of this new repellent response.

Evidence for a gene encoding a photoactive yellow protein homologue in *Rb. sphaeroides* led to the prediction that a negatively phototactic response to blue light, similar to the repellent response as described above, may also occur in this species (Kort *et al.*, 1996*a*). *Rb. sphaeroides* shows typical swimming behaviour: its flagellum rotates only in a clockwise direction; reorientation of swimming cells occurs as a result of Brownian motion during stop periods of several seconds (Armitage & Macnab, 1987). Recently, it was shown that *Rb. sphaeroides* cells show a repellent response towards a *decrease* in photosynthetic light (Grishanin *et al.*, 1997). This response was only observed in phototrophically grown cells, and the magnitude of the response decreased when the light intensity during growth was lowered. The photosynthetic apparatus is most probably the primary photoreceptor for this response (Grishanin *et al.*, 1997). Recent data, using computer-assisted motion analysis using the methods of Khan *et al.* (1993) and Zacks *et al.* (1993), indicated that motile *Rhodobacter* cells also respond by a transient stop, followed by adaptation, to an *increase* in blue light (Kort, Hellingwerf & Spudich, unpublished observations). The observed adaptation suggests the involvement of a methyl-accepting taxis protein as one of the components of the downstream signal transduction pathway to the flagellar motor. Indeed, blue-light induced release of volatile methanol has subsequently been observed, which is generated when tactic adaptation occurs (Kort, Hellingwerf & Spudich, unpublished observations).

This newly discovered photoresponse, which may involve a xanthopsin as photoreceptor, is only observable in phototrophically grown cells. The physiological roles of the two responses may be that in the positive one will result in migration of the cells towards light conditions that are most favourable for photosynthesis, while the second photophobic one may function to avoid high light intensities, being harmful to the cell. If so, their function would be strikingly analogous to the primary function of the two sensory rhodopsins in halobacteria (see, for example, Hoff, Jung & Spudich, 1997). One might even speculate further that both in archaeal and in bacterial 'vision' there is a correlation with vision in mammals, for instance based on the abundance of the pigments; the first response would then correlate with the function of rods and the second with the one of cones.

Consistent with a photosensory role for xanthopsins in the second response, it has been possible to detect the chromophoric group of PYP, 4-hydroxycinnamic acid, in intact cells of *Rb. sphaeroides*, though it was only possible to isolate from phototrophically grown cells (Haker & Kort, unpublished observations). These latter findings for *Rb. sphaeroides* are strongly reminiscent of the characteristics of the purple sulphur bacterium *Rhodospirillum salexigens*, in which protein-attached chromophore, as well as the PYP homologue, could only be identified in cells grown anaerobically in the light (Kort *et al.*, 1996*a*).

The formal name for all representatives of the Domain of Bacteria that carry out anoxygenic photosynthesis is anoxyphotobacteria; it distinguishes them from the cyanobacteria, which perform oxygenic photosynthesis (see, for example, Hellingwerf *et al.*, 1994). Traditionally, these organisms were grouped in such diverse families as the green- or purple, non-sulphur- or sulphur bacteria, in the more recently discovered heliobacteria or even among the azorhizobia (Hellingwerf *et al.*, 1994; Stackebrandt, Rainey & Ward-Rainey, 1996). The development of a phylogenetic classification scheme, however, has led to a more rational classification of these organisms and to the realization that some non-phototrophic prokaryotes may actually be closely related to their phototrophic counterparts (Schlegel, 1986; Woese, 1987). It may therefore be anticipated that the negative phototactic response observed in *E. halophilia* and *Rb. sphaeroides* also occurs in non-phototrophic members of the Bacteria, in particular among the proteobacteria. The negative phototactic response that has been observed in *Beggiatoa* may be an example, since it also shows a wavelength dependence that peaks at 450 nm (Nelson & Castenholz, 1982). Significantly, screening of a large number of organisms from the Domain of the Bacteria showed that many of these organisms possess a single protein of a size similar to the size of PYP that cross-reacts with a specific polyclonal antiserum against PYP (Hoff *et al.*, 1994*b*). Many non-photosynthetic organisms also reacted positively. The conclusion from this experiment is blurred by the fact that even some non-motile organisms reacted positively. The finding of a PYP homologue in the non-sulphur bacterium *Rb. sphaeroides*, however, is not in agreement with the suggestion that the distribution of PYP homologues is limited to only a very few members of the phylogenetic tree (Thiemann & Imhoff, 1995).

At an early stage in *E. coli* chemotaxis research, it was observed that non-physiologically high levels of blue light induced a repellent response (Macnab & Koshland, 1974). Initially these effects were ascribed to artefactual light-dependent oxidation of a flavoprotein of the electron transfer chain. More recently, however, it was speculated that a specific chemotaxis receptor for oxygen (that is Aer, which happens to be a flavoprotein) is responsible for this artefact (Bibikov *et al.*, 1997). When the pathway for haem biosynthesis is impaired in *E. coli*, a blue-light induced repellent response can be observed at much lower light intensities, which is due to the interaction of light-activated haem precursors or their products with the chemotaxis machinery (Yang, Inokuchi & Adler, 1995). It is evident that this response is not of a physiological nature either.

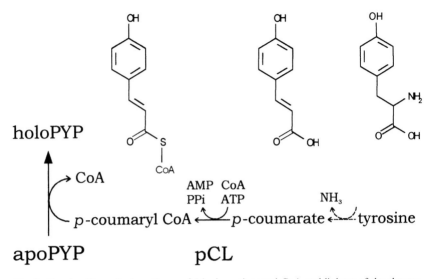

Fig. 2. Putative biosynthesis pathway of 4-hydroxycinnamyl-CoA and linkage of the chromophore to apoPYP. This model is based on the observation that the genes encoding apoPYP and a putative CoA-ligase are adjacent in the chromosomes of E. halophila and Rb. sphaeroides.

MOLECULAR GENETICS OF PHOTOACTIVE YELLOW PROTEINS IN PHOTOSYNTHETIC BACTERIA

Studies of the genes possibly involved in negative phototaxis were initiated through the cloning of pyp gene of E. halophila (Baca et al., 1994; Kort et al., 1996a). Subsequent Southern analyses revealed the presence of homologous genes in additional members of the purple bacteria. Cloning and sequencing of the gene encoding PYP of Rsp. salexigens has been described by Kort et al. (1996a). The encoding protein also possesses the chromophore trans 4-hydroxycinnamic acid, as was recently demonstrated with high performance capillary zone electrophoresis (Kort et al., 1996a). Additionally, evidence was presented for the presence of a gene encoding a PYP homologue in Rb. sphaeroides, the first genetically well-characterized bacterium in which this photoreceptor has been identified (Kort et al., 1996a). The genetic region encoding this gene has now been cloned and sequenced (Kort & Phillips-Jones, unpublished observations).

Analysis of the flanking regions of the pyp gene in E. halophila revealed an open reading frame (ORF) encoding a CoA-ligase homologue, suggesting a plant-like conversion of 4-hydroxycinnamic acid to its CoA derivative before linkage to apoPYP (Fig. 2). The pathway of biosynthesis of 4-hydroxycinnamic acid has been studied extensively in higher plants, but no information is available on the conservation of this pathway in E. halophila or other members of the Bacteria (Hahlbrock & Scheel, 1989). In higher plants, the two enzymes of central importance in the metabolic conversions relevant for

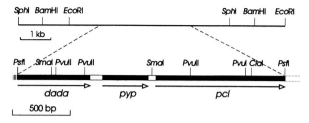

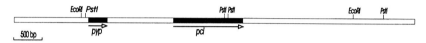

Fig. 3. Schematic representation of the regions of the genomes in *Rb. sphaeroides* and *E. halophila* containing the *pyp* genes. Standard abbreviations for restriction sites have been used. The *dada* and *pcl* genes refer to a D-amino acid dehydrogenase a CoA-ligase homologue, respectively.

4-hydroxycinnamic acid Are: phenylalanine ammonia lyase (PAL), which catalyses the reaction from either phenylalanine or tyrosine to 4-hydroxycinnamic acid, and *p*-coumaryl:CoA ligase (pCL, or 4-hydroxycinnamyl:CoA ligase), which activates 4-hydroxycinnamic acid through a covalent coupling to CoA, via a thiol ester bond (Hahlbrock & Scheel, 1989). A similar *pcl* homologue has been identified approximately 1 kb downstream of *pyp* in *Rb. sphaeroides* (Fig. 3).

The *pyp* gene of *E. halophila* has been expressed in both *E. coli* and *Rhodobacter*. In both organisms the protein could be immunologically detected, but it yellow colour was not observed (Kort *et al.*, 1996*a*; Devanathan *et al.*, 1997). Subsequent molecular genetic construction of a histidine-tagged version of PYP led to its 2500-fold over-production in *E. coli* and allowed the rapid purification of the heterologously produced apoprotein. Native as well as recombinant holoPYP can be reconstituted by the addition of 4-hydroxycinnamic acid anhydride to apoPYP (Imamoto *et al.*, 1995; Kort *et al.*, 1996*a*).

From the preliminary analysis of the *pyp*-gene cluster of *Rb. sphaeroides*, there is now firm evidence for the presence of a *pyp* gene and a possible pCL-based activation of the chromophore in a member of the α-group of the proteobacteria. Genetic characterization of the putative signal transduction pathway involving PYP remains to be performed, but the identification of a negative phototactic response in the genetically well-characterized *Rb. sphaeroides*, will significantly facilitate this work. Additional analyses of the flanking regions of the different *pyp* genes may also be useful for this purpose.

STRUCTURE AND FUNCTION OF PHOTOACTIVE YELLOW PROTEINS

Photoactive yellow protein (PYP) was discovered in 1985 as part of a study on the characterization of coloured proteins from *E. halophila* (Meyer, 1985). Subsequently, PYP was also isolated from two other halophilic phototrophic purple sulphur bacteria, *Rsp. salexigens* and *Chromatium salexigens* (Meyer *et al.*, 1990; Koh *et al.*, 1996). Since extensive detail of the structure and function of the first of these proteins is available, its properties will be discussed below. PYP is a 14 kDa (125 amino acid) water-soluble protein with a main absorption band at 446 nm (ε_{max} = 45.5 mM^{-1} cm^{-1}), as a result of the presence of a thiol-ester linked 4-hydroxycinnamate chromophore (Meyer, 1985; Hoff *et al.*, 1994*a*; Baca *et al.*, 1994; Hoff *et al.*, 1996). Its correct three-dimensional structure is now available at 1.4 Å resolution (Borgstahl, Williams & Getzoff, 1995). The structure of the protein in solution has been obtained at high resolution with ^{1}H-NMR (Düx & Kaptein, unpublished observations).

After absorption of a blue photon, PYP enters a cyclic chain of reactions. In this photocycle the ground state (λ_{max}446 nm, pG) is converted into a red-shifted intermediate (λ_{max}465 nm, pR), followed by the formation of a relatively long-lived intermediate (λ_{max}355 nm, pB), and recovery of the ground state, as indicated in Fig. 4 (Meyer *et al.*, 1987; Hoff *et al.*, 1994*c*). In the pB state the chromophore is presumably protonated and isomerized to the *cis* configuration (Kort *et al.*, 1996*b*).

The photocycle of PYP strongly resembles that of the archaeal sensory rhodopsins, although the chromophore in PYP has a completely different chemical structure (see, for example, Hellingwerf *et al.*, 1996). In pG, the anionic chromophore is in hydrogen bonding contact with the buried and protonated E46 and with Y42 (Borgstahl *et al.*, 1995; see also Fig. 5A). The pK_a of both the chromophore and of E46 are strongly shifted by the protein environment, to a lower and higher value, respectively.

Photo-excitation of PYP results in the formation of the red-shifted intermediate pR. FTIR analyses revealed that the hydrogen bond between the chromophore and E46 remains intact in pR (Xie *et al.*, 1996; Imamoto *et al.*, 1997). Therefore, photo-isomerization of the double bond of the chromophore most likely takes place by rotation across both the double bond and the C–S bond that links the chromophore to the apoprotein, in other words, a two-bond isomerization process (Fig. 5C). Initially most models to explain chromophore photoisomerization in PYP were based on the assumption that isomerization would only involve the double bond, which would imply that the aromatic ring of the chromophore would have to flip (Fig. 5B). However, existence of a hydrogen bond between the chromophore and E46 in pR, in which the chromophore supposedly has undergone photoisomerization, now invalidates such a model.

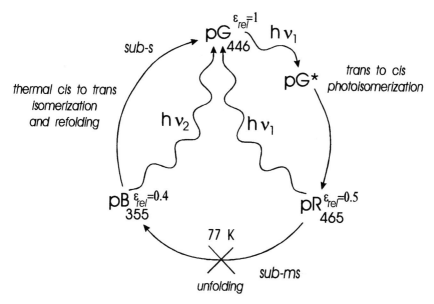

Fig. 4. The photocycle of PYP. PG, pR and pB are the ground state, the red-shifted and the blue-shifted intermediate of PYP, respectively. For each, the wavelength of maximal absorbance is indicated by a subscript; the relative extinction coefficient by a superscript. pG* is the excited state of PYP, formed after absorption of a blue photon. These intermediates interconvert through thermal (straight lines) and light-induced (wavy lines) reactions. The approximate time scale of the thermal reactions is indicated. The subscripts 1 and 2 for hv refer to a blue and a UV photon, respectively. The photocycle of PYP is blocked at a temperature of 77 K or below, after formation of pR. The formation of pB is paralleled by a partial unfolding of the protein.

Subsequently, the long-lived intermediate pB is formed. To achieve this, the chromophore must take up a proton, while E46 most likely becomes deprotonated. One possibility is that a direct proton transfer occurs between these two groups (Xie et al., 1996; Imamoto et al., 1997). However, it has been reported that PYP takes up a proton from solution, in parallel with the formation of pB (Meyer, Cusanovich & Tollin, 1993). Therefore, the chromophore may also be directly protonated from the solution. Further experiments are required to resolve this. In the pB state, the protein presumably must significantly change its conformation, to generate a signal for the initiation of a phototactic response. Different techniques have yielded different estimates of the extent of this conformational transition. Time-resolved X-ray diffraction experiments with PYP crystals have led to the conclusion that this conformational transition is mainly confined to the chromophore-binding pocket (Genick et al., 1997). Spectroscopic and thermodynamic analyses of the photocycle transitions of PYP in solution, on the other hand, have led to a model that describes formation of the pB

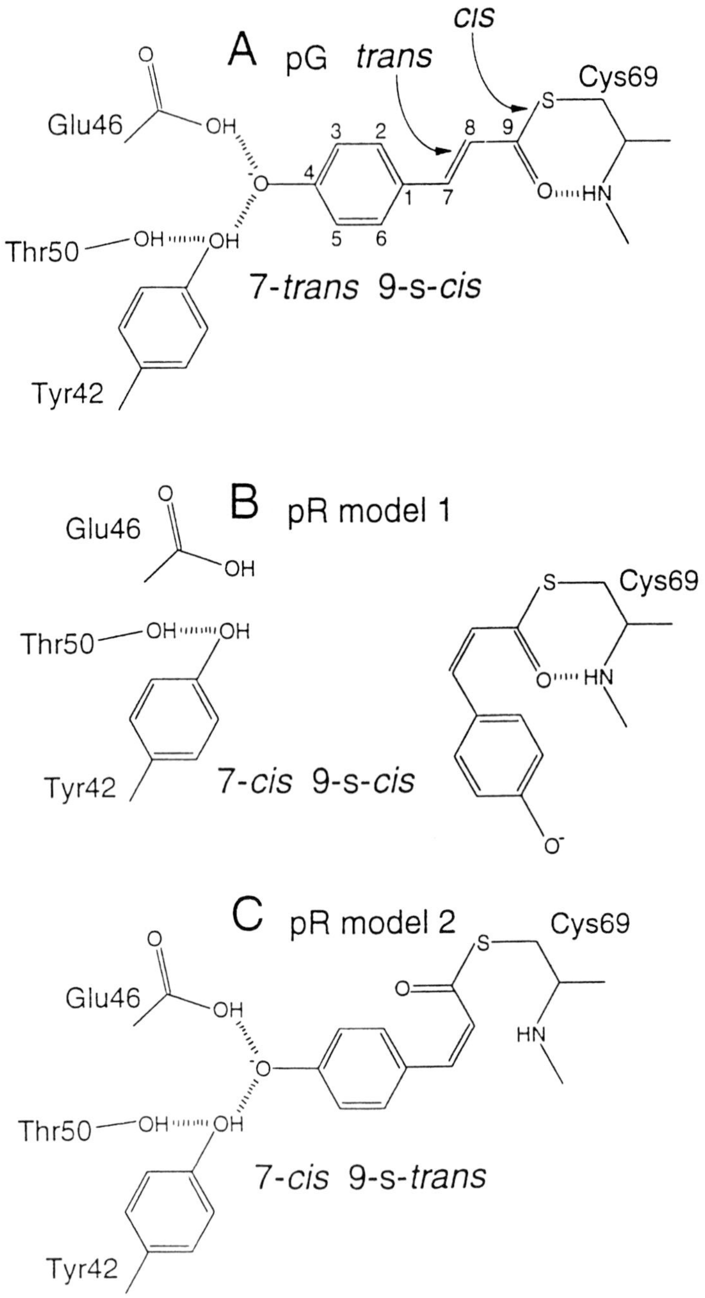

Fig. 5. Model of the structure of the chromophore binding pocket before (A) and after (B and C) photoisomerization. At the bottom of each part of the figure the isomeric configuration of the chromophore is given. For explanation see text. From Xie *et al.* (1996), with permission.

state as a partial unfolding of the protein, in which a much larger part of the protein is involved (Van Brederode *et al.*, 1996).

Finally, the ground state of PYP (that is pG) is recovered after re-isomerization and deprotonation of the chromophore and conformational relaxation. Thus, in common with the photocycle of the sensory rhodopsins, photo-isomerization and proton transfer are essential features in the PYP photocycle as well. This may explain at least part of the strong similarity of their photochemical properties. It remains to be established whether the partial reactions in the recovery process of the pG state of PYP proceed independently or in one concerted reaction. The detailed characterization of the spatial structure of the signalling state of PYP forms a major challenge for future research. Recently initiated time-resolved X-ray experiments with nanosecond time resolution, as well as ^{1}H-NMR experiments, in combination with proper illumination are promising approaches towards this goal. The biophysical characterization of PYP is facilitated greatly by the possibility to heterologously express and reconstitute PYP (see above). The reconstitution has also been carried out with a large range of chromophore analogues (Kroon *et al.*, 1996). These experiments revealed that, by varying the chemical structure of the chromophore of PYP, a considerable (more than 40 nm) red shift of its absorbance maximum can be obtained. However, such shifts are not sufficient to mimic the photoreceptor that mediates the negative phototactic response in *Rsp. centenum* (see above). Consequently, it is not likely that a PYP homologue plays a role in this latter process in the organism.

If it is demonstrated conclusively that PYP is a photoreceptor involved in negative phototaxis, then the signal transduction pathway from PYP to the flagellar motor, which presumably regulates the direction of rotation, remains to be identified. It is most likely that the blue-shifted intermediate pB would be the signalling state of PYP, since this is by far the most stable intermediate, and it has a characteristic conformation, allowing specific interactions with a hypothetical transducer (for a review, see Hellingwerf, Hoff & Crielaard, 1996). The amino acid residues important for signalling may be localised in a region of PYP, homologous to the PAS domain. This latter domain has also been identified in phytochrome and a number of proteins involved in the regulation of circadian rhythmicity and may play a role in (photo)receptor dimerisation (Lagarias, Wu & Lagarias, 1995). The homologous region ranges from residues 24 to 66, and contains a number of conserved residues including three glycines (G37, G51 and G59), which may act as hinges in the dynamics of the protein in the processes of formation of the signalling state (Van Aalten & Crielaard, unpublished observations). The PYP transducer might also be a homologue of the methyl-accepting chemotaxis proteins, like the transducers of SRI and SRII (HtrI and HtrII; Spudich, this volume). This would provide a straightforward mechanism for adaptation, as observed for free swimming *E. halophila* cells in their blue light response. It would also be consistent with the observation of blue-light mediated methanol release in *Rb. sphaeroides* (Kort *et al.*, unpublished observations). Further downstream, a two-component regulatory system may be involved, consisting of homologues of the kinase CheA and the response regulators CheY and CheB, just like in chemotaxis in *E. coli*.

In addition to the Photoactive Yellow Protein, new types of photoreceptors have been detected in several classes of organisms. Examples are a presumed rhodopsin and a phytochrome in cyanobacteria (Geerdink *et al.*, 1995; Hughes *et al.*, 1997), the rhodopsins and flavin-containing photoreceptors in unicellular eukaryotes (Campuzano *et al.*, 1996; Saranak & Foster, 1997), and even a human blue-light photoreceptor (Hsu *et al.*, 1996). There will be a lot of excitement in the detailed characterization of the structure of these photoreceptors and the elucidation of their role in cellular physiology. It will be even more exciting to characterize those photoreceptors, for instance, with respect to roles in tactic migration in gliding motility and for circadian synchronization, of which so far only their existence are known.

REFERENCES

Armitage, J. P. (1992). Behavioural responses in bacteria. *Annual Review of Physiology*, **54**, 683–714.

Armitage, J. P. & Macnab, R. M. (1987). Unidirectional, intermittent rotation of the flagellum of *Rhodobacter sphaeroides*. *Journal of Bacteriology*, **169**, 514–18.

Armitage, J. P. & Evans, M. C. W. (1981). The reaction centre in the phototactic and chemotactic responses of photosynthetic bacteria. *FEMS Microbiology Letters*, **11**, 89–92.

Baca, M., Borgstahl, G. E. O., Boissinot, M., Burke, P. M., Williams, D. R., Slater, K. A. & Getzoff, E. D. (1994). Complete chemical structure of photoactive yellow protein: novel thioester-linked 4-hydroxycinnamyl chromophore and photocycle chemistry. *Biochemistry*, **33**, 14369–77.

Bibikov, S. I., Biran, R., Rudd, K. E. & Parkinson, J. S. (1997). A signal transducer for aerotaxis in *Escherichia coli*. *Journal of Bacteriology*, **179**, 4075–9.

Borgstahl, G. E. O., Williams, D. R. & Getzoff, E. D. (1995). 1.4 Å structure of photoactive yellow protein, a cytosolic photoreceptor: unusual fold, active site and chromophore. *Biochemistry*, **34**, 6278–87.

Campuzano, V., Galland, P., Alvarez, M. I. & Eslava, A. P. (1996). Blue-light receptor requirement for gravitropism, autochemotropism and ethylene response in *Phycomyces*. *Photochemistry and Photobiology*, **63**, 686–94.

Clayton, R. K. (1953). Studies in the phototaxis of *Rhodospirillum rubrum*. II. The relation between phototaxis and photosynthesis. *Archives of Microbiology*, **19**, 125–40.

Devanathan, S., Genick, U. K., Getzoff, E. D., Meyer, T. E., Cusanovich, M. A. & Tollin. G. (1997). Preparation and properties of a 3,4-dihydroxycinnamic acid chromophore variant of the photoactive yellow protein. *Archives of Biochemistry and Biophysics*, **340**, 83–9.

Engelmann, T. W. (1883). *Bacterium photometricum*. Ein Beitrag zur vergleichenden Physiologie des Licht- und Farbensinnes. *Archives of Physiology*, **30**, 95–124.

Gauden, D. E. & Armitage, J. P. (1995). Electron transport-dependent taxis in *Rhodobacter sphaeroides*. *Journal of Bacteriology*, **177**, 5853–9.

Geerdink, J. H., Haker, A., Matthijs, H. C. P., Hoff, W. D., Hellingwerf, K. J. & Mur, L. R. (1995). A rhodopsin as photoreceptor in chromatic adaptation of the cyanobacterium *Calothrix* sp. Photosynthesis: from light to biosphere, ed. P. Mathies, vol. I, pp. 303–306, Dordrecht, The Netherlands: Kluwer Academic Publishers.

Genick, U. K., Borgstahl, G. E. O., Ng, K., Ren, Z., Pradervand, C., Burke, P. M., Srajer, V., Teng, T-Y., Schildkamp, W., McRee, D. E., Moffat, K. & Getzoff, E. D. (1997). Structure of a protein photocycle intermediate by millisecond time-resolved crystallography. *Science*, **275**, 1471–5.

Gest, H. (1995). Phototaxis and other sensory phenomena in purple photosynthetic bacteria. *FEMS Microbiological Reviews*, **16**, 287–94

Glagolev, A. N. & Skulachev, V. P. (1978). The proton pump is a molecular engine of motile bacteria. *Nature*, **272**, 280–2.

Grishanin, R. N., Gauden, D. E. & Armitage, J. P. (1997). Photoresponses in *Rhodobacter sphaeroides*: role of photosynthetic electron transport. *Journal of Bacteriology*, **179**, 24–30.

Häder, D-P. (1987). Photosensory behaviour in prokaryotes. *Microbiological Reviews*, **51**, 1–21.

Hahlbrock, K. & Scheel, D. (1989). Physiology and molecular biology of phenylpropanoid metabolism. *Annual Review of Plant Physiology and Plant Molecular Biology*, **40**, 347–69.

Hellingwerf, K. J., Crielaard, W., Hoff, W. D., Matthijs, H. C. P., Mur, L. R. & van Rotterdam, B. J. (1994). Photobiology of Bacteria, *Antonie van Leeuwenhoek*, **65**, 331–47.

Hellingwerf, K. J., Hoff, W. D. & Crielaard, W. (1996). Photobiology of microorganisms: how photosensors catch a photon and use it to initialize signalling. *Molecular Microbiology*, **21**, 683–93.

Henrichsen, J. (1983). Twitching motility. *Annual Review of Microbiology*, **37**, 81–93.

Hoff, W. D., Düx, P., Hård, K., Devreese, B., Nugteren-Roodzant, I. M., Crielaard, W., Boelens, R., Kaptein, R., Van Beeumen, J. & Hellingwerf, K. J. (1994a). Thiol ester-linked *p*-coumaric acid as a new photoactive prosthetic group in a protein with rhodopsin-like photochemistry. *Biochemistry*, **33**, 13959–62.

Hoff, W. D., Sprenger, W. W., Postma, P. W., Meyer, T. E., Veenhuis, M., Leguijt, T. & Hellingwerf, K. J. (1994b). The photoactive yellow protein from *Ectothiorhodospira halophila* as studied with a highly specific polyclonal antiserum: (intra)cellular localization, regulation of expression, and taxonomic distribution of cross-reacting proteins. *Journal of Bacteriology*, **176**, 3920–7.

Hoff, W. D., Van Stokkum, I. H. M., Van Ramesdonk, H. J., Van Brederode, M. E., Brouwer, A. M., Fitch, J. C., Meyer, T. E., Van Grondelle, R. & Hellingwerf, K. J. (1994c). Measurement and global analysis of the absorbance changes in the photocycle of the photoactive yellow protein from *Ectothiorhodospira halophila*. *Biophysical Journal*, **67**, 1691–705.

Hoff, W. D., Devreese, B., Fokkens, R., Nugteren-Roodzant, I. M., Van Beeumen, J., Nibbering, N. & Hellingwerf, K. J. (1996). Chemical reactivity and spectroscopy of the thiol ester-linked *p*-coumaric acid chromophore in the photoactive yellow protein from *Ectothiorhodospira halophila*. *Biochemistry*, **35**, 1274–81.

Hoff, W. D., Jung, K-H. & Spudich, J. L. (1997). Molecular mechanism of photosignaling by archaeal sensory rhodopsins. *Annual Review of Biophysics and Biomolecular Structure*, **26**, 221–56.

Hsu, D. S., Zhao, X., Zhao, S., Kazantsev, A., Wang, R-P., Todo, T., Wei, Y-F. & Sancar, A. (1996). Putative human blue-light photoreceptors hCRY1 and hCRY2 are flavoproteins. *Biochemistry*, **35**, 13781–7.

Hughes, J., Lamparter, T., Mittmann, F., Hartmann, E., Gärtner, W., Wilde, A. & Börner, T. (1997). A prokaryotic phytochrome. *Nature*, **386**, 663.

Hustede, E., Liebergesell, M. & Schlegel, H. C. (1989). The photophobic response of various sulphur and nonsulphur purple bacteria. *Photochemistry and Photobiology*, **50**, 809–15.

Imamoto, Y., Ito, T., Kataoka, M. & Tokunaga, F. (1995). Reconstitution photoactive yellow protein from apoprotein and *p*-coumaric acid derivates. *FEBS Letters*, **374**, 157–60.

Imamoto, Y., Mihara, K., Hisatomi, O., Kataoka, M., Tokunaga, F., Bojkova, N. & Yoshihara, K. (1997). Evidence for proton transfer from Glu-46 to the chromophore during the photocycle of photoactive yellow protein. *Journal of Biological Chemistry*, **272**, 12905–8.

Imhoff, J. F. (1992). Taxonomy, phylogeny and general ecology of anoxygenic phototrophic bacteria. In *Photosynthetic Prokaryotes*, ed. N. H. Mann & N. G. Carr, pp. 53–92. New York: Plenum Press.

Khan, S., Castellano, F., Spudich, J. L., McCray, J. A., Goody, R. S., Reid, G. P. & Trentham, D. R. (1993). Excitatory signaling in bacteria probed by caged chemoeffectors. *Biophysical Journal*, **65**, 2368–82.

Koh, M., Van Driessche, G., Samyn, B., Hoff, W. D., Meyer, T. E., Cusanovich, M. A. & Van Beeumen, J. J. (1996). Sequence evidence for strong conservation of the photoactive yellow proteins from the halophilic phototrophic bacteria *Chromatium salexigens* and *Rhodospirillum salexigens*. *Biochemistry*, **35**, 2526–34.

Kort, R., Hoff, W. D., van West, M., Kroon, A. R., Hoffer, S. M., Vlieg, K. H., Crielaard, W., Van Beeumen, J. J. & Hellingwerf, K. J. (1996*a*). The Xanthopsins: a new family of eubacterial blue-light photoreceptors. *EMBO Journal*, **15**, 3209–18.

Kort, R., Vonk, H., Xu, X., Hoff, W. D., Crielaard, W. & Hellingwerf, K. J. (1996*b*). Evidence for *trans–cis* isomerization of the *p*-coumaric acid chromophore as the photochemical basis of the photocycle of photoactive yellow protein. *FEBS Letters*, **382**, 73–8.

Kroon, A., Hoff, W. D., Fennema, H., Gijzen, J., Koomen, G-J., Verhoeven, J. W., Crielaard, W. & Hellingwerf, K. J. (1996). Spectral tuning, fluorescence and photoactivity in hybrids of photoactive yellow protein, reconstituted with native and modified chromophores. *Journal of Biological Chemistry*, **271**, 31949–56.

Lagarias, D. M., Wu, S-H. & Lagarias, J. C. (1995). Atypical phytochrome gene structure in the green alga *Mesotaenium caldariorum*. *Plant Molecular Biology*, **29**, 1127–42.

Macnab, R. & Koshland, D. E. Jr. (1974). Bacterial motility and chemotaxis: light-induced tumbling response and visualization of individual flagella. *Journal of Molecular Biology*, **85**, 399–406.

Mattick, J. S., Whitchurch, C. B. & Alm, R. A. (1996). The molecular genetics of type-4 fimbriae in *Pseudomonas aeruginosa* – a review. *Gene*, **179**, 147–55.

Meyer, T. E. (1985). Isolation and characterization of soluble cytochromes, ferredoxins and other chromophoric proteins from the halophilic bacterium *Ectothiorhodospira halophila*. *Biochimica et Biophysica Acta*, **806**, 175–83.

Meyer, T. E., Yakali, E., Cusanovich, M. A. & Tollin, G. (1987). Properties of a water soluble, yellow protein isolated from a halophilic phototrophic bacterium that has photochemical activity analogous to sensory rhodopsin. *Biochemistry*, **26**, 418–23.

Meyer, T. E., Fitch, J. C., Bartsch, R. G., Tollin, G. & Cusanovich, M. A. (1990). Soluble cytochromes and a photoactive yellow protein from the moderately halophilic purple phototrophic bacterium *Rhodospirillum salexigens*. *Biochimica et Biophysica Acta*, **1016**, 364–70.

Meyer, T. E., Cusanovich, M. A. & Tollin, G. (1993). Transient proton uptake and release is associated with the photocycle of the photoactive yellow protein from the purple phototrophic bacterium *Ectothiorhodospira halophila*. *Archives of Biochemistry and Biophysics*, **306**, 515–17.

Nelson, D. C. & Castenholz, R. W. (1982). Light responses of *Beggiatoa*. *Archives of Microbiology*, **131**, 146–55.

Parkinson, J. S. (1993). Signal transduction schemes of bacteria. *Cell*, **73**, 857–71.

Ragatz, L., Jiang, Z-Y., Bauer, C. E. & Gest, H. (1994). Phototactic purple bacteria. *Nature*, **370**, 104.

Ragatz, L., Jiang, Z-Y., Bauer, C. E. & Gest, H. (1995). Macroscopic phototactic behaviour of the purple photosynthetic bacterium *Rhodospirillum centenum*. *Archives of Microbiology*, **163**, 1–6.

Saranak, J. & Foster, K. W. (1997). Rhodopsin guides fungal phototaxis. *Nature*, **387**, 465–6.

Schlegel, H. G. (1986). *General Microbiology*. Cambridge, UK: Cambridge University Press.

Sprenger, W. W., Hoff, W. D., Armitage, J. P. & Hellingwerf, K. J. (1993). The eubacterium *Ectothiorhodospira halophila* is negatively phototactic, with a wavelength dependence that fits the absorption spectrum of the photoactive yellow protein. *Journal of Bacteriology*, **175**, 3096–104.

Stackebrandt, E., Rainey, F. A. & Ward-Rainey, N. (1996). Anoxygenic phototrophy across the phylogenetic spectrum: current understanding and future perspectives. *Archives of Microbiology*, **166**, 211–23.

Stock, A. M. & Mowbray, S. L. (1995). Bacterial chemotaxis: a field in motion. *Current Opinions in Structural Biology*, **5**, 744–51.

Thiemann, B. & Imhoff, J. F. (1995). Occurrence and purification of the photoactive yellow protein of *Ectothiorhodospira halophila* (PYP) and immunologically related proteins from *Rhodospirillum salexigens* and *Chromatium salexigens* and intracellular localization of PYP. *Biochimica et Biophysica Acta*, **1253**, 181–8.

Van Brederode, M. E., Hoff, W. D., Van Stokkum, I. H. M., Groot, M-L. & Hellingwerf, K. J. (1996). Protein folding thermodynamics applied to the photocycle of the photoactive yellow protein. *Biophysical Journal*, **71**, 365–80.

Woese, C. R. (1987). Bacterial evolution. *Microbiological Reviews*, **51**, 221–71.

Xie, A., Hoff, W. D., Kroon, A. R. & Hellingwerf, K. J. (1996). Glu46 donates a proton to the 4-hydroxycinnamate anion chromophore during the photocycle of photoactive yellow protein. *Biochemistry*, **35**, 14671–8.

Yang, H., Inokuchi, H. & Adler, J. (1995). Phototaxis away from blue light by an *Escherichia coli* mutant accumulating protoporphyrin IX. *Proceedings of the National Academy of Sciences, USA*, **92**, 7332–6.

Zacks, D. N., Derguini, F., Nakanishi, K. & Spudich, J. L. (1993). Comparative study of phototactic and photophobic receptor chromophore properties in *Chlamydomonas reinhardtii*. *Biophysical Journal*, **65**, 508–18.

THE ROLE OF LIGHT IN THE REGULATION OF MOSS DEVELOPMENT

DAVID J. COVE[1] AND TILMAN LAMPARTER[2]

[1]*School of Biology, University of Leeds, Leeds LS2 9JT UK*
[2]*Institut für Pflanzenphysiologie, Free University Berlin, Königin Luise Strasse 12–16, D-14195, Berlin, Germany*

INTRODUCTION

From a technical standpoint, mosses are micro-organisms. They can be cultured axenically under controlled conditions, on solid or in liquid media. Physiologically, mosses are higher plants containing the same hormones, using similar control circuitry for signal transduction and having gene sequences that often show remarkable homology to those of flowering plants. Genetically, they combine features of both microbes and higher plants, but the predominance of the haploid phase gives to mosses many of the technical features that commend fungi for genetic analysis. Mosses therefore provide ideal model systems for studying plant development using classical and molecular genetic techniques. In addition, developmental responses can be observed in living tissue at the level of the individual cell and so it is possible to study directly the role of light in morphogenetic programming. Consequently, their potential for bringing together molecular and cell biological analyses of the role of light in the regulation of development is considerable. These same qualities also allow the mechanisms regulating the timing of developmental events to be studied, although the full potential of mosses for the study of this aspect of development have yet to be realised (but see below).

Reviews have appeared earlier of the effect of light on moss development (Hartmann & Jenkins, 1984) and of the use of mosses as model systems (Cove, Knight & Lamparter, 1997).

MOSS DEVELOPMENT

Mosses show alternation of generations. The familiar moss plant, the predominant phase of the life cycle, is haploid, and is called the gametophyte generation because it produces gametes directly by mitosis. Zygotes are produced following fertilization of female gametes by free-swimming male gametes, and develop into the diploid sporophyte generation. This is, at least in part, nutritionally independent on the gametophyte (Courtice, Ashton & Cove, 1978), and comprises a spore capsule born on a stalk or seta. Spores, produced within the spore capsule by meiosis, germinate to produce a new generation of gametophytes.

Most developmental studies have been carried out on the gametophyte generation. In the majority of moss species, including *Funaria hygrometrica*, *Physcomitrella patens* and *Ceratodon purpureus*, the three species in which development has been studied most extensively, gametophyte development occurs in two stages. Spore germination leads to the production of a filamentous stage, the protonemata, which subsequently produces buds which develop into the leafy shoots that bear the gametangia. Protonemal filaments grow by the serial division of their apical cells. The sub-apical cells produced by the division of apical cells may also divide, but usually only once or twice. Divisions of the sub-apical cells of a filament do not lead to extension of that filament, but instead lead to the production of side branches. These side branches may either develop into further filaments or into buds that in turn develop into leafy shoots.

In *F. hygrometrica*, two types of protonemal filaments can be clearly distinguished. Chloronemata are slower growing and contain many large chloroplasts, whereas caulonemata grow more rapidly and contain fewer smaller chloroplasts. The same two cell types can be recognized in the protonemata of *P. patens*, although the degree of differentiation is somewhat less. In this species, the pattern of protonemal development has been studied in detail. Spore germination gives rise to primary chloronemal filaments. The apical cells of some primary chloronemal filaments can undergo differentiation to give rise to the second filament type, caulonema. Caulonemal apical cells extend more rapidly and divide more often than the apical cells of chloronemal filaments. Whereas the side branches produced by the division of sub-apical cells of primary chloronemal filaments develop into further chloronemal filaments, the side branches from caulonemal sub-apical cells have a number of developmental fates. Some remain as single-cell side branch initials, some develop into either further caulonemal filaments or into buds, but the majority develop into secondary chloronemal filaments. Both hormones and light quality play important roles in controlling the choice between these alternative developmental fates. The role of light is reviewed below. Figure 1 summarizes these developmental relationships.

Studies of the mechanisms controlling the timing of developmental events, particularly during protonemal development in *P. patens*, have recently been initiated. Although protonemal development is not rigorously determined, pattern can be described by a number of simple statements relating to the probability of developmental changes occurring in time and space (A. Russell & D. J. Cove, unpublished data). Examples of these statements, for development of *P. patens* under standard culture conditions, which utilize white light intensities of $5-20 \, \mathrm{W \, m^{-2}}$ and a temperature of $25\,^{\circ}\mathrm{C}$, are:

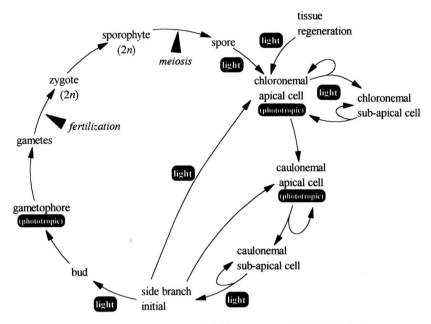

Fig. 1. Life cycle of *Physcomitrella patens*. Diploid stages are marked '(2n)'. All other stages are haploid. Steps marked 'light' require light.

- chloronemal apical cells divide about every 24 h and extend at a rate of about 4 μm h^{-1}.
- caulonemal apical cells have a cell cycle time of about 6 h and extend at about 40 μm h^{-1}.
- almost all caulonemal sub-apical cells divide after about 21 h, to produce a side-branch initial.
- 40% of caulonemal sub-apical cells divide again after a further 21 h, to produce a second side-branch initial.
- fewer than 1% of caulonemal sub-apical cells divide for a third time to produce a third side-branch initial, but the cell cycle time for such cells is again about 21 h.

Nothing is yet known of the mechanisms controlling the timing of the division cycles of protonemal cells but it is now possible to analyse the effect of mutation on the developmental parameters that have been established, to assess the complexity of their genetic control.

THE EFFECT OF LIGHT ON THE PATTERN OF GAMETOPHYTE DEVELOPMENT

Moss protonemata offer an unrivaled opportunity to analyse the effect of light on morphogenesis at the level of the individual cell. Filament apical cells grow by extension of their most apical region, and so protonemal filaments are comparable to pollen tubes, for which numerous studies have unravelled the signal transduction chain that mediates regulation of growth rate and direction. Protonemal filaments grow slower than pollen tubes but they grow continuously, allowing prolonged observation and allowing mutant isolation (see below). In addition, moss protonemata show numerous responses to light, often similar to responses shown by flowering plants, making mosses ideal subjects for studying the effect of light on pattern formation and cell morphogenesis. Most studies of the effect of light in mosses have been concentrated on the protonemal stage in which growth is essentially two-dimensional, and much less is known about photomorphogenesis in the three-dimensional leafy shoot. Despite their simple morphology, numerous effects of light have been identified and there is evidence that mosses possess the same set of photoreceptors as are found in flowering plants, namely UV-B, blue-light photoreceptors and phytochromes.

The pattern of protonemal development is dependent on light conditions. The effect of light periodicity has not been studied extensively, but an effect of photoperiod on the overall time of the cell division cycle of protonemal apical cells of *C. purpureus* has been reported (Larpent-Gourgaud & Aumaître, 1980). In *P. patens* no major effects of day length can be detected (D. J. Cove, unpublished data). The pattern of development in white light at intensities of 5 W m^{-2} or more is essentially similar in continuous white light or in a 16 h light/8 h dark cycle. Development is however slower in intermittent light conditions. The time taken to reach developmental landmarks such as the formation of the first buds, is dependent only on the total number of hours of light received, even in cycles as extreme as 4 h light/20 h dark (D. J. Cove, unpublished data). Thus it seems that, provided high-intensity white light is perceived by the developing gametophyte for some period every 24 h, development follows the pattern described for standard conditions. However, development in darkness or in lower intensities of white light shows major differences. Chloronemal cells *P. patens* do not divide in darkness, even when an exogenous source of reduced carbon is supplied (Jenkins & Cove, 1983a). Chloronemal cells of *F. hygrometrica* also show a requirement for light for growth, but this appears to be confined to the need for light for photosynthesis, since the action spectrum for the light requirement closely parallels that for photosynthesis, and chloronemal extension and cell division both occur in darkness in this species when a source of reduced carbon is supplied (Berthier, Larpent & Larpent-Gourgaud, 1976; Brière, Buis & Larpent, 1979). In both *F. hygrometrica* and *P. patens*, the division of caulonemal apical cells is essentially unaffected in darkness. The protonemal cells of *C. purpureus* are not differentiated into chloronema and caulonema, and most filament apical cells continue to divide in darkness. Tissue regeneration, which in *P. patens* occurs via chloronemal formation and therefore in that species does not occur in darkness, does not require light in *C. purpureus*.

In *P. patens*, light intensity affects the division of caulonemal sub-apical cells and the developmental fates of the side branches that are formed from them. In darkness, sub-apical cell division is reduced and after prolonged culture in darkness, side branches cease to be produced. Only very low light intensities or brief exposures to higher light intensities are required to induce side branch formation (Cove & Ashton, 1988). At low light intensities, the principal developmental fate of side branch initials is the production of further caulonemal filaments. As light levels increase, caulonemal filament production falls and at intermediate light levels, almost all side branch initials remain undivided. With increasing light levels, the production of chloronemal filaments becomes more prevalent (Cove *et al.*, 1978; G. I. Jenkins & D. J. Cove, unpublished data). Bud production, which can be enhanced considerably by the application of exogenous cytokinins, is also dependent on light in both *F. hygrometrica* (Simon & Naef, 1981) and *P. patens* (Cove & Ashton, 1988). In the latter species, maximum bud production requires white light intensities greater than about 1 W m^{-2}, which lead to about 3% of side branch initials developing into buds (A. Russell & D. J. Cove, unpublished data). The development of leafy shoots is also affected by light intensity, with higher light levels needed both to prevent etiolation and to induce leaf expansion (Cove *et al.*, 1978).

C. purpureus filaments form almost no side branches when grown in darkness and in this species branch formation is also light dependent (Kagawa *et al.*, 1997). Blue and red light both initiate the formation of side branches, the red light effect being controlled by the photoreceptor phytochrome.

THE EFFECT OF LIGHT ON CELL POLARITY

Light not only effects the developmental fate of gametophyte cells, but also affects the polarity of their growth. In darkness, caulonemal apical cells of *F. hygrometrica* and *P. patens*, and protonemal apical cells of *C. purpureus*, grow negatively gravitropically (Jenkins, Courtice & Cove, 1986; Walker & Sack, 1990). Unidirectional light treatments lead to a phototropic response from these cells by changing the polarity of the apical tip cell. In *P. patens* the mode of response is dependent on the light intensity. At low intensities, filaments grow positively phototropically, that is towards the light. In higher light intensities, caulonemal apical cells of *P. patens* grow perpendicular to the light direction. The apical cells of primary chloronemal filaments (i.e. filaments arising directly following spore germination or tissue regeneration) will not divide or grow in darkness, but show phototropic responses to unidirectional light (Jenkins & Cove, 1983*b*). In low intensities of red light, these cells grow positively phototropically whereas in high levels of red light, growth is lateral to the light direction. There is evidence that the switch from the low level response to the high level response is mediated via phytochrome

(Jenkins & Cove, 1983*b*). *C. purpureus* tip cells grow positively phototropically under low and high intensity of red light. The phototropic response of apical cells of *C. purpureus* filaments has been shown to be mediated solely by phytochrome (Hartmann, Klingenberg & Bauer, 1983) and there is similar but less complete evidence for the involvement of phytochrome in the phototropic response of caulonemal apical cells of *P. patens* (Cove *et al.*, 1978). In high levels of blue light, growth of primary chloronemal apical cells of *P. patens* and of protonemal apical cells of *C. purpureus* is away from the light direction, suggesting that a blue-light receptor may also be involved. Filament apical cells also show tropic responses to polarized light, polarotropism (Nebel, 1969; Jenkins & Cove, 1983*b,c*; Hartmann & Jenkins, 1984), providing evidence for the anisotropic distribution of phytochrome.

Protoplasts can be readily isolated from moss protonemal tissue. Cultured in osmotically buffered medium, protoplasts regenerate directly to form protonemal filaments. Protoplasts isolated from chloronemal tissue of *P. patens* show a requirement for light in order to regenerate (Jenkins & Cove, 1983*a*). In darkness, protoplasts from this species form cell walls and become osmotically resistant, but remain spherical. At low light levels, cell division occurs but polar outgrowth and filament formation only occurs at high levels of white light.

Protoplasts from protonemal tissue of *C. purpureus* regenerate and show polar outgrowth in darkness, but the orientation of the protoplast regeneration axis can be influenced by light (Cove, Quatrano & Hartmann, 1996). Because light is not essential for polar axis formation in this species, it is possible to investigate the effects of a broad range of light treatments on polar axis formation. As a result of these studies, it has become evident that the role of light in establishing the orientation of the regeneration axis is complex. Before polar outgrowth can be observed, two morphogenetic decisions are taken. Under standard conditions, protoplasts cultured in unidirectional light fix the alignment of their regeneration axis about nine hours before polar outgrowth occurs. Consequently, re-orientation of protoplasts with respect to the light direction does not affect the alignment of the regeneration axis of those protoplasts that regenerate in the nine hours after re-orientation. Axis alignment once fixed, is initially apolar and potentially outgrowth can occur either towards the light source or away from it. The decision on the polarity of the axis is taken about five hours before polar outgrowth is observed. Different wavelengths of light affect axis alignment and axis polarity differentially. Thus in red light alignment is strongly determined but polarity is only determined weakly, whereas in blue light, alignment is weak but polarity is strong (see Fig. 2). To explain these findings, it has been proposed that two distinct morphogenetic gradients are established, one determining axis alignment, and the other subsequently determining polarity along the aligned axis (Cove *et al.*, 1996).

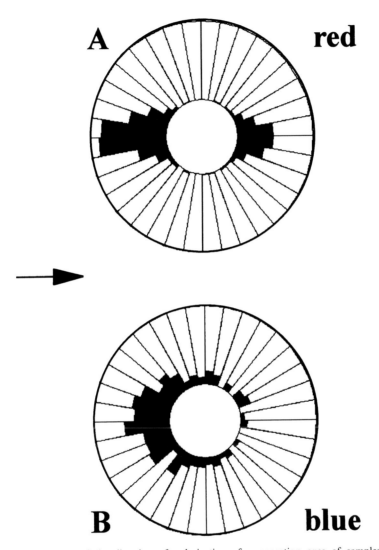

Fig. 2. Distribution of the direction of polarization of regeneration axes of samples of protoplasts cultured in uni-directional red light (A) and uni-directional blue light (B). Light direction is indicated by the arrow. In red light, alignment of axes with respect to the light source is good, but only 70% of protoplast axes are polarized towards the light source, the remainder showing the opposite polarity. In blue light, almost all protoplast polarize their axes towards the light source, but the axis alignment is less precisely determined. For details of the experimental procedures used to generate these distributions see Cove et al., 1996.

OTHER EFFECTS OF LIGHT

Chlorophyll synthesis, chloroplast division and chloroplast movement

Effects of light on chlorophyll accumulation and plastid development show similarities to the de-etiolation process of flowering plants. However, aetioplast-like organelles have not been found in mosses, indicating that the adaptation to darkness is not as strong as in flowering plants. Chlorophyll can still be detected in plastids of *C. purpureus* protonemata grown for 4 weeks in darkness and indirect evidence suggests that chlorophyll synthesis also occurs in darkness (Lamparter *et al.*, 1997). This is possible because, in contrast to angiosperms, the last step in chlorophyll synthesis, the proto-chlorophyllide-chlorophyll oxidoreductase, is not light-requiring in this species. It has been shown that chlorophyll biosynthesis is up-regulated in light both by way of phytochrome and of a blue light photoreceptor; the chlorophyll content of light-adapted cells being up to 30 times higher that of dark-grown cells. Plastids of dark-grown *C. purpureus* tip cells show a typical zonation (see Fig. 3A). Few plastids are located in the apical dome, this region is followed by a plastid-free zone and then by a zone with higher plastid density. This zonation is lost in light grown cells (see Fig. 3B). Here, plastids are more or less evenly distributed over the entire tip cell, they are also more numerous than in dark-adapted cells. Analysis of a mutant defective in the biosynthesis of the phytochrome chromophore has highlighted the role of phytochrome in chlorophyll accumulation (Lamparter *et al.*, 1997). Chlorophyll data of this mutant indicate that chlorophyll synthesis is also affected in a phytochrome-independent manner in a similar way to that found in the corresponding mutants of flowering plants (Terry, 1997). This suggests that a block in the synthesis of the phytochrome chromophore leads to haem accumulation and to feedback-inhibition of δ-aminolaevulinic acid synthesis and protoporphyrin synthesis.

A blue light effect has been found for chloroplast division of *P. patens*. The mutant PC22 contains only one large chloroplast per cell, as a result of impaired chloroplast division. Chloroplast division in this mutant can partially be restored by additional blue light or by an exogenous supply of cytokinin (Reski, 1994), implying that both agents play a role in moss chloroplast division.

Light-dependent chloroplast movement can be detected in almost any plant species (Haupt & Häder, 1994). Whereas phytochrome plays an important role in the green algae, *Mesotaenium* sp. and *Mougeotia scalaris* (Kraml, 1994) and in ferns (Wada & Sugai, 1994), this does not seem to be the case in mosses. Voerkel (1933) had already shown that chloroplast movement in leaves of *F. hygrometrica* is induced by blue light, whereas red light is inactive. In *C. purpureus* protonemata, chloroplast movement is also only induced by blue and not by red light. Chloroplast movement was induced in a phytochrome-chromophore deficient mutant (Kagawa *et al.*, 1997).

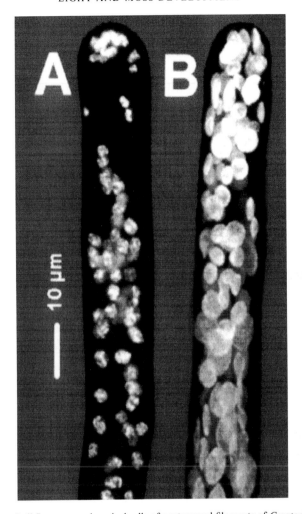

Fig. 3. Chlorophyll fluorescence in apical cells of protonemal filaments of *Ceratodon purpureus* grown for 6 days in darkness (A) or for 5 days in darkness followed by 1 day red light (B). The images were obtained using a confocal laser-scanning microscope. Chlorophyll fluorescence is displayed in white/grey (corresponding to the intensity) on a black background (see Lamparter *et al.*, 1997 for further details). To show the contour of the cells, a software-processed transmission image was superimposed, so that the background outside the cell is displayed in grey. Exposure to red light has led to a breakdown in the zonation of plastids as well as to an increase in chlorophyll and in plastid size.

Gravitropism

Caulonemal apical cells of *P. patens* and of protonemal apical cells of *C. purpureus* show negative gravitropism in darkness. Aphototropic mutants (see below) have been used to demonstrate that there is active suppression of the gravitropic response by light. Mutants of both species in which the phenotypic abnormalities are specifically limited to phototropism, show no gravitropic response in the light. However, in mutants of *C. purpureus* that are affected in the synthesis of the phytochrome chromophore, protonemal apical cells show a negative gravitropic response in red light (Lamparter *et al.*, 1996). It therefore appears that the gravitropic response of apical cells is actively turned off in the light by way of the phytochrome photoreceptor. In *C. purpureus*, gravitropism is also affected by a blue light photoreceptor; blue light irradiation can reverse the gravitropic response so that filaments grow positively gravitropically (T. Lamparter, J. Hughes & E. Hartmann, unpublished data).

Spore germination

It has been shown in a number of moss species, that light stimulates the level of spore germination, and in some, including *P. patens*, no spore germination occurs in darkness even when an exogenous source of reduced carbon is supplied. In various studies, red light has been demonstrated to be the most effective wavelength for the induction of germination, and far-red reversibility of the effect of red light has been demonstrated for *F. hygrometrica* (Bauer & Mohr, 1959), *C. purpureus* and *Dicranum scoparium* (Valanne, 1966) and *P. patens* (Schild, 1981), indicating that phytochrome is the photoreceptor concerned.

MUTANT STUDIES

Mutants of *P. patens* which have abnormal phototropic responses have been known for some time (Cove *et al.*, 1978) and are designated *ptr*. Detailed phenotypic and genetic analyses have been carried out on *ptr* mutants (Knight & Cove, 1989). The primary screen used to identify mutants was an inability of caulonemal apical cells to respond to a unilateral light source. This screen has identified a number of mutants all of which have essentially similar phenotypes. The caulonemal apical cells of *ptr* mutants have not only lost their phototropic response but are also insensitive to polarized light, confirming that phototropism and polarotropism are closely related. The caulonemal filaments grow much straighter than those of wild-type strains, which tend to spiral. These mutants have also lost the phototropic response of their gametophores, suggesting that there must be common components for the signalling pathways used in the single cell response of filament apical cells and in the multicellular response of shoots. Surprisingly, such mutants have not lost the phototropic response of their primary chloronemal apical cells, but switch from the low light response (towards the light source) to the high level response (at right angles to a source of red light, away from a source of blue light) at a lower light intensity than the wild type (Jenkins & Cove, 1983*c*). Although *ptr* mutants show pleiotropic effects on phototropism and polarotropism, no other abnormalities of light-mediated developmental processes have been detected. Thus spore germination, chloronemal development, light reversal of gravitropism, light induction of bud formation, light suppression of shoot etiolation and light induction of leaf expansion are all similar to the responses of wild-type strains. Genetic analysis of *ptr* mutants has been carried out using both conventional sexual crosses and complementation tests using somatic hybrids obtained following protoplast fusion (Courtice, 1979). These analyses reveal that there are at least three *ptr* genes, and that *ptr*A and *ptr*B are unlinked. There is no molecular genetic analysis of *ptr* mutants in *P. patens*.

More recently, phototropically abnormal mutants of *C. purpureus* have also been isolated (Lamparter *et al.*, 1996). These mutants have been shown to fall into two distinct classes. One mutant class is essentially similar to *P. patens ptr* mutants, being affected in the phototropic and polarotropic responses of protonemal apical cells, but appearing to be otherwise unaltered. The other class has further pleiotropic effects including reduced chlorophyll synthesis, and a gravitropic response that is insensitive to light. The phototropic response of this second class of mutant can be restored by supplying the tetrapyrroles biliverdin, the proposed precursor of the phytochrome chromophore, or phycocyanobilin, which may replace the phytochrome chromophore (Lamparter *et al.*, 1997). These observations support the conclusion that this class of mutants is impaired in the synthesis of the phytochrome chromophore. Their phenotype therefore confirms that phytochrome is used in mosses to detect light direction in phototropism, but also provides evidence that the gravitropic response of caulonemal/protonemal apical cells is actively suppressed in light, by way of phytochrome.

It is possible to use a selective screen to isolate *ptr* mutants in *C. purpureus*, and so many more mutants have already been identified in this species than in *P. patens*. However, complementation analyses have not yet been carried out and so the genetic complexity of the two classes of *ptr* mutants has yet to be determined.

PHOTORECEPTORS

Blue light photoreceptors

Numerous blue light responses have been found in mosses (see above). Since the molecular nature of most blue light photoreceptors has remained obscure for many decades, their action has mostly been defined via physiological experimentation. Their effects need to be distinguished from those of phytochromes which also absorb in the blue region of the spectrum. Mutants with a defect in the biosynthesis of the phytochrome chromophore help to distinguish between effects mediated by a separate blue light photoreceptor and phytochrome. The recent identification of *CRY* genes (Ahmad & Cashmore, 1996) and the *NPH1* gene (Briggs & Liscum, 1997) in flowering plants, which encode putative blue light photoreceptors, has stimulated PCR searches to be carried out for similar genes in the lower plants. *CRY* genes have been found in *P. patens* as well as in some other non-vascular plants (H. Ü. Kosukisaoglu & H. A. W. Schneider-Poetsch, unpublished data).

Phytochromes

The photochromic nature of phytochrome allows a more detailed analysis of this photoreceptor. Spectrophotometric detection of phytochrome was described for gametophytes of the moss *Atrichum undulatum* (Lindemann, Braslavsky & Schaffner, 1989) and for protonemata of *C. purpureus* (Lamparter *et al.*, 1995) and *P. patens* (T. Lamparter, unpublished data). Searches for moss DNA sequences homologous to those coding for the phytochrome apoprotein in flowering plants revealed one gene in *P. patens* (Schneider-Poetsch *et al.*, 1994) and two in *C. purpureus*, designated *CpPHY1* and *CpPHY2* (Hughes, Lamparter & Mittmann, 1996; Pasentsis *et al.*, 1997). Recently, evidence has been presented that there may also be a second phytochrome gene in *P. patens* (Pasentsis *et al.*, 1997). At least in *C. purpureus*, it is clear that phytochrome abundance is light regulated with kinetics similar to other lower plants (Lamparter *et al.*, 1995; Pasentsis *et al.*, 1997). Light-adapted filaments have relatively low levels of phytochrome but following incubation in darkness, the phytochrome content increases steadily, and is depleted again if protonemata are transferred back to light (Lamparter *et al.*, 1995). Two different modes of regulation have been suggested. Since the photosynthesis inhibitor, DCMU (3-(3,4-dichlorophenyl)-1,1-dimethylyurea) blocks the light-induced decrease of phytochrome, an effect of photosynthesis on the regulation of phytochrome abundance seems to be involved (Pasentsis *et al.*, 1997). However, mRNA levels show very little dependence on light conditions, implying that the regulation of phytochrome levels occurs post-transcriptionally (Pasentsis *et al.*, 1997). Data from phytochrome chromophore-deficient mutants imply that phytochrome is degraded in its Pfr form. Some mutants contain around 15% of the wild-type level of photoactive phytochrome. It has been shown that this holophytochrome is specifically lost if mutants are kept in red light, whereas the abundance of apoprotein (lacking chromophore) was not affected by the light treatment (H. Esch & T. Lamparter, unpublished data). The finding that regulation of phytochrome abundance occurs post-transcriptionally indicates that, in this respect, regulation is similar to that for phytochrome A in angiosperms; however, other aspects of regulation are in contrast to those in angiosperms.

Mutants lacking the phytochrome chromophore allow the identification of responses mediated by phytochrome. However, no mutant with a defect in the synthesis of a phytochrome apoprotein has yet been isolated. Thus it is not clear if either or both of the two phytochrome genes of *C. purpureus* encode the active photoreceptor, and it is even possible that an as yet unidentified phytochrome might play this role. Immunoblots, using an anti-apophytochrome antibody that does not discriminate between *CpPHY1* and *CpPHY2*, reveal only one band corresponding to a molecular weight of around 130 kD. The protein precipitated with this antibody is also stained with the monoclonal Z-3B1 antibody, the epitope of which is present in *CpPHY2* but not in *CpPHY1* (Lamparter *et al.*, 1995). In addition, antibodies which specifically stain *CpPHY1* failed to give a signal (Algarra, Linder & Thümmler, 1993; J. Hughes & T. Lamparter, unpublished data). Thus *CpPHY1* seems either not to be expressed in *C. purpureus* filaments or to be expressed at very low levels, whereas *CpPHY2* most likely encodes the bulk of photoreversible phytochrome in that species. These protein data are consistent with the finding that mRNA levels of *CpPHY1* are very low, around 1/50 of the *CpPHY2* mRNA levels. It has recently been found that two different splice products from *CpPHY2* scan be detected in *C. purpureus*, the relative abundance of which is dependent on light conditions (Pasentsis *et al.*, 1997).

SIGNAL TRANSDUCTION

The ease of cytological observation of moss tissue has also allowed Ca^{2+} and actin, cell components that may play a part in the phototropic response, to be monitored. Chlorotetracycline-stained Ca^{2+} shows a tip to base gradient in the apical cells of dark-grown protonemata of *C. purpureus*. Following unilateral irradiation, a shift of this signal towards the site of prospective growth is observed, before the outgrowth itself can be seen (Hartmann & Weber, 1988; Meske & Hartmann, 1995). This suggests that Ca^{2+} may act as a second messenger in the phytochrome-mediated phototropic response, even though the precise role of Ca^{2+} has yet to be determined. A phytochrome effect on membrane potential and possibly on plasmalemma calcium channels has been described for *P. patens* by Ermolayeva *et al.* (1996) and Ermolayeva, Sanders & Johannes (1997).

P. patens strains, transgenic for apo-aquorin, have also been used to measure transient changes in cytoplasmic Ca^{2+} levels (Russell *et al.*, 1996). This technique has been used recently to show that irradiation of dark-adapted protonemal tissue with blue light leads to a transient increase in Ca^{2+} level in an intensity-dependent manner (A. Russell, T. L. Wang & D. J. Cove, unpublished data). Irradiation of the same tissue with red light has no effect on Ca^{2+} levels. The effect of blue light is attenuated by the calmodulin inhibitor W7, suggesting that calmodulin is involved in the transduction pathway. Although it is not yet possible to establish the final developmental outcome that results from these treatments, the principal result of brief exposure to light of dark-adapted protonemata of *P. patens* is the induction of cell division in filament sub-apical cells, to produce side-branch initials.

A red light-induced effect on the organization of the actin cytoskeleton can be observed in *C. purpureus* (Meske & Hartmann, 1995). As with changes in Ca^{2+} levels, the rearrangement of the actin cytoskeleton also precedes outgrowth towards the new direction. The microtubule cytoskeleton does not seem to play a part in the phototropic response, but is involved in gravitropism; inhibitors of microtubule assembly eliminate the gravitropic response but the phototropic response is not impaired (Meske, Ruppert & Hartmann, 1996).

Agents like the ionophore monensin that transiently inhibit growth of *C. purpureus*, show that the information about light direction can be stored for several hours. Monensin-treated filaments, unilaterally irradiated with red, show no bending because growth is inhibited. Growth is resumed after washing out the agent. The tip cell now grows towards the direction from which it was irradiated, even if the light has been switched off (Hartmann & Weber, 1988). The information about light direction can be stored for several hours, possibly by phytochrome itself (the red light effect remains far-red reversible) or an early intermediate of the signal transduction cascade. In contrast, filaments have not been treated with monensin, do not seem to store the phytochrome-mediated signal; they require a continuous unilateral irradiation to show the typical phototropic response. Regenerating proto-plasts of both *C. purpureus* (Cove, Quatrano & Hartmann, 1996) and *P. patens* (D. J. Cove, unpublished data) also retain a memory of light direction for long periods after transfer to darkness. As described above, regenerating protoplasts of *C. purpureus* fix the alignment of the regeneration axis about 9 hours before protoplasts become visibly asymmetrical. However, protoplasts regenerating more than nine hours after exposure to unidirectional red light, continue to orient their axis to the direction of this light, even if kept in darkness or in omni-directional white light (Cove, Quatrano & Hartmann, 1996). These observations and those involving monensin suggest that there is active control over the gradients controlling polar outgrowth, such that they are revised rapidly in changed light conditions but retained when growth is not possible or when no light input is available.

REFERENCES

Ahmad, M. & Cashmore, A. R. (1996). Seeing blue: the discovery of cryptochrome. *Plant Molecular Biology*, **30**, 851–61.
Algarra, P., Linder, S. & Thümmler, F. (1993). Biochemical evidence that phyto-chrome of the moss *Ceratodon purpureus* is a light-regulated protein kinase. *FEBS Letters*, **315**, 69–73.

Bauer, L. & Mohr, H. (1959). Der Nachweis des reversiblen Hellrot-Dunkelrot Reaktionssystems bei Laubmoosen. *Planta*, **54**, 68–73.

Berthier, J., Larpent, J. P. & Larpent-Gourgaud, M. (1976). Light action on vegetative propagation in bryophytes. *Journal of the Hattori Botanical Laboratory*, **41**, 193–203.

Brière, C., Buis, R. & Larpent, J. P. (1979). Cellular growth and cellular division in relation to age and illumination in the protonemata of *Ceratodon purpureus* Brid. *Zeitschrift für Pflanzenphysiologie*, **95**, 315–22.

Briggs, W. & Liscum, E. (1997). The role of mutants in the search for the photoreceptor for phototropism in higher plants. *Plant, Cell and Environment*, **20**, 766–72.

Courtice, G. R. M. (1979). Developmental genetic studies of *Physcomitrella patens*. PhD thesis, University of Cambridge, UK.

Courtice, G. R. M., Ashton, N. W. & Cove, D. J. (1978). Evidence for the restricted passage of metabolites into the sporophyte of the moss, *Physcomitrella patens* (Hedw.) Br. Eur. *Journal of Bryology*, **10**, 191–8.

Cove, D. J. & Ashton, N. W. (1988). Growth regulation and development in *Physcomitrella patens*: an insight into growth regulation and development in bryophytes. *Botanical Journal of the Linnean Society*, **98**, 247–52.

Cove, D. J., Schild, A., Ashton, N. W. & Hartmann, E. (1978). Genetic and physiological studies of the effect of light on the development of the moss *Physcomitrella patens*. *Photochemistry and Photobiology*, **27**, 249–54.

Cove, D. J., Quatrano, R. S. & Hartmann, E. (1996). The alignment of the axis of asymmetry in regenerating protoplasts of the moss *Ceratodon purpureus*, is determined independently of axis of polarity. *Development*, **122**, 371–9.

Cove, D. J., Knight, C. D. & Lamparter, T. (1997). Mosses as model systems. *Trends in Plant Science*, **2**, 99–105.

Ermolayeva, E., Hohmeyer, H., Johannes, E. & Sanders, D. (1996). Calcium-dependent membrane depolarisation activated by phytochrome in the moss *Physcomitrella patens*. *Planta*, **199**, 352–8.

Ermolayeva, E., Sanders, D. & Johannes, E. (1997). Ionic mechanism and role of phytochrome-mediated membrane depolarisation in caulonemal side branch initial formation in the moss *Physcomitrella patens*. *Planta*, **201**, 109–18.

Hartmann, E. & Jenkins, G. I. (1984). Photomorphogenesis of Mosses and Liverworts. In *The Experimental Biology of Bryophytes*, ed. A. F. Dyer & J. G. Duckett, pp. 203–28. London: Academic Press.

Hartmann, E. & Weber, M. (1988). Storage of the phytochrome-mediated phototropic stimulus of moss protonemal tip cells. *Planta*, **175**, 39–49.

Hartmann, E., Klingenberg, B. & Bauer, L. (1983). Phytochrome-mediated phototropism in protonemata of the moss. *Ceratodon purpureus* Brid. *Photochemistry and Photobiology*, **38**, 599–603.

Haupt, W. & Häder, D. P. (1994). Photomovement. In *Photomorphogenesis in Plants*, 2nd edn. ed. R. E. Kendrick & G. H. M. Kronenberg, pp. 707–31. Dordrecht: Kluwer Academic Publishers.

Hughes, J. E., Lamparter, T. & Mittmann, F. (1996). *CpPHY2* (*PHYCER2*), a 'normal' phytochrome in *Ceratodon* (Accession No. U56698). *Plant Physiology*, **112**, 446.

Jenkins, G. I. & Cove, D. J. (1983*a*). Light requirements for regeneration of protoplasts of the moss *Physcomitrella patens*. *Planta*, **157**, 39–45.

Jenkins, G. I. & Cove, D. J. (1983*b*). Phototropism and polarotropism of primary chloronemata of the moss *Physcomitrella patens*: responses of the wild type. *Planta*, **158**, 357–64.

Jenkins, G. I. & Cove, D. J. (1983*c*). Phototropism and polarotropism of primary chloronemata of the moss *Physcomitrella patens*: responses of mutant strains. *Planta*, **159**, 432–8.

Jenkins, G. I., Courtice, G. R. M. & Cove, D. J. (1986). Gravitropic responses of wild-type and mutant strains of the moss, *Physcomitrella patens*. *Plant, Cell and Environment*, **9**, 637–44.

Kagawa, T., Lamparter, T., Hartmann, E. & Wada, M. (1997). Phytochrome-mediated branch formation in protonemata of the moss *Ceratodon purpureus*. *Journal of Plant Research* (in press).

Knight, C. D. & Cove, D. J. (1989). The genetic analysis of tropic responses. *Environmental and Experimental Botany*, **29**, 57–70.

Kolukisaoglu, H. U., Braun, B., Martin, W. F. & Schneider-Poetsch, H. A. (1993). Mosses do express conventional, distantly B-type-related phytochromes. Phytochrome of *Physcomitrella patens* (Hedw.). *FEBS Letters*, **334**, 95–100.

Kraml, M. (1994). Light direction and polarization. In *Photomorphogenesis in Plants*, 2nd edn. eds R. E. Kendrick & G. H. M. Kronenberg, pp. 417–46. Dordrecht: Kluwer Academic Publishers.

Lamparter, T., Podlowski, S., Mittmann, F., Schneider-Poetsch, H. J., Hartmann, E. & Hughes, J. (1995). Phytochrome from protonemal tissue of the moss *Ceratodon purpureus*. *Journal of Plant Physiology*, **147**, 426–34.

Lamparter, T., Esch, H., Cove, D. J., Hughes, J. & Hartmann, E. (1996). Aphototropic mutants of the moss *Ceratodon purpureus* with spectrally normal and with dysfunctional phytochrome. *Plant, Cell and Environment*, **19**, 560–8.

Lamparter, T., Esch, H., Cove, D. J. & Hartmann, E. (1997). Phytochrome control of phototropism and chlorophyll accumulation in the apical cells of protonemal filaments of wild-type and an aphototropic mutant of the moss *Ceratodon purpureus*. *Plant and Cell Physiology*, **38**, 51–8.

Larpent-Gourgaud, M. & Aumaître, M. P. (1980). Photoperiodism and morphogenesis of the protonema of *Ceratodon purpureus* (Hedw.) Brid. *Experientia*, **36**, 1366.

Lin, C., Ahmad, M. & Cashmore, A. R. (1996). *Arabidopsis* cryptochrome 1 is a soluble protein mediating blue light-dependent regulation of plant growth and development. *Plant Journal*, **10**, 893–902.

Lindemann, P., Braslavsky, S. E. & Schaffner, K. (1989). Partial purification and initial characterisation of phytochrome from the moss *Atrichum undulatum* P. Beauv. grown in the light. *Planta*, **178**, 436–42.

Meske, V. & Hartmann, V. (1995). Reorganization of microfilaments in protonemal tip cells of the moss *Ceratodon purpureus* during the phototropic response. *Protoplasma*, **188**, 58–68.

Meske, V., Ruppert, V. & Hartmann, V. (1996). Structural basis for the red light induced repolarization of tip growth in caulonema cells of *Ceratodon purpureus*. *Protoplasma*, **192**, 189–98.

Nebel, B. J. (1969). Response of moss protonemata to red and far-red polarized light: evidence for disc-shaped phytochrome photoreceptors. *Planta*, **87**, 170–9.

Pasentsis, K., Paulo, N., Algarra, P., Dittrich, P. & Thümmler, F. (1997). Characterization and expression of the phytochrome gene family in the moss *Ceratodon purpureus*. *Plant Journal*, in press.

Reski, R. (1994). Plastid genes and chloroplast biogenesis. In *Cytokinins, Chemistry, Activity and Function*, eds D. W. S. Mok & M. C. Mok, pp. 179–95. Ann Arbor, London, Tokyo: CRC Press.

Russell, A. J., Knight, M. R., Cove, D. J., Knight, C. D., Trewavas, A. J. & Wang, T. L. (1996). The moss, *Physcomitrella patens*, transformed with apoaequorin cDNA responds to cold shock, mechanical perturbation and pH with transient increases in cytoplasmic calcium. *Transgenic Research*, **5**, 167–70.

Schild, A. (1981). Untersuchungen zur Sporenkeimung and Protonemaentwicklung bei dem Laubmoos *Physcomitrella patens*. PhD thesis, University of Mainz, Germany.

Schneider-Poetsch, H. A. W., Marx, S., Kolukisaoglu, H. Ü., Hanelt, S. & Braun, B. (1994). Phytochrome evolution: phytochrome genes in ferns and mosses. *Physiologia Plantarum*, **91**, 241–50.

Schumaker, K. S. & Gizinski, M. J. (1995). 1,4-Dihydropyridine binding sites in moss plasma membranes. Properties of receptors for a calcium channel antagonist. *Journal of Biological Chemistry*, **270**, 23461–7.

Simon, P. E. & Naef, J. B. (1981). Light dependency of cytokinin-induced bud initiation in protonemata of the moss, *Funaria hygrometrica*. *Physiologia Plantarum*, **53**, 13–18.

Terry, M. (1997). Phytochrome chromophore-deficient mutants. *Plant, Cell and Environment*, **20**, 740–5.

Vallanne, N. (1966). The germination phases of moss spores and their control by light. *Annales Botanici Fennici*, **3**, 1–60.

Voerkel, S. H. (1933). Untersuchungen über die Phototaxis der Chloroplasten. *Planta*, **21**, 156–205.

Wada, M. & Sugai, M. (1994). Photobiology of ferns. In *Photomorphogenesis in Plants*, 2nd edn. ed. R. E. Kendrick & G. H. M. Kronenberg, pp. 783–802. Dordrecht: Kluwer Academic Publishers.

Walker, L. M. & Sack, F. D. (1990). Amyloplasts as possible statoliths in gravitropic protonemata of the moss *Ceratodon purpureus*. *Planta*, **181**, 71–7.

PHOTOSYNTHETIC LIGHT HARVESTING

RICHARD J. COGDELL[1], PAUL FYFE[1], NIALL FRASER[1], CHRIS LAW[1], TINA HOWARD[1], KAREN MCLUSKEY[2], STEVEN PRINCE[2], ANDY FREER[2] AND NEIL ISAACS[2]

Department of Biochemistry and Molecular Biology[1] and Department of Chemistry[2], University of Glasgow, Glasgow G12 8QQ, UK

INTRODUCTION

Photosynthesis usually begins with the absorption of light energy by the light-harvesting or antenna system. This is then followed by a fast, efficient energy transfer to the reaction centres, where this absorbed solar energy is 'trapped' and used to 'power' the primary charge separation reactions. In purple bacteria, for example, it has been estimated that about 90% of the absorbed photons are used productively by the reaction centres (Sundström & van Grondelle, 1995). However, the overall efficiency of the primary reactions is a lot less when calculated on a 'per energy basis' and depends strongly on the wavelength of the absorbed photon.

Antenna complexes have evolved to increase the effective cross-sectional area for light absorption of each reaction centre. In the absence of an antenna system, photosynthesis could still occur but only in 'high' light conditions. The light-harvesting complexes allow each reaction centre to be supplied with photons at a suitable rate permitting photosynthesis to occur over a wide range of incidence light intensities. Photosynthetic organisms could, in principle, react to changes in light intensity by making more and more reaction centres, thereby trying to absorb a greater fraction of the incoming photons to sustain their growth. However, since each reaction centre needs an electron transport chain to utilize the products of the primary reactions to produce ATP and reduced pyridine nucleotides, this option is very expensive in terms of metabolic energy (it requires a lot of extra protein synthesis). It has proved advantageous to evolve specific antenna complexes and to respond to changes in incident light intensity by elaborating a larger photosynthetic unit.

Now is a very exciting time for people who are interested in the molecular details of the structure and function of antenna complexes. There is almost an embarrassment of riches with high-resolution structures now available for two types of water soluble Bchla-proteins (FMO) from green sulphur bacteria (Fenna & Matthews, 1975; Tronrud & Matthews, 1986; Li *et al.*, 1997), a number of water soluble phycobiliproteins (for example, see

Schrimer, Bode & Huber, 1987 and, for a review, Sidler, 1994), the major light-harvesting complex from spinach (LHC2) (Kühlbrandt, Wang & Fujiyoshi, 1994), two types of purple bacterial antenna complexes (LH2) (McDermott *et al.*, 1995; Koepke *et al.*, 1996) and the water-soluble peridinin–chlorophyll *a* complex from a dinoflagellate (Hoffmann *et al.*, 1996). The main features of those antenna complexes which contain chlorophylls are summarized in Table 1. It is now timely therefore, to compare these and to ask whether they share any general features? In other words, are there any clear structural patterns or motifs that characterize an antenna complex?

This type of approach has been very successful when applied to reaction centres (Michel & Deisenhofer, 1988; Krauss *et al.*, 1996). High-resolution structures are currently only available for two species of purple bacteria, *Rhodobacter sphaeroides* (Allen *et al.*, 1987) and *Rhodopseudomonas viridis* (Deisenhofer *et al.*, 1984). However, sequence comparisons between the purple bacterial reaction centre subunits M and L and the D1 and D2 polypeptides from plant PSII reaction centres have suggested strong structural homologies. The currently available structure of the PSI reaction centre from the cyanobacterium *Synechococcus elongatus* (Krauss *et al.*, 1996) also shows that the 'core' of this reaction centre contains a set of transmembrane α-helices whose arrangement is also rather homologous to that seen in the purple bacterial reaction centres. It is very clear that all these reaction centres are constructed on a common structural principle. This is manifestly not the case with the antenna complexes.

A COMPARISON OF THE STRUCTURES OF ANTENNA COMPLEXES

The antenna complexes whose structures have been determined can be divided into two groups, the three water-soluble ones (FMO, phycobiliproteins and the peridinin–chl*a* complex) and the integral membrane ones (LHC2 and LH2). However, even this categorization reveals no strong common structural motifs. The protein folding in these complexes is different. The FMO proteins are largely β-sheet structures. In this case the protein scaffold is like a 'string-bag' in which seven bacteriochlorophyll *a* molecules per monomer are rather randomly arranged (Fig. 1). The phycobiliproteins and the peridinin-chl*a* protein are largely α-helical structures, but the arrangement of these helices and the chromophores (phycobilins and, peridinins and chlorophyll *a* respectively) are completely different (Fig. 2). The protein folds in the phycobiliproteins are rather similar to those found in the globins (Schrimer *et al.*, 1987), while the disposition of the α-helices in the peridinin-chl*a* protein resemble a wide open, right-handed 'jelly roll' (Hoffman *et al.*, 1996). The peridinin–chl*a* complex is also unique since it has four molecules of the carotenoid peridinin for every molecule of chl*a*. The two integral membrane antenna complexes for which there are structures are

Table 1. *A comparison of the structure and composition of the major chlorophyll-containing antenna complexes for which there are high resolution structures*

Type of complex (organism)	Method of structural determination (resolution)	Molecular weight of apoprotein(s) (native molecular weight)	Type and number of pigments in the holocomplex	Density of chls (chl/protein w/w)	Major type of secondary structure	References
LH2 (*Rps. acidophila*)	X-ray crystallography (currently 2.0 Å)[a]	α – 5.8 kD, β –4.8 kD (93.6 kD)	27 Bchl*a* and 18 carotenoids	~1:3.4	α-helices	McDermott *et al.* (1995)
LHC2 (Spinach)	Electron diffraction[c] on 2D crystals (nominal 3.4 Å)	28 kD, 27 kD, 25 kD[b] (app. 80 kD)	39 Chls (21 chl*a* + 18 chl*b*) and ~9 carotenoids	~1:2.1	α-helices	Kühlbrandt *et al.* (1994)
FMO (*P. aestuarii*)	X-ray crystallography (1.9 Å)	~39 kD (120 kD)	21 Bchl*a*	~1:5.7	β-sheet	Tronrud & Matthews (1986)
Peridinin-chl*a* protein (*Am. carterae*)	X-ray crystallography (2.0 Å)	30.2 kD (90.6 kD)	6 Chl*a* and 24 carotenoids	~1:15	α-helices	Hoffmann *et al.* (1996)

[a] Prince *et al.* unpublished observations

[b] The exact polypeptide composition of the LHC2 complex is not known.

[c] This resolution is only for that part of the structure parallel to the membrane plane. Normal to the membrane plane the resolution is lower.

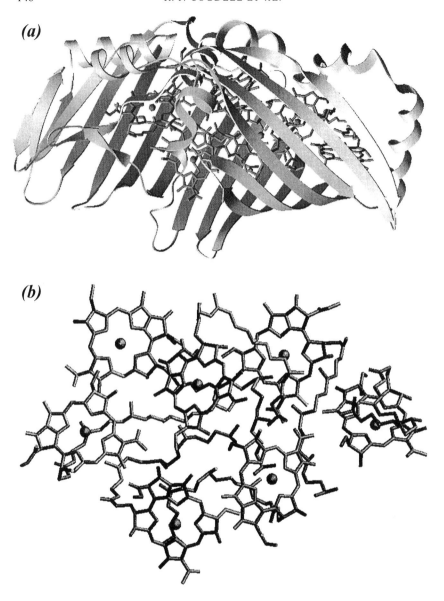

Fig. 1. The structure of the Bacteriochlorophyll *a*–protein complex from *Prosthecochloris aestuarii. (a)* A ribbon diagram of the monomer of this antenna complex. The repeating, strands of β-sheets are clearly seen. Seven Bchl*a* molecules are contained in a space with dimensions of 45 × 35 × 15 Å. Five of the Bchls are liganded to His residues, one to water and one to a backbone carbonyl residue. The average centre-to-centre distance of the bacteriochlorin rings is 12 Å. The closest distance between Bchls in adjacent monomers within the intact dimer is about 30 Å. Figures 1 and 2 were prepared from the coordinates deposited in the Brookhaven Database using SETOR (Evans, 1993). *(b)* The distribution of the seven Bchls in the monomer with the protein removed for clarity.

(a)

(b)

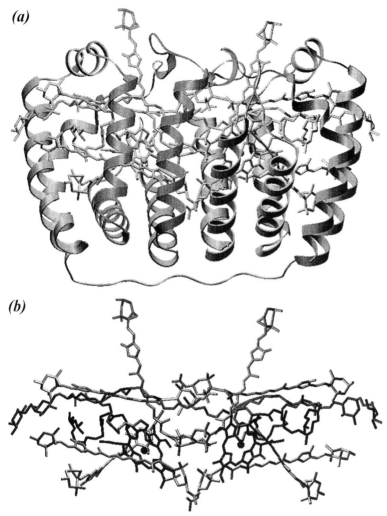

Fig. 2. The structure of the peridinin–chlorophyll *a* complex from *Amphidinium carterae* and allophycocyanin from *Spirulina platensis* (Hoffmann *et al.*, 1996; Brejc *et al.*, 1995). *(a)* A ribbon diagram of a monomer of the peridinin–chlorophyll *a* complex. The crystal structure shows that the monomer contains two almost identical domains, each of which contains eight α-helices and binds one molecule of chlorophyll *a* and four molecules of peridinin. The whole monomer contains a central hydrophobic space of about $23 \times 23 \times 53$ Å which contains the pigments. The centre-to-centre distance between the two chlorophylls in the monomer is 17.4 Å, and between nearest chlorophylls from monomer to monomer is 44 Å. *(b)* The arrangement of the chromophores from a monomer of the peridinin–chlorophyll *a* protein, with the protein removed for clarity. Each polyene chain from the four peridinin molecules in each cluster comes within <4 Å of the chlorophyll *a*. Both chlorophylls are liganded to the protein via histidine residues. *(c)* A ribbon diagram of the complete allophycocyanin molecule looking down on to the trimer. *(d)* A ribbon diagram of the α-subunit of allophycocyanin, which shows the folding pattern of the α-helices and the location of the two covalently bound bilin pigments. Notice that, although the bilins are 'linear tetrapyrroles', they assume very twisted configurations in the complex. This is important for their photochemical properties.

(c)

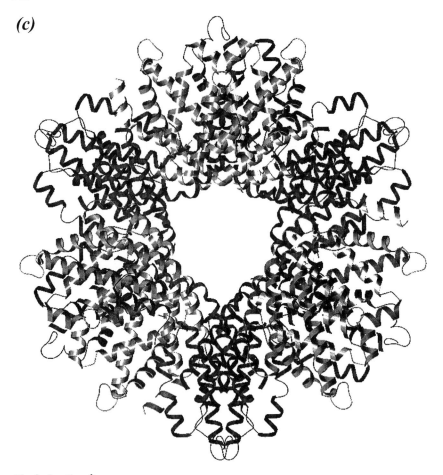

Fig. 2. (*continued*).

both α-helical; however, again the organization of these helices are very different in each (Fig. 3). The LH2 complexes are highly symmetric (nine-fold or eight-fold symmetry) with all the transmembrane helices arranged in two concentric circles. LHC2 shows no such symmetry.

In the chlorophyll-containing antenna complexes for which there are high-resolution structures the pigments are all non-covalently bound to the apoproteins. In contrast, the phycobilins are covalently linked to their apoprotein via thio-ether linkages (see, for example, Wedemayer, Kidd & Glazer, 1996). All of the chlorophylls are penta-coordinated, with a variety of amino acid side chains and/or water providing the fifth ligand to the central Mg^{2+}. In modelling antenna structures prior to the availability of crystallographic data, there has been a tendency to over-emphasize the

(d)

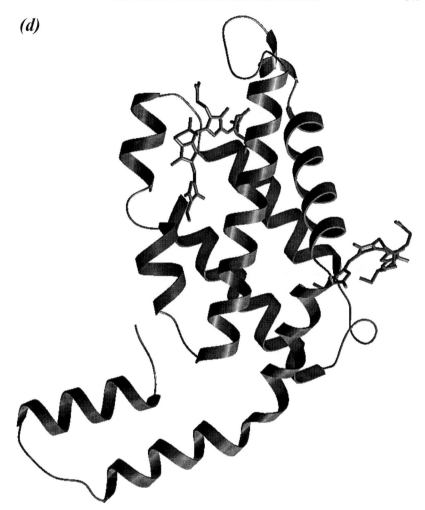

Fig. 2. (*continued*).

possible role of histidine residues as prospective Mg^{2+} ligands. It is therefore, interesting to note that the range of ligands actually used in nature is quite broad and as well as histidine includes, for example, glutamate, aspartate and H_2O.

All these types of antenna complexes work well yet they appear to show no obvious structural similarities. Does this mean that the structural requirements for light-harvesting are less stringent that those for reaction centres? Are there any common features of light-harvesting systems at all?

In order to begin to answer those questions, a more fundamental problem needs to be considered. What properties must an antenna complex have in

(a)

(b)

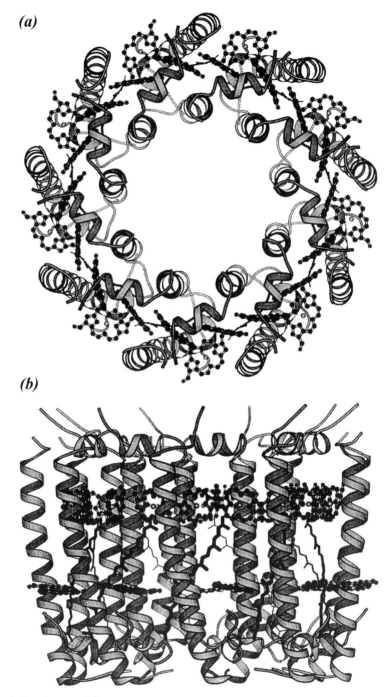

Fig. 3. (*caption opposite*).

(c)

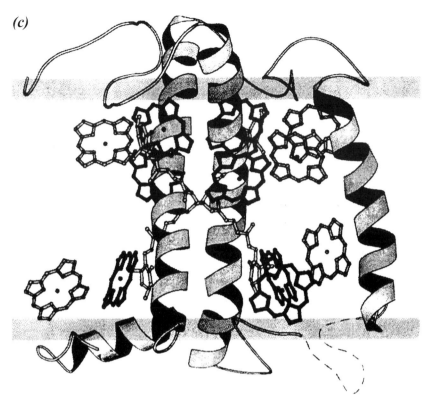

Fig. 3. The structure of the two integral-membrane antenna complexes, LH2 (McDermott *et al.*, 1995) and LHC2 (Kühlbrandt *et al.*, 1994). *(a)* A schematic view of the LH2 nonameric complex viewed from the periplasmic side of the membrane and *(b)* from within the membrane. The arrangement of the two groups of Bchls within the walls of transmembrane helices are shown: the ring of nine B800 Bchls parallel to the membrane: and the second ring of 18 B850 Bchls perpendicular to the membrane. The 'well-resolved' carotenoid is also shown. The phytol chains have been removed from the Bchls for clarity. These two figures were produced using Molscript (Kraulis, 1991). *(c)* A view of a monomer of the LHC2 antenna complex from within the membrane. The protein is depicted as a ribbon and the distribution of the pigment (seven chl*a*s, six chl*b*s and the two lutein carotenoids) are clearly seen. Notice the more irregular organization of the chls in this complex as compared with LH2. This diagram was reproduced and modified from Kühlbrandt *et al.* (1994) with permission.

order to function as an efficient light harvester? First of all, the complex must be able to absorb light in the 'correct' region of the spectrum. What is 'correct' will depend entirely upon the ecological niche in which the photosynthetic organism involved finds itself. The dinoflagellates, such as *Amphidinium carterae*, live in the sea at depths in the water column where most of the incident solar energy is in the blue region of the spectrum. They have therefore, developed an antenna complex where most of the light absorption is carried out by carotenoids. The cyanobacteria and the red algae, which compete for light energy with other algae and plants using only chl*a* and chl*b*, are able to use phycobilins (linear tetrapyrroles) which absorb light energy in the green, orange and red regions of the spectrum where chl*a* and chl*b* do not. In these two cases, where non-chlorophyll pigments are used, it might be expected that rather different protein folding would be needed to accommodate them in addition to the chlorophylls. This is the case. However, when one compares the antenna systems which are based mainly on the chlorophylls (FMO, LH2, LHC2), it might be expected that the structural similarities would be more evident than they actually are.

The second feature of an antenna complex is that its light-absorbing pigments must be arranged to balance maximum light-absorption capability with the need for rapid and efficient transfer of absorbed energy from pigment to pigment. In other words, the 'pigment density' (i.e. interpigment distance and orientations) must be carefully controlled to maximize light absorption and yet not so close as to produce 'quenching' centres which would prevent the light energy being transferred on to the reaction centre (i.e. there must be no concentration quenching). It is worthwhile at this point digressing slightly to consider the type of energy transfer that might be expected to take place in the antenna systems and to discuss what structural constraints these put upon the pigment organisation.

Most classical descriptions of singlet–singlet energy transfer in photosynthesis have dealt mainly with the so-called Förster resonance transfer mechanism (for example, see Borisov, 1978). This is a mechanism typically thought to apply when two pigments are rather weakly coupled, i.e. separated by distances of 15 Å up to even 100 Å (Förster, 1948). Although this type of energy transfer takes place via a non-radiative resonance process, it can be considered to be equivalent to the donor molecule emitting its absorbed photons as fluorescence and the acceptor molecule re-absorbing this emitted photon. Therefore, the rate of this process depends on the spectral overlap of the fluorescence of the donor with the absorption of the acceptor, the distance (R) between the donor and the acceptor (rate $\propto 1/R^6$), the orientation of the transition dipole moment of the donor and acceptor molecules (the more parallel these are the faster the energy transfer) and various parameters that are dependent on the medium between the two pigments. In light-harvesting complexes with chlorophylls, where there is good spectral overlap between pigments, then if the protein packs the chlorophylls close together with their transition dipoles moments reasonably well aligned, there is no apparent advantage in having a highly ordered, symmetrical LH2-type arrangement of the pigments as compared with a less regular arrangement in either the LHC2 or the FMO protein. Once an antenna chlorophyll has absorbed a photon, a 'clock' starts to count down. When sufficient time has elapsed, the absorbed energy will be lost in a series of wasteful reactions such as fluorescence. Thus, it is important that pigments in the antenna system are organized such that the time taken for energy transfer from pigment to pigment, both within the antenna complex itself and to reach the reaction centre, is short compared with the time taken for the 'loss' reactions to occur. Typically, this means that the antenna system must deliver the energy to the reaction centre in less than about 100 ps (Sundström & van Grondelle, 1996) to compete successfully with typical fluorescence lifetimes (in the absence of other antenna complexes or reaction centres) of several hundred picoseconds or even a few nanoseconds.

If pigment molecules are strongly coupled (that is very close to each other), mechanisms other than the weak Förster interaction can apply (for example, see Davydov, 1962; Pearlstein, 1996). For the purposes of this short review, this can be illustrated for the triplet–triplet energy transfer which occurs between triplet chlorophylls and carotenoids (Frank & Cogdell, 1993). Carotenoids are essential photoprotective pigments in photosynthesis. When chlorophylls are over-excited, then chlorophyll triplet states are formed. If these last long enough to react with oxygen, then they will sensitize the formation of singlet oxygen (Foote, 1968). This is a very powerful oxidizing agent and will rapidly, irreversibly destroy the chlorophyll and kill the organism. Carotenoids prevent this harmful reaction by quenching the chlorophyll triplet states before they have a chance to react with oxygen. The energy level of the triplet state of the carotenoids concerned is lower than that of singlet oxygen and so now it cannot be formed.

$$\text{Chla(Bchla)} \xrightarrow{h\nu} {}^1\text{Chl}a^*$$
$${}^1\text{Chl}a^* \longrightarrow {}^3\text{Chl}a^*$$
$${}^3\text{Chl}a + O_2 \longrightarrow \text{Chl}a + {}^1\Delta g\ O_2^*$$
$${}^3\text{Chl}a^* + \text{car} \longrightarrow {}^3\text{car}^* + \text{Chl}a$$

This triplet–triplet energy transfer proceeds via an electron exchange mechanism (Dexter, 1953). For this to occur in proteins, as it does even at cryogenic temperatures, the carotenoid and the chlorophylls must be in van der Waal's contact. In those antenna complexes which contain carotenoids, e.g. peridinin–chlorophyll a complex, LH2 and LHC2 and the carotenoids are all in van der Waal's contact with at least one chlorophyll.

It should be clear from this brief discussion of the physical mechanisms of energy transfer that the mechanism which operates in the antenna complexes is dependent upon which pigments are involved and how 'close' they are to each other (that is strongly or weakly interacting, or an intermediate case). In the phycobiliproteins the measured rates of energy transfer can be rather successfully modelled just using the classical Förster weak interaction case (Sauer & Scheer, 1988). On the other hand, within the tightly coupled ring of B850 Bch$a$s in LH2 (centre to centre distances about 9 Å) other mechanisms involving delocalized excited states are more appropriate (for example, see Novoderezhkin & Razjivin, 1995; Sauer et al., 1996; Pullerits & Sundström, 1996; Alden et al., 1997). However, the strong take home message from this section is that the structural requirements for rapid and efficient energy transfer with antenna complexes are such that a wide variety of antenna structures are possible while still retaining high efficiency.

The final requirement of an antenna complex is that its structure, as far as possible, is well matched to the reaction centre or to the next antenna complex in the energy transfer pathway between it and the reaction centre. By well 'matched' the meaning is that there can be fast and therefore, efficient energy transfer between the complexes. The purple bacterial photosynthetic unit provides a beautiful example of this (Fig. 4).

The structure of LH1 from *Rhodospirillum rubrum*, as determined from a projection map at a resolution of 8.5 Å (Karrasch, Bullough & Ghosh, 1995), reveals two concentric rings of apoproteins forming transmembrane helices and with 16-fold symmetry (that is, very similar to the LH2 structure but $\alpha_{16}\beta_{16}$ rather than $\alpha_9\beta_9$ or $\alpha_8\beta_8$ as found with LH2). Since there is clear homology between the LH2 and the LH1 apoproteins in the transmembrane helical regions, it is possible to use the LH2 structure to produce a 'high resolution' model of LH1 (Papiz *et al.*, 1996; Hu *et al.*, 1997 but see also Hunter, van Grondelle & Olsen, 1989). Now, since the reaction centre is located in the centre of the larger LH1 ring (for example, see Stark *et al.*, 1984) it is possible to produce a persuasive model for the photosynthetic unit. In this model when the LH2 rings and LH1 rings are brought into close contact, the closely coupled B850 Bchl*a*s (LH2) and B875 Bchl*a*s (LH1) line up next to each other and the same depth in the membrane. Moreover, so do the 'special pair' Bchl*a*s in the reaction centre (the ultimate destination of the energy). This then is a beautiful example of 'matching' the antenna complex structure to ensure ultra-fast and very efficient energy transfer.

FINAL COMMENTS

A comparison of the available structures of photosynthetic antenna complexes reveals a diverse group of structures, both from the protein folding viewpoint and from the arrangement of pigments. It is clear that the constraints placed upon these systems by the basic mechanisms of single-singlet energy transfer are far less stringent than those seen for the electron transport reactions in the reaction centres. The exact mechanism of energy transfer involved in the antenna complex varies, depending upon which pigments are being considered and how strongly coupled they are.

The antenna systems in the purple bacteria and the phycobilisomes show a strong 'wavelength programming'. In these systems the more peripheral an antenna complex is, the shorter the wavelength maximum of its long wavelength absorption band. This energy gradient (that is phycoerythrin 540–550 nm, phycocyanin 610–620 nm and allophycocyanin approximately 650 nm) imparts directionality on the energy transfer processes so that it 'funnels' downhill towards the reaction centre (Glazer, 1984). This phenomenon is much less clearcut in plants and algae, indeed under mainly different

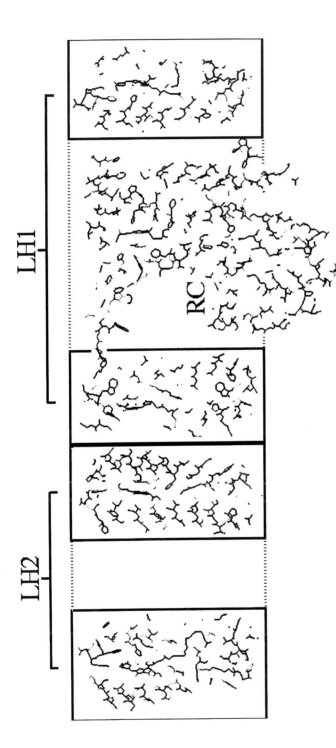

Fig. 4. A model of the purple bacterial photosynthetic unit. A section of the model presented in Papiz *et al.* (1996) viewed from within the membrane. The reaction centre structure used is that of *Rhodobacter sphaeroides* R-26 (Yeates *et al.*, 1988). This view highlights the similarity in depth in which the B850 Bchls from LH2, and B875 Bchls from LH1 and the 'special pair' Bchls in the reaction centre are positioned in the membrane. The distance between the ring of B850 Bchls and the ring of B875 Bchls is ≃ 24 Å, and from the ring of B875 Bchls to the 'special pair' ≃ 43 Å. This figure was produced using 'O' (Jones *et al.*, 1991). A colour version of this is available at the following website: http://www.chem.gla.ac.uk./protein/LH2/lh2.html.

environmental conditions the problem for many oxygenic eukaryotic photo-synthetic organisms is not too little light but too much. It is beyond the scope of this brief review of light harvesting to discuss the various regulation mechanisms antenna complexes exhibit in order to prevent over-excitation of preformed reaction centres but, for a recent review on this topic, the reader can consult Horton, Ruban and Walters, 1996. Also, antenna systems have not been considered, such as are seen in the chlorosomes of green photo-synthetic bacteria, where the antenna complexes are largely devoid of protein (Blankenship, Olsen & Miller, 1995).

ACKNOWLEDGEMENTS

This review was prepared with support from the BBSRC and the Gatsby Foundation.

REFERENCES

Alden, R. G., Johnson, E., Nagarajan, V., Parson, W. W., Law, C. J. & Cogdell, R. J. (1997). Calculations of spectroscopic properties of the LH2 bacteriochlorophyll–protein antenna complex from *Rhodopseudomonas acidophila*. *Journal of Physical Chemistry B*, **101**, 4667–80.

Allen, J. P., Feher, G., Yeates, T. O., Komiya, H. & Rees, D. C. (1987). Structure of the reaction centre from *Rhodobacter sphaeroides* R-26: the co-factors. *Proceedings of the National Academy of Sciences, USA*, **84**, 5730–4.

Blankenship, R. E., Olson, J. M. & Miller, M. (1995). Antenna complexes from green photosynthetic bacteria. In *Anoxygenic Photosynthetic Bacteria*, ed. R. E. Blankenship, M. T. Madigan & C. E. Bauer, pp. 399–435. The Netherlands: Kluwer Academic Press.

Borisov, A. Y. (1978). Energy migration mechanisms in antenna chlorophylls. In *The Photosynthetic Bacteria*, ed. R. K. Clayton & W. R. Sistrom, pp. 323–32. New York and London: Plenum Press.

Brejc, K., Ficner, R., Huber, R. & Steinbacher, S. (1995). Isolation, crystallisation, crystal structure analysis and definement of allophycocyanin from cyanobacterium *Spiralina platensis* of 2.3 Å resolution. *Journal of Molecular Biology*, **249**, 424–40.

Davydov, A. S. (1962). *Theory of Molecular Excitons* (translated by Kasa, M. & Oppengeimer, M. Jr.). New York: McGraw-Hill.

Deisenhofer, J., Epp, O., Miki, R., Huber, R. & Michel, H. (1984). The structure of the protein subunits in the photosynthetic reaction centre of *Rhodopseudomonas viridis* at 3 Å resolution. *Nature*, **318**, 618–24.

Dexter, D. L. (1953). A theory of sensitised luminescence in solids. *Journal of Chemical Physics*, **21**, 836–60.

Evans, S. V. (1993). SETOR: hardware three-dimensional solid model representations of macromolecules. *Journal of Molecular Graphics*, **11**, 134–8.

Fenna, R. E. & Matthews, B. W. (1975). Chlorophyll arrangement in a bacterio-chlorophyll protein from *Chlorobium limicola*. *Nature*, **258**, 573–7.

Foote, C. S. (1968). Mechanisms of photosensitised oxidation. *Science*, **162**, 963–70.

Förster, T. (1948). Zwischenmolecular Eneryiewanderung und Fluorescenz. *Annals of Physics*, **6**, 55–75.

Frank, H. A. & Cogdell, R. J. (1993). Photochemistry and function of carotenoids in photosynthesis. In *Carotenoids in Photosynthesis*, ed. A. Young & G. Britton, pp. 52–326. London: Chapman & Hall.

Glazer, A. N. (1984). Phycobilisome. A macromolecular complex optimised for energy transfer. *Biochimica et Biophysica Acta*, **768**, 29–51.

Hoffmann, E., Wrench, P. M., Sharples, F. P., Hiller, R. G., Welte, W. & Diederichs, K. (1996). Structural basis of light-harvesting by carotenoids: peridinin-chlorophyll-protein from *Amphidinium carterae*. *Science*, **272**, 1788–91.

Horton, P., Ruban, A. V. & Walters, R. G. (1996). Reulation of light-harvesting in green plants. *Annual Review of Plant Physiology and Plant Molecular Biology*, **47**, 655–84.

Hu, X., Ritz, T., Damjanovic, A. & Schulten, K. (1997). Pigment organisation and transfer of electronic excitation in the photosynthetic unit of purple bacteria. *Journal of Physical Chemistry B*, **101**, 3854–71.

Hunter, C. N., van Grondelle, R. & Olsen, J. D. (1989). Photosynthetic antenna proteins: 100ps before photochemistry starts. *Trends in Biochemical Sciences*, **14**, 72–6.

Jones, T. A., Zou, J. Y., Cowan, S. W. & Kjeldgaard, M. (1991). Improved methods for building protein models in electron density maps and the location of errors in these maps. *Acta Crystallography*, **A47**, 110–19.

Karrasch, S., Bullough, P. A. & Ghosh, R. (1993). 8.5 of projection map of the light-harvesting complex 1 from *Rhodospirillum rubrum* reveals a ring composed of 16 subunits. *The EMBO Journal*, **14**, 631–8.

Koepke, J., Hu, X., Muenke, C., Schulten, K. & Michel, H. (1996). The crystal structure of the light-harvesting complex II (B800-85) from *Rhodospirillum molischianum*. *Structure*, **4**, 581–97.

Kraulis, P. J. (1991). Molescript – a program to produce both detailed and schematic plots of protein structures. *Journal of Applied Crystallography*, **24**, 946–50.

Krauss, N., Schubert, W. D., Klukas, O., Fromme, P., Witk, H. T. & Saenger, W. (1996). Photosystem I at 4 Å resolution represents the first structural model of a joint photosynthetic reaction centre and core antenna system. *Nature Structural Biology*, **3**, 965–73.

Kühlbrandt, W., Wang, D. N. & Fujiyoshi, Y. (1994). Atomic model of plant light-harvesting complex by electron crystallography. *Nature*, **367**, 614–21.

Li, Y-F., Zhou, W., Blankenship, R. E. & Allen, J. P. (1997). Crystal structure of the bacteriochlorophyll *a* protein from *Chlorobium tepidum*. *Journal of Molecular Biology*, **271**, 456–71.

McDermott, G., Prince, S. M., Freer, A. A., Hawthornthwaite-Lawless, A. M., Papiz, M. Z., Cogdell, R. J. & Isaacs, N. W. (1995). Crystal structure of an integral membrane light-harvesting complex from photosynthetic bacteria. *Nature*, **374**, 517–21.

Michel, H. & Deisenhofer, J. (1988). Relevance of the photosynthetic reaction centre from purple bacteria to the structure of photosystem II. *Biochemistry*, **27**, 1–7.

Novoderezhkin, V. L. & Razjivin, A. P. (1995). Exciton dynamics of circular aggregates: application to antenna of photosynthetic purple bacteria. *Biophysical Journal*, **68**, 1089.

Papiz, M. Z., Prince, S. M., Hawthornthwaite-Lawless, A. M., McDermott, G., Freer, A. A., Isaacs, N. W. & Cogdell, R. J. (1996). A model for the photosynthetic apparatus of purple bacteria. *Trends in Plant Science*, **1**, 198–206.

Pearlstein, R. M. (1996). Coupling of exciton motion in the core antenna and primary change separation in the reaction centre. *Photochemistry and Photobiology*, **48**, 75–82.

Pullerits, T. & Sundström, V. (1996). Photosynthetic light-harvesting pigment-protein

158   R. J. COGDELL *ET AL.*

complexes: toward understanding how and why. *Account of Chemical Research*, **29**, 381–9.

Sauer, K. & Scheer, H. (1988). Excitation transfer in C-phycocyanin – Förster transfer rate and exciton calculations based on new crystal structure data for C-phycocyanins from *Agmenellum quadruplicatum* and *Masticocladus laminosus*. *Biochimica et Biophysica Acta*, **936**, 157–70.

Sauer, K., Cogdell, R. J., Prince, S. M., Freer, A. A., Isaacs, N. W. & Scheer, H. (1996). Structure-based calculations of the optical spectra of the LH2 bacteriochlorophyll–protein complex from *Rhodopseudomonas acidophila*. *Photochemistry and Photobiology*, **64**, 564–76.

Schrimer, T., Bode, W. & Huber, R. (1987). Refined three-dimensional structures of two cyanobacterial C-phycocyanins at 2.1 and 2.5 Å resolution: a common principle of phycobilin–protein interaction. *Journal of Molecular Biology*, **196**, 677–96.

Sidler, W. (1994). Phycobilisome and phycobiliprotein structures. In *The Molecular Biology of Cyanobacteria*, ed. D. A. Bryant, pp. 139–216. The Netherlands: Kluwer Academic Publishers.

Stark, W., Kühlbrandt, W., Wildhaber, I., Wehrli, E. & Mühlethaler, K. (1984). The structure of the photoreceptor unit of *Rps. viridis*. *The EMBO Journal*, **3**, 777–83.

Sundström, V. & van Grondelle, R. (1995). The kinetics of excitation transfer and trapping in purple bacteria. In *Anoxygenic Photosynthetic Bacteria*, ed. R. E. Blankenship, M. T. Madigan & C. E. Bauer, pp. 349–372. The Netherlands: Kluwer Academic Publishers.

Tronrud, D. E. & Matthews, B. W. (1986). Structure and X-ray amino acid sequence of the bacteriochlorophyll *a* protein from *Prosthecochloris aestuarii* refined at 1.9 Å resolution. *Journal of Molecular Biology*, **188**, 443–54.

Wedemayer, G. J., Kidd, D. G. & Glazer, A. N. (1996). Cryptomonad biliproteins: Bilintypes and locations. *Photosynthesis Research*, **48**, 163–70.

Yeates, T. O., Komiya, H., Chorino, A., Rees, D. C., Allen, J. P. & Feher, G. (1988). Structure of the reaction centre from *Rhodobacter sphaeroides* R-26 and 2.4.1: Protein–cofactor (bacteriochlorophyll, bacteriopheophytin and carotenoid) interactions. *Proceedings of the National Academy of Sciences, USA*, **85**, 7993–7.

LIGHT REGULATION OF PIGMENT–PROTEIN GENE EXPRESSION IN *RHODOBACTER* SPECIES

MARY K. PHILLIPS-JONES

Department of Microbiology, University of Leeds, Woodhouse Lane, Leeds LS2 9JT, UK

INTRODUCTION

For those micro-organisms which are able to utilize light as an energy source, capture of light energy and conversion to a cellular chemical form, requires the synthesis of a specialized photosystem. One of the most intensively studied photosystems, particularly in terms of structure and function, is that of the purple photosynthetic bacteria (see Cogdell *et al.*, this volume). In such bacteria, the photosystem is composed of three membrane-bound pigment-protein complexes: the light harvesting I (LHI), light harvesting II (LHII) and the reaction centre (RC) complexes (Fig. 1) (Deisenhofer *et al.*, 1985; McDermott *et al.*, 1995). Light energy is mainly captured by bacteriochlorophyll (Bchl) and carotenoid (Crt) pigments which are non-covalently bound to the light harvesting proteins (for example, see Hunter, 1995). It is then 'funnelled down' to the 'special pair' of Bchl molecules located in the reaction centre and used to initiate the first photochemical steps of photosynthesis and energy conservation (Fig. 1).

REGULATION OF PHOTOSYNTHETIC COMPLEXES BY OXYGEN AND LIGHT

It has been known for many years that, in photosynthetic species such as *Rhodobacter sphaeroides* and *R. capsulatus*, synthesis of photosystem components and the intracytoplasmic membrane that houses them, is profoundly influenced by external environmental factors, particularly light and oxygen (Cohen-Bazire, Sistrom & Stanier, 1957; Aagaard & Sistrom, 1972; Drews, 1986; Kiley & Kaplan, 1988). These environmental signals trigger a widespread change in the amount, morphology and composition of the cell membrane. Oxygen has the most profound effect, repressing synthesis by up to 30-fold. Given sufficient aeration, *R. sphaeroides* cells can be grown almost free of pigment. Oxygen and redox regulation of photosynthetic complex formation has been, and continues to be, an intensively studied area. A description of the molecular mechanisms of regulation by oxygen is beyond the scope of this review and the reader is referred to the work of McGlynn and Hunter (1992); Lee *et al.* (1993); Pollich, Jock and Klug (1993); Penfold

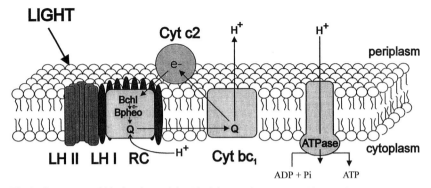

Fig. 1. Structure and biochemistry of the *Rhodobacter* photosystem. The reaction centre (RC) is shown as a square, surrounded by light harvesting I polypeptides (LHI). To the left is shown the peripheral or 'variable' light harvesting II complex. Also shown are cytochrome c2 (cyt c2), the cytochrome bc1 complex (cyt bc1), bacteriochlorophyll (Bchl), bacteriopheophytin (Bpheo) and ubiquinone (Q). Components are embedded in, or associated with, the intracytoplasmic membrane. (Adapted and redrawn from Bauer & Bird, 1996, with permission.)

and Pemberton (1994); Phillips-Jones and Hunter (1994); Ponnampalam, Buggy and Bauer (1995); Zeilstra-Ryalls and Kaplan (1995, 1996); Bauer and Bird (1996); Eraso and Kaplan (1996); Gomelsky and Kaplan (1997); O'Gara and Kaplan (1997) (and reference cited therein); Yeliseev, Krueger and Kaplan (1997) and M. Kirndörfer, M. Hebermehl, S. K. Hemschemeier and G. Klug (unpublished data), which describe the oxygen-dependent regulators of photosynthesis gene transcription identified to date in *R. sphaeroides* or *R. capsulatus*, to the work of Klug (1993a, 1995) (and to references cited therein) on oxygen-mediated effects on mRNA stability and to the work of Biel and Marrs (1985); Klug, Liebetanz and Drews (1986); Neidle and Kaplan (1993); Varga and Kaplan (1993) and Yeliseev, Eraso and Kaplan (1996) in which possible translational and/or post-translational effects of oxygen are reported. From these studies it has become clear that oxygen exerts its effects at many levels in the cell, but particularly at the level of gene transcription.

Synthesis of photosystem components is also influenced by light intensity (Cohen-Bazire *et al.*, 1957; Aagaard & Sistrom, 1972; Drews, 1986). High light levels result in relatively low levels of photosynthetic components, whereas a change to low light conditions results in increased synthesis of these components, presumably to enable the cell to capture as much of the available light as possible. Light intensity exerts a particularly marked effect on the intracytoplasmic membrane (ICM) in which the pigment–protein complexes are housed and also on the relative abundance of the LHII complex (Drews & Oelze, 1981; Kiley & Kaplan, 1988; also see the work of Zucconi & Beatty (1988) described below). In *R. capsulatus* cells adapted to low light conditions, the area of the ICM is approximately six-fold higher

Table 1. *Adaptation of* Rhodobacter capsulatus *to low light intensity*[a]

	High light	Low light
Cells		
nmol RC (mg cell protein)$^{-1}$	0.03	0.145
nmol Bchl (mg cell protein)$^{-1}$	2.14	18.3
doubling time for cell protein (min)	190	190
area of ICM (μm^2)	0.4	2.5
photophosphorylation:		
nmol ATP (mg cell protein)$^{-1}$ min^{-1}	140	180
Membranes		
nmol RC (mg membrane protein)$^{-1}$	0.052	0.205
nmol Bchl (mg membrane protein)$^{-1}$	3.4	25.9
size of photosynthetic unit:		
mol Bchl (mol RC)$^{-1}$	65	126
photophosphorylation:		
nmol ATP (nmol RC)$^{-1}$ min^{-1}	4.3	1.3
nmol ATP (mg membrane protein)$^{-1}$ min^{-1}	220	260

[a] Taken from Drews (1986), with permission.

than that of high light-grown cells, the number of photosynthetic units (measured as the number of RCs per cell) increases five-fold and the size of the photosynthetic unit (that is, the number of antenna Bchls per RC, reflecting LHI and LHII levels) doubles (Table 1 (taken from Drews, 1986; further data reviewed in Drews, 1986, 1996).

Other photosynthetic species such as *Rhodopseudomonas palustris* and *Rps. acidophila*, respond to variations in light levels by making different types of LHII complex (Tadros & Waterkamp, 1989; Hawthornthwaite & Cogdell, 1991). It is known that, whilst *R. capsulatus* and *R. sphaeroides* possess one pair of genes encoding the LHII polypeptides (*pucB* and *pucA*), the *pucBA* genes in *Rps. palustris* and *Rps. acidophila* are present in multiple copies in these bacteria (Gardiner, Cogdell & Takaichi, 1993; Tadros *et al.*, 1993 and references cited therein). Studies to determine the precise roles of individual gene pairs, and to compare the structural features of the proteins they encode under different intensities of light, are most conveniently pursued by expressing these gene pairs in species such as *R. sphaeroides*, which provide suitable genetic (*pucBA*-minus) backgrounds for such studies. One such study recently compared two *puc* gene-pair products from *Rps. palustris*, one associated with high light growth conditions and the other associated with low light. From these comparisons it was possible to identify potentially important features of the LHII structure, such as residues involved in energy transfer and those involved in assembly and stability of LHII complexes in the membrane (Fowler & Hunter, 1996).

Regulation by light in *R. sphaeroides* and *R. capsulatus* is less pronounced than that by oxygen. Whereas changes in oxygen concentration can exert a

30-fold effect on levels of photosynthetic complexes in the membrane, the overall effects of light are only two- to five-fold (Buggy, Sganga & Bauer, 1994*a*; Drews, 1986). Nevertheless, in common with oxygen regulation, regulation by light also appears to be exerted at the levels of photosynthesis gene transcription, post-transcription, and translation/post-translation. It should be noted, however, that the effects of these regulatory mechanisms are greater on some photosynthetic components than on others. For example, when light intensity is reduced, mRNA levels for light harvesting and RC polypeptides increase very significantly over those for Bchl synthesis (Zhu & Hearst, 1986). One study, which used reporter genes to measure transcription, failed to find any significant change in *bch* (Bchl) gene transcription in response to changes in light intensity (Biel & Marrs, 1983). Hence, transcriptional and post-transcriptional regulation is exerted most strongly on expression of LHI, LHII and RC polypeptides, rather than on pigment gene expression. Light regulation of Crt and Bchl biosynthesis appears to be exerted primarily through the activities of bacteriochlorophyll biosynthetic enzymes, that is, post-translationally. This aspect is discussed in more detail in Armstrong (1995) and Biel (1995).

Further evidence for post-translational control of Bchl synthesis comes from the finding that light intensity appears to affect the stability and accumulation of Bchl, with increases in light intensity resulting in increased degradation of Bchl (Biel, 1986, 1995). The precise mechanism by which this occurs is not fully understood, but it has been shown that Bchl regulation by light occurs independently of photosynthesis, the photosynthetic membrane system, photosynthesis electron transport and photosynthetic ATP generation, since *R. capsulatus* mutants which lack RC/LHI or LHII complexes exhibit normal light regulation of Bchl accumulation (Biel, 1986; Yurkova & Beatty, 1996). For accounts of pigment synthesis regulation through the activities of Bchl synthesizing enzymes, see Armstrong (1995) and Biel (1995). This aspect is important, since levels of cellular Bchl influence the stabilization of apoproteins into the membrane and hence determine the final levels of membrane-bound complexes (Dierstein, 1983; Biel, 1986; Klug, Liebetanz & Drews, 1986). Pigment and apoprotein levels appear to be finely tuned in the cell, suggesting coordinated regulation, since in wild-type cells that are adapted to synthesize the photosynthetic apparatus, the levels of Bchl precursors free in the cell are invariably low (Beck & Drews, 1982).

In *Rhodobacter*, the two major carotenoid pigments present are spheroidene and its keto derivative spheroidenone. These derivatives are differentially synthesized in *R. sphaeroides*, depending on light and other growth conditions. Spheroidene is the major form present under anaerobic, low light conditions, whereas spheroidenone is predominant under semi-aerobic or anaerobic/high light conditions. One recent study has demonstrated that spheroidene is associated mostly with the LHII complex (which is synthesized in greater amounts under low light conditions), whereas spheroidenone occurs more abundantly in the LHI complex (Yeliseev *et al.*, 1996). Using mutants defective in either the LHI or LHII complexes, an association was observed between the formation of either LHI or LHII complexes and the accumulation of spheroidene or spheroidenone, suggesting the possible involvement of these carotenoid forms in adaptation to changes in light intensity and oxygen concentration (Yeliseev *et al.*, 1996).

Some of the practical aspects associated with studies of light regulation of photosynthesis gene expression are worth mentioning. One of the difficulties associated with these studies is the effects of self-shading as bacterial cultures reach significant cell densities. Thus, as a batch culture grows, the amount of light to which cells are exposed becomes less and less. This is usually (though not necessarily wholly) overcome by studying cells grown only to low culture densities. An alternative strategy is to subject cells to higher levels of light. These two methods produce cells which have been growing at two different rates and are in two different growth phases prior to harvesting. It is therefore important to consider the conditions under which cells are grown and exposed to light, particularly when comparing data which use these two different approaches to solving this problem.

The rest of this chapter is primarily concerned with the transcriptional and post-transcriptional control mechanisms governing expression of photosynthetic complexes in direct response to changes in light intensity. As mentioned above, these mechanisms are largely responsible for regulating expression of apoproteins, rather than pigment expression directly (though HvrB, described below, is an example of a transcriptional regulator affecting levels of a Bchl precursor). Regulation of pigment biosynthesis through the activities of Bchl synthesizing enzymes is not discussed here (see Armstrong, 1995; Biel, 1995). In addition, one study has implicated translation or post-translation as a possible mechanism(s) for light regulation of apoprotein (LHII) synthesis (Zucconi & Beatty, 1988). Studies have considered the possibility of specific post-translational effects (Klug, 1995), although none has so far been linked specifically with light. Similarly, no light-specific translational regulation has yet been demonstrated; however, it is relevant to include some discussion of the recently emerging reports of translation processes, since is not yet understood which signals (for example, light, oxygen or intracellular signals) these processes are responding to.

ARRANGEMENT OF THE PHOTOSYNTHESIS GENES

Before an account of transcriptional regulation can be given, a description of the photosynthesis genes involved is required. Figure 2 shows the clustered arrangement of genes in *R. sphaeroides*, which occupies approximately 46 kb of the chromosome. Genes encoding the biosynthetic enzymes for Bchl (*bch*) and Crt (*crt*) pigments are flanked by those encoding the LHI (*pufB* and *pufA*) and RC (*puhA, pufL* and *pufM*) polypeptides. The genes required for

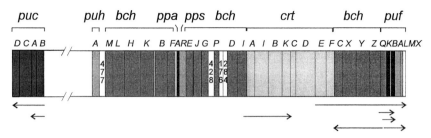

Fig. 2. Arrangement of genes in the 45 kb photosynthesis gene cluster of *Rhodobacter sphaeroides*, showing *pucBACD* (encoding the LHII polypeptides and located outside the main cluster) (and excluding *hemF*), *puhA* and *pufLM* (encoding reaction centre polypeptides), *pufBA* (encoding the LHI polypeptides), *bch*, (encoding Bchl biosynthetic enzymes) and *crt* (encoding carotenoid biosynthetic enzymes). Different shades of box hatching are used to differentiate between the adjacent groupings of *crt* and *bch* genes and also to differentiate *ppa* and *ppsR* genes more easily within the *bch* clusters. Numbers shown inside unshaded boxes denote open reading frames: *orf477* ('F1696') (adjacent to *puhA*), *orf428* (located between the *bchG* and *bchP* genes), and *orf176* and *orf284* (located between *bchP* and *bchD*) (C. N. Hunter, unpublished data). The *ppa* and *ppsR* genes are shown as the darker and lighter boxes respectively, within the *bch* cluster (Gomelsky & Kaplan, 1995*b*). Basis of the genetic map is redrawn and updated with permission (H. Lang, University of Sheffield) and from Hunter & Coomber (1988), Coomber *et al.* (1990), Hunter *et al.* (1991), Gibson *et al.* (1992), McGlynn & Hunter (1993), Lang *et al.* (1994), Hunter *et al.* (1994), Gong & Kaplan (1996). *cycA* and *hem* genes not shown. Arrows indicate identified transcripts.

LHII expression (*pucB, pucA, pucC* and *pucD* (*pucE*)) are not found in this cluster; they are located some distance from these genes (Fig. 2). The arrangement of photosynthesis genes is generally similar in *R. capsulatus*, although one notable difference is the presence of an additional gene in the *puc* operon of this organism.

Studies of both *R. sphaeroides* and *R. capsulatus* have revealed that many of the photosynthesis genes are arranged in operons. The transcripts identified to date in *R. sphaeroides* are shown in Fig. 2. Work on *R. capsulatus* genes has demonstrated that some of the genes are further arranged into larger units, or 'superoperons', since expression of the *bchCXYZ* genes is influenced by readthrough transcription from the upstream *crtE* and *crtF* genes, and the *bchCXYZ* genes contain both an oxygen-regulated and a constitutive promoter for the downstream *puf* operon (Young *et al.*, 1989). This is also the case in *R. sphaeroides* (Hunter *et al.*, 1991). This arrangement of genes into operons and superoperons has consequences for the coordinated and differential expression of these genes (Belasco *et al.*, 1985).

LIGHT REGULATION AT THE LEVELS OF PHOTOSYNTHESIS GENE TRANSCRIPTION AND mRNA STABILITY

Evidence that light regulates expression of photosynthetic components at the level of gene transcription first emerged when studies of steady-state levels of

mRNA in *R. sphaeroides* showed that cells contained higher levels of *pufBA*-
(LHI β and α polypeptides), *pufL*- (RC L subunit) and *pufM*- (RC M
subunit) specific mRNAS when grown under low light (620 to 1100 nm),
compared with those grown under high light (620 to 1100 nm) (Zhu &
Kaplan, 1985). Studies by Zhu & Hearst (1986) showed that this was also
the case for LHI- and RC-specific transcripts in *R. capsulatus* and, in
addition, that LHII, RC-H (*puhA*) and Bchl (*bchJGD, bchHKF*) transcript
levels were also affected by light. Furthermore, this study demonstrated that
the opposite was true for levels of Crt mRNAs; high light levels resulted in
increased levels of these transcripts (Zhu & Hearst, 1986). This might be
expected, since carotenoids not only serve as accessory light harvesting
pigments, but also have the additional function of protecting cells from
photooxidative damage, which occurs under conditions of high light and
oxygen (Zhu & Hearst, 1986). Since the studies of Zhu and Kaplan (1985)
and Zhu and Hearst (1986) examined steady-state mRNA levels, it was not
possible to conclude whether any light-dependent mRNA processing events
also contributed to the final mRNA levels measured in these studies. Earlier
studies by Biel and Marrs (1983) used reporter constructs, in which the *bch*
promoter regions were fused to *lacZ* reporter genes, to measure levels of *bch*
gene transcription initiation. They found no detectable effect of light on *bch*
transcription. Taken with the studies of steady-state mRNA levels, this
would imply that light regulation was exerted through transcript processing.
However, other interpretations are also possible. For example, Biel and
Marrs suggested that the lack of change in *bch-lacZ* transcription initiation
may be the result of the inability of the fusion strains to synthesize Bchl and
thus pigment–protein complexes (Biel & Marrs, 1983).

Zucconi and Beatty (1988) have demonstrated that the final steady-state
levels of LHII transcripts are regulated by at least one light-dependent post-
transcriptional process. *R. capsulatus* cells grown anaerobically under low
light conditions $(2 \, \mathrm{W \, m^{-2}})$ possessed levels of LHII which were four times
greater than those found in cells grown under high light intensity
$(140 \, \mathrm{W \, m^{-2}})$. However, when LHII transcript levels were measured, it was
found that levels were four times higher in the high light-grown cells, rather
than in the low light cells. This difference was shown to be due, in part, to
light-dependent rates of mRNA degradation, since transcript half-lives were
about 10 minutes in cells grown with high light and approximately 19 minutes
in cells grown under low light conditions. It was also concluded that there
must be more frequent initiations of transcription of LHII genes in cells
grown under high light conditions in order to account for both the recorded
steady-state levels and half-lives of these mRNAs (Zucconi & Beatty, 1988).
One further suggestion from this study was the presence of light-dependent
translational or post-translational regulation. Zucconi & Beatty (1988)
considered the possibility that the observed lower levels of photosynthetic
complexes in the membrane may occur through inhibitory effects on

bacteriochlorophyll biosynthesis since, as mentioned above, bacteriochlorophyll limitation is known to reduce the amount of complexes integrated in the membrane (Dierstein, 1983; Biel, 1986; Klug et al., 1986). One possible mechanism for this might be an effect on pufQ gene expression, since the pufQ gene product was suggested to be required for activity of bacteriochlorophyll biosynthesis enzymes; a reduction in pufQ transcription would therefore limit bacteriochlorophyll levels in the cell and hence reduce apoprotein and complex formation in the membrane (Bauer & Marrs, 1988). However, another study has reported that, at least in R. sphaeroides, the Q protein appears to be involved in the ultimate use of Bchl, perhaps assembly (Gong, Lee & Kaplan, 1994).

In studies examining the effects of different wavelengths of light on mRNA levels, it was found that blue light with a wavelength optimum of 450 nm exerted a strong repressive effect on levels of puc and puf transcripts when an intensity of $23 \, \mu mol \, m^{-2} \, s^{-1}$ was used (Shimada, Iba & Takamiya, 1992). In these studies too, it was not possible to differentiate between possible effects on rates of transcript initiation and degradation, but later studies by this group confirmed an effect on initiation when they reported the isolation of a blue-light dependent transcription factor, Spb (see below).

Evidence that light can regulate transcription initiation of photosynthesis genes in R. capsulatus and R. sphaeroides also comes from the work of Klug et al. (1991) and Lee and Kaplan (1992), respectively. Klug et al. (1991) found that a single base change within a region of dyad symmetry located upstream of the puf promoter affected light regulation (as well as oxygen regulation) of RC and LHI transcripts. Lee and Kaplan (1992) reported studies in which different sequences located upstream of the puc genes (encoding LHII) were fused to a lacZ reporter gene in order to identify a light-regulated upstream region located − 629 to − 150 base pairs upstream from the pucBAC transcription start point. Additional evidence also comes from a study by Lee et al. (1993) which reported an integration host factor (IHF) binding site located upstream of the puc operon that could be involved in light-mediated (as well as oxygen-mediated) transcriptional regulation.

Transcription factors

Although significant progress has been made in our understanding of the oxygen-dependent transcription factors and associated signalling pathways involved in regulating photosynthesis gene expression, less information is available concerning light-dependent factors. This is mainly because oxygen exerts the greatest influence on expression of photosynthesis components. The effects of oxygen are sufficiently great to facilitate the use of molecular genetic selection procedures for the isolation of mutants affected in oxygen regulation of photosynthetic complexes. This does not appear to be the case for light regulation; no reports using such techniques have yet appeared.

Nevertheless, three light-dependent factors which affect the transcription of genes involved in apoprotein or bacteriochlorophyll biosynthesis have been identified to date in *R. sphaeroides* or *R. capsulatus*. Furthermore, one recent study has shown that high light repression of *puc* (and probably *crt* and *bch*) genes is mediated through the aerobic transcription factor PpsR (Gomelsky & Kaplan, 1997). The following is a brief description of these four light-regulated factors.

HvrA is a light-dependent transcriptional activator of *puf* and *puh* operons in *R. capsulatus* (Buggy, Sganga & Bauer, 1994a). It was first identified just downstream of the *regA* gene, during genetic analyses of the *reg* gene region which contains the genes encoding components of the Reg oxygen-regulatory pathway. When mutations were introduced into this region, some mutants were isolated which possessed a reduced ability to grow photosynthetically under dim light conditions and a significantly reduced ability to increase photopigment biosynthesis as light conditions became limiting. The 11.5 kDa HvrA protein exhibits homology with another light-responsive transcription factor, Spb (described below), and possesses a 20-residue region which is likely to be a helix–turn–helix motif involved in DNA binding (Fig. 3(b)). Reporter studies showed that, whereas wild-type cells exhibited a two-fold increase in photosynthesis gene expression in response to a reduction from near saturating light conditions to low light, a *hvrA* mutant failed to increase *puf* and *puh* gene expression under the same conditions. The effect appeared to be specific for *puf* and *puh*, which encode LHI and RC polypeptides, since *puc* and Bchl gene expression was unaffected by a reduction in light conditions. Footprinting experiments confirmed that HvrA is able to bind to the *puf* and *puh* promoter regions (Buggy *et al.*, 1994a). Thus, *hvrA* appears to be responsible for regulating LHI and RC gene expression, but not LHII gene expression, in response to changes in light intensity. This finding is consistent with what is currently understood about the discoordinated expression of *puf* and *puh* genes from the *puc* genes (Schumacher & Drews, 1979; Klug, Kaufmann & Drews, 1985; Kiley & Kaplan, 1988). As Buggy *et al.* (1994a) reason, presumably discoordinate expression provides the cell with greater flexibility with regard to adjustment of complex levels in response to changes in light intensity. Northern hybridization experiments demonstrated that the *hvrA* gene is co-transcribed with the *senC* and *regA* genes involved in oxygen regulation of photosynthesis genes, in the order *senC-regA-hvrA*, and that *in vivo* mRNA processing occurs which may have consequences for final levels of HvrA relative to RegA and SenC components. The mechanism by which HvrA activity is controlled has yet to be elucidated. But whether this is by the direct interaction of light with HvrA (though no chromophore has yet been identified), or the presence of a more complex light signal transduction pathway, the findings that *hvrA* transcription may possibly be co-regulated with that of *regA* and *senC* suggests the possibility of some coordination with the Reg regulatory network.

(a)

-729   -689
GGGTGCGGCGATCCGGCGCGCTACCGGAACGCCCGTTATGG
       -722   -709

(b)

Hvra   RKKKEAFAELDEIARKMGYPLAEILTMVETKPRKTVAAK
                     helix-turn-helix motif

Spb   RRKRDALAELEEKARELGFSLSELTGTAATKRRAAAQPK

c-JUN   KRKLERIARLEEKVKTLKAQNSELASTANMLREQVAQLK
                     leucine-zipper motif

Fig. 3. (*a*) The Spb binding site in the *puf* promoter region of *R. sphaeroides*. The horizontal line shows the sequence protected in the DNAse I footprinting experiments of Shimada *et al.* (1993). The sequence is numbered with respect to the start codon of *pufB*, with the first base of the start codon assigned as + 1 and the base preceding it as − 1. (*b*) Amino acid sequence alignments of the DNA binding regions of the Spb protein of *R. sphaeroides*, the HvrA protein of *R. capsulatus* and c-JUN from chicken. The solid line shows the leucine-zipper motifs whilst the unfilled line shows the extent of the helix–turn–helix motif identified in HvrA (Buggy *et al*, 1994; Shimada *et al.*, 1996). The leucine residues are in bold type. (Redrawn from Shimada *et al.*, 1993 and Shimada *et al.*, 1996.)

The only light-dependent transcription factor identified in *R. sphaeroides* to date is Spb. This protein was isolated and characterised following studies which showed that blue light of 450 nm reduced the expression of the *puf* and *puc* operons under semi-aerobic conditions in *R. sphaeroides* (Shimada *et al.*, 1992). It was proposed that, under these conditions, a blue pigment is present which regulates (represses) the expression of these operons. Bandshift experiments demonstrated the presence of a putative transcription factor, which behaved as a repressor of *puf* transcription under blue light or high oxygen conditions (Shimada *et al.*, 1993). The protein was shown to bind at a position between − 709 and − 722 base pairs upstream of the *puf* transcription start site, and is located within a 37 base pair region identified by Hunter *et al.* (1991) as the oxygen-regulated promoter for *pufQ*, and probably the entire *puf* operon (Shimada *et al.*, 1993) (Fig. 3(a)). The palindromic sequence present in this region may indicate that, in common with other bacterial transcription factors, the protein may bind to the DNA target in a dimeric form. The protein was isolated and sequenced in order to facilitate the cloning of the gene (*spb*) (Shimada *et al.*, 1996). The *spb* gene encodes a protein of 11.5 kDa (104 amino acids) which exhibits 53% identity (71% similarity) with HvrA (see above), and also possesses a leucine-zipper motif typical of eukaryotic transcription factors (Fig. 3(b)). In a sequence comparison of Spb with HvrA, it is clear that the leucine-zipper motif identified in Spb overlaps the proposed helix–turn–helix of HvrA (Shimada *et al.*, 1996) (Fig. 3(b)). This sequence may therefore be the DNA-binding domain. In view of the high level of homology between Spb and HvrA, it is possible that Spb may be the *R. sphaeroides* homologue of HvrA. During recent analyses of the *spb* gene region, it has been shown that the *spb* gene is a positional counterpart of *hvrA* (Mizoguchi *et al.*, 1997). However, Spb functions as a repressor of *puf* transcription, whereas HvrA serves as a *trans*-acting activator or *puf* and *puhA* genes. Whether or not Spb turns out to be the only HvrA homologue, these two transcription factors nevertheless provide an interesting example of structurally similar proteins functioning to bring about opposite effects at the promoter region of the *puf* operon. The *spb* gene is expressed constitutively, irrespective of growth conditions, but binding to the *puf* promoter region only occurs when blue light is present under semi-aerobic growth conditions. No binding occurred when cells were grown semi-aerobically in the dark, confirming that binding was blue light dependent (Shimada *et al.*, 1996). However, binding could occur under these conditions if cell extracts were pretreated with alkaline phosphatase, suggesting that DNA-binding activity might be regulated by a phosphorylation–dephosphorylation mechanism. Perhaps a more complex light signal transduction pathway (possibly similar to that in which HvrA might function), may be involved.

In addition to Spb, another blue light-responsive protein has been identified in the purple bacteria. The photoactive yellow protein (PYP) was first characterized in *Ectothiorhodospira halophila*, an obligately anaerobic halophilic bacterium (Meyer, 1985) and has recently been further characterized at the molecular genetic and biochemical levels by Kort *et al.*, 1996 (and references therein). Recently, it was also shown that homologous proteins also occur in *Rhodobacter* species (Hoff *et al.*, 1994; Kort *et al.*, 1996; R. Kort, S. M. Hoffer, M. K. Phillips-Jones, P. S. Duggan, W. D. Hoff, W. Crielaard & K. J. Hellingwerf, unpublished data). The precise function of the protein has not yet been fully established, although current evidence suggests that it may be a photoreceptor involved in negative phototaxis in *E. halophila* (Sprenger *et al.*, 1993). This is discussed in more detail in Hellingwerf, Kort & Crielaard, this volume.

The third light-dependent transcription factor identified to date is HvrB, which regulates the expression of at least one gene involved in Bchl synthesis in *R. capsulatus* (Buggy, Sganga & Bauer, 1994*b*). HvrB is a 32 kDa protein

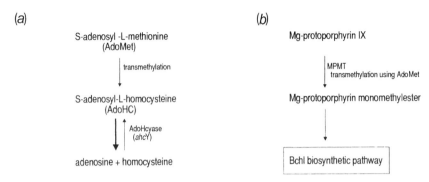

Fig. 4. Two reactions involving S-adenosyl-L-methionine (AdoMet)-mediated transmethylation steps in *Rhodobacter* species. (*a*) S-adenosyl-L-homocysteine (AdoHC) is synthesized as a direct product of AdoMet-mediated methyl group transfer reactions. AdoHC is, in turn, hydrolysed by S-adenosyl-L-homocysteine hydrolase (AdoHcyase) (product of the *ahc*Y gene), yielding adenosine and homocysteine. Disruption of *ahc*Y results in a build-up of AdoHC, which is an inhibitor of transmethylation reactions involving AdoMet. (*b*) The transmethylation of Mg-protoporphyrin IX to Mg-protoporphyrin monomethylester by S-adenosyl-L-methionine:magnesium protoporphyrin methyltransferase (MPMT) using AdoMet as methyl group donor. The monomethylester is the first stable intermediate which is unique to the Bchl biosynthetic pathway.

resembling the LysR family of transcriptional regulators and is required for the low-light activation of the *ahc*Y gene and the high-light activation of another gene, *orf5* (Buggy *et al.*, 1994*b*). The *ahc*Y gene encodes S-adenosyl-L-homocysteine hydrolase (AdoHcyase), which catalyses the reversible hydrolysis of S-adenosyl-L-homocysteine (AdoHC) to adenosine and homocysteine (Fig. 4(a)). Strains lacking AdoHcyase activity accumulate AdoHC, which is a potent inhibitor of transmethylation reactions involving S-adenosyl-L-methionine (AdoMet), some of which are required for the first steps in Bchl biosynthesis (Fig. 4(b)). The function of the *orf5* gene product is unknown. Buggy *et al.* (1994*b*) showed that, whilst HvrB appeared to be a low-light transcriptional activator of *ahc*Y, mutant strains disrupted in *hvr*B nevertheless exhibited normal *in vivo* light regulation of Bchl biosynthesis. Thus, it was suggested that regulation of *ahc*Y expression may be just one mechanism by which Bchl biosynthesis is regulated by light. This is consistent with evidence that regulation of Bchl biosynthesis is primarily exerted at the level of activities of Bchl biosynthetic enzymes, rather than directly at the genetic level.

Recently, an additional photosynthesis-associated transcription factor has been shown to function in response to light (Gomelsky & Kaplan, 1997). The PpsR protein was first described by Penfold and Pemberton (1994) who showed that it was involved in repression of carotenoid and bacteriochlorophyll synthesis under aerobic conditions in *R. sphaeroides*. Later, it was shown that PpsR serves as an aerobic repressor of *puc* and *bchF* genes in *R. sphaeroides* (Gomelsky & Kaplan, 1995*b*) and of *puc, crtI, bchH, bchC* and *bchD* genes in *R. capsulatus* (in which the protein is known as CrtJ) (Ponnampalam *et al.*, 1995). The 464 residue protein (469 in *R. capsulatus*) possesses a helix–turn–helix motif and was shown to bind to a proximal TGT-N_{12}-ACA palindromic sequence located upstream of *puc* (Gomelsky & Kaplan, 1995*b*; Ponnampalam *et al.*, 1995). Gomelsky and Kaplan (1997) have shown that PpsR can also repress *puc* (and presumably *crt* and *bch* genes) under anaerobic/high light conditions. Comparisons of the levels of photosynthetic complexes in wild-type and PpsR null mutant *R. sphaeroides* strains grown under anaerobic/high light conditions revealed that, whereas levels of LHI complexes in the two strains were similar, the levels of LHII complexes remained higher in the mutant, suggesting the loss of LHII-specific repressor function. The effect was not observed under low light conditions. These findings may well account, at least in part, for the well-known repression of LHII expression under high light conditions. The precise mechanism by which PpsR functions is not yet fully understood. Gomelsky and Kaplan (1997) suggest that, since PpsR expression levels appear generally unaffected by growth conditions, the extent of PpsR-mediated repression may depend on the activity of the PpsR protein itself, which in turn could be regulated by interaction with another protein. One possible candidate is AppA, which has also been shown to have a role in regulating photosynthetic complex levels (Gomelsky & Kaplan, 1995*a*). These workers used molecular genetic methods to demonstrate a possible interaction between PpsR and AppA proteins, with AppA acting 'upstream' of the PpsR protein (Gomelsky & Kaplan, 1997).

An overview of transcriptional activation and repression by oxygen and light

Figure 5 summarizes what is currently known about the transcriptional events which occur within the photosynthetic gene clusters of *R. sphaeroides* and *R. capsulatus* in response to a change from chemoheterotrophic (aerobically in the dark) to phototrophic (anaerobically in the light) growth. A reduction in oxygen tension coupled with the presence of light, induces events which reduce the activities, function and/or amounts of transcriptional repressors at their target sites in the promoter regions (Fig. 5*(a)*), and events which promote the binding and activities of transcriptional activators (Fig. 5*(b)*). Those factors which affect Bchl precursor formation are not shown here. For more details of the oxygen-responsive factors and their associated signalling pathways see Bauer and Bird (1996), Phillips-Jones and Hunter (1994) and Eraso and Kaplan (1996) (and references cited therein) for details of the Reg circuit, Gomelsky and Kaplan (1995*a*, 1997) for AppA (which affects PpsR activity), and Penfold and Pemberton (1994), Gomelsky and Kaplan (1995*b*, 1997) (and references therein) and Ponnampalam *et al.* (1995) for PpsR (CrtJ).

(a) Aerobic conditions (chemoheterotrophic growth)
Rhodobacter sphaeroides

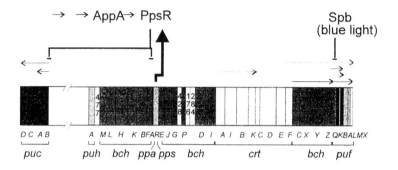

Rhodobacter capsulatus

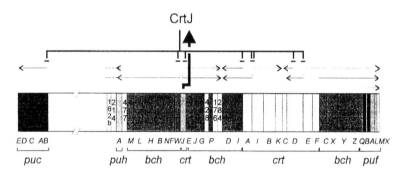

Fig. 5. Transcriptional regulation of genes in the photosynthetic cluster under (*a*) aerobic chemoheterotrophic, and (*b*) anaerobic/light conditions in *R. sphaeroides* and *R. capsulatus*. The genes/operons are *puc* (encoding the LHII polypeptides), *puhA* (encoding one of the RC polypeptides), *bch* (encoding BchI biosynthetic enzymes), *crt* (encoding carotenoid biosynthetic enzymes) and *puf* (encoding L and M RC and LHI polypeptides). Different shades of box hatching are used to differentiate between the adjacent groupings of *crt* and *bch* genes and also to differentiate *ppa*, *ppsR* and *crtJ* regulatory genes more easily within the *bch* clusters. Numbers shown inside unshaded boxes denote open reading frames: *orf477* ('F1696') (adjacent to *puhA*), *orf428* (located between the *bchG* and *bchP* genes), *orf176* and *orf284* (located between *bchP* and *bchD*) and *orf214* and *orf162b* (located adjacent to *puhA*). Thin arrows drawn parallel to the genes show the transcripts identified. Thick arrows indicate the genes located within the clusters encoding either PpsR (in *R. sphaeroides*) or CrtJ (in *R. capsulatus*). *cycA* and *hem* genes are not shown. The oxygen-dependent factors include PpsR in *R. sphaeroides* and CrtJ in *R. capsulatus*,

(b) Anaerobic/light conditions (phototrophic growth)
Rhodobacter sphaeroides

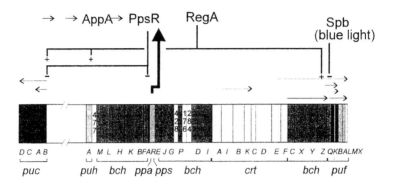

Rhodobacter capsulatus

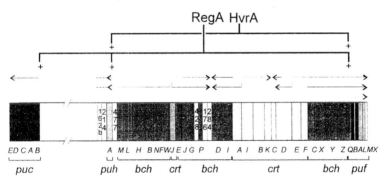

which serve as aerobic repressors (with PpsR additionally involved in light regulation under aerobic light conditions through interactions with AppA in *R. sphaeroides* (see text)), and RegA which serves as an activator of transcription in both species under anaerobic conditions. The activator of *hemA* (FnrL) is not shown here (Zeilstra-Ryalls & Kaplan, 1995). The light-dependent factors identified to date include Spb which represses *puf* transcription under blue light conditions in *R. sphaeroides*, and HvrA (which shows homology to Spb) which activates *puf* and *puh* transcription under low light conditions in *R. capsulatus*. The low light transcriptional activator of *ahcY*, HvrB (described in text) is not shown here. Gene activation is shown by a + symbol and repression by a − symbol. For further details, see text. (Adapted from drawings by H. Lang, University of Sheffield, and from data in references cited in Fig. 2 legend (*R. sphaeroides*), and Bollivar *et al.*, 1994, Bauer & Bird, 1996; Drews, 1996 and Wong *et al.*, 1996 (*R. capsulatus*).)

Under aerobic conditions, oxygen-dependent PpsR serves as a repressor of the *puc* genes and some pigment genes in both *R. sphaeroides* (Penfold & Pemberton, 1994; Gomelsky & Kaplan, 1995*b*) and *R. capsulatus* (CrtJ; Ponnampalam *et al.*, 1995). It does not appear to repress *puf* and *puhA* genes encoding RC and LHI components. This is consistent with the observation that levels of LHII are much more variable compared with LHI or RC in response to environmental changes. Presumably, further repressors have yet to be identified which repress *puf* and *puh* transcription under aerobic conditions.

Upon a shift to anaerobic/light conditions, transcription of *puf*, *puc* and *puhA* genes is activated by the oxygen-sensitive Reg pathway in both *R. sphaeroides* and *R. capsulatus* (Phillips-Jones & Hunter, 1994; Bauer & Bird, 1996; Eraso & Kaplan, 1996). In addition, HvrA activates *puf* and *puhA* transcription in *R. capsulatus* under low light conditions, thus functioning as a regulator of LHI and RC expression (Buggy, Sganga & Bauer, 1994*a*), whilst in *R. sphaeroides* the only other factor identified to date is the oxygen- and light-sensitive PpsR protein which regulates LHII and pigment expression (Penfold & Pemberton, 1994; Gomelsky & Kaplan, 1995*b*, 1997). Note that the aerobic repressor PpsR serves as a repressor under both aerobic and anaerobic/high light conditions but not under anaerobic/low light conditions. As discussed above, this finding is entirely consistent with the finding that the levels of membrane-bound LHII complex (encoded by the *puc* genes), are high under low light conditions and low under high light conditions.

Some of the factors identified so far have been reported either only in *R. sphaeroides* (Spb) or only in *R. capsulatus* (HvrA). It should be noted that this does not necessarily mean that they do not occur in both species. Rather, this illustrates, that in some cases, there is a need for further work to establish whether they are indeed common to both species. This is important since it is known that there are differences in oxygen regulation displayed by *R. sphaeroides* and *R. capsulatus* (Klug, 1993*b*) and it is possible that the underlying mechanisms responsible for these differences may be closely associated with, or cross-connect with, light-dependent mechanisms.

REGULATION AT THE LEVEL OF TRANSLATION

Several studies indicate the possibility that regulation of photosynthesis gene expression is also exerted at the level of translation (e.g. Zucconi & Beatty, 1988; Gong & Kaplan, 1996; Babic *et al.*, in press), and that translational regulation may be one possible mechanism by which light affects the final level and composition of photosynthetic complexes synthesized in the cell (Zucconi & Beatty, 1988; Klug, 1995). Although these studies indicate a possible role for translation, to date there have been no reports of any light-dependent translational control. However, it is of relevance to present a short summary of what is currently understood about the photosynthetically relevant translational processes; further studies in this area will be required to elucidate whether any such processes are subject to regulation by light or oxygen.

The first study to demonstrate translational control was that of Gong & Kaplan (1996) who were studying regulation of *puf* operon expression in *R. sphaeroides* 2.4.1. The *puf* operon of this strain contains five genes: the *pufBA* genes encoding the LHI polypeptides, *pufLM* encoding two of the reaction centre polypeptides and *pufX* encoding the PufX assembly protein. Two major transcripts are produced from the *puf* operon; a small 0.5 kb (*pufBA*) transcript coding for the two LHI polypeptides PufB and PufA and a large 2.7 kb transcript (*pufBALMX*) coding for all five proteins. Just upstream of the *pufB* gene (but co-transcribed with the *pufBA* and *pufBALMX* genes), is a small gene known as *pufK* which encodes a 20-residue polypeptide. Gong and Kaplan (1996) demonstrated that the small *pufK* gene is translated and that the rate of translation is attributable to the occurrence of rare codons in the *pufK* message. Most significantly, it appeared that translation of *pufB* was dependent on translation of *pufK*. This finding led Gong and Kaplan to suggest that the abundance and distribution of rare codons within *pufK* may serve as a 'gateway' for the entry of ribosomes to *pufB* and hence exert control over the rate of translation from *pufB*.

Another recent study examined translation initiation; in particular, the role of a translation initiation factor (Babic *et al.*, 1997). The initiation of translation in prokaryotes requires several protein factors including translation initiation factor 3 (IF3). Studies in *E. coli* have demonstrated that IF3 functions in the proofreading of the initiation AUG codon-initiator tRNA^{fMet} complex on the 30S ribosomal subunit (Gualerzi & Pon, 1990; Hartz, McPheeters & Gold, 1990), in the dissociation of the 70S ribosomes to ensure a plentiful supply of 30S subunits for new rounds of translation initiation (Godefroy-Colburn *et al.*, 1975) and in the discrimination against atypical start codons (Sacerdot *et al.*, 1996; Sussman, Simons & Simons, 1996). With the multiple important roles played by this protein during translation initiation, it is perhaps not surprising that this protein is essential for cell viability. Interestingly, in *E. coli* and *Myxococcus xanthus*, IF3 appears to have an additional role as a pleiotropic regulation factor controlling expression of certain developmental genes or groups of genes. In *E. coli*, IF3 differentially affects recombination genes by suppressing the *recJ* gene, whilst in *M. xanthus*, the IF3 protein (Dsg) exerts a selective effect on expression of sporulation and fruiting body development genes (Haggerty & Lovett, 1993; Cheng, Kalman & Kaiser, 1994; Kalman, Cheng & Kaiser, 1994). The mechanisms by which these selective effects are exerted are not well understood, but they appear to be different in these two species, since the selective function exhibited by Dsg was shown to be associated with a unique C-terminal peptide of 66 residues, which is not present in the *E. coli* protein

(a)

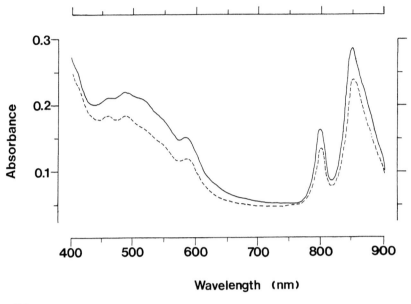

(b)

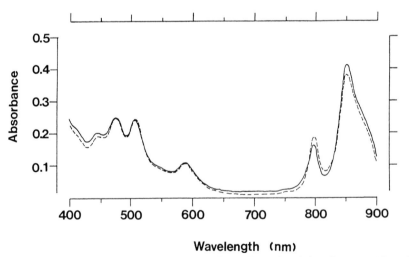

Fig. 6. Room temperature absorbance spectra of *R. sphaeroides* whole cells grown (*a*) semi-aerobically or (*b*) anaerobically in the light, showing peaks associated with the LHII complex (absorbing in the 800–850 nm range), the LHI complex (absorbing in the 875 nm range), and free pigments (in the 420–600 nm range). Wild-type cells (solid line) and *pifC* mutant cells (dashed line). (Taken from Babic *et al.*, 1997, with permission.)

(Kalman *et al.*, 1994). These studies raised the possibility that other developmental processes such as photosystem synthesis in photosynthetic bacteria may also be subject to selective effects by IF3. The work of Babic *et al.* (1997) reveals that this may possibly be the case. Following the cloning and sequencing of the *R. sphaeroides* gene encoding the IF3 homologue PifC, it has been shown that partial inactivation of the *pifC* gene resulted in reduced levels of Bchl and overall photosynthetic complex levels under semi-aerobic conditions but had no effect on cells grown anaerobically in the light (Fig. 6). It is not yet known whether any signals such as light, oxygen or other intracellular factor may be involved in these observed effects. Although the effect of the mutation was not as large as that observed for some photosynthesis-specific regulator genes, it should be noted that the mutant still possessed significant (but reduced) PifC IF3 activity, attributable to the synthesis of an altered, partially active form or reduced levels of this essential protein (Babic *et al.*, 1997).

It is well established that the synthesis of the photosynthetic apparatus in photosynthetic bacteria such as *R. sphaeroides* and *R. capsulatus* is subject to strict regulation by environmental factors, especially oxygen and light. Oxygen exerts the greatest effect, particularly at the levels of transcription of photosynthesis genes, post-transcriptional effects on the stability of the resulting mRNAs and also translation and/or post-translation. Light also appears to regulate expression at these levels, although fewer light-dependent protein factors have been identified, to date. This may be because there are fewer light-dependent factors present, or because not all factors have yet been identified. Expression of the components of the photosynthetic apparatus, the final stoichiometry of components and assembly of those components in the membranes of photosynthetic bacteria are all complex and finely coordinated processes. The mechanisms by which these processes (particularly expression), are affected by changes in levels of external oxygen and light are also complex and more work is required to obtain a full understanding of how these mechanisms work.

ACKNOWLEDGEMENTS

I would like to thank Professor C. N. Hunter for discussions, and for making data available prior to publication. I would also like to thank Professor G. Klug and Professor S. Kaplan for reading the manuscript and for their suggestions and comments.

REFERENCES

Aagaard, J. & Sistrom, W. R. (1972). Control of synthesis of reaction centre bacteriochlorophyll in photosynthetic bacteria. *Photochemistry and Photobiology*, **15**, 209–25.

Armstrong, G. A. (1995). Genetic analysis and regulation of carotenoid biosynthesis: structure and function of the *crt* genes and gene products. In *Anoxygenic Photosynthetic Bacteria*, ed. R. E. Blankenship, M. L. T. Madigan & C. E. Bauer, pp. 1135–57. Dordrecht, The Netherlands: Kluwer Academic Publishers.

Babic, S., Hunter, C. N., Rakhlin, N. J., Simons, R. W. & Phillips-Jones, M. K. Molecular characterisation of the *pifC* gene encoding translation initiation factor 3, which is required for normal photosynthetic complex formation in *Rhodobacter sphaeroides* NCIB 8253. *European Journal of Biochemistry*, **249**, 564–75.

Bauer, C. E. & Bird, T. H. (1996). Regulatory circuits controlling photosynthesis gene expression. *Cell*, **85**, 5–8.

Bauer, C. E. & Marrs, B. L. (1988). *Rhodobacter capsulatus puf* operon encodes a regulatory protein (*pufQ*) for bacteriochlorophyll biosynthesis. *Proceedings of the National Academy of Sciences, USA*, **85**, 7074–8.

Beck, J. & Drews, G. (1982). Tetrapyrrole derivatives shown by fluorescence emission and excitation spectroscopy in cells of *Rhodobacter capsulatus* adapting to phototrophic conditions. *Zeitschrift Naturforschung*, **37c**, 199–204.

Belasco, J. G., Beatty, J. T., Adams, C. W., von Gabain, A. & Cohen, S. N. (1985). Differential expression of photosynthesis genes in *R. capsulatus* results from segmental differences in stability within the polycistronic *rxcA* transcript. *Cell*, **40**, 171–81.

Biel, A. J. (1986). Control of bacteriochlorophyll accumulation by light in *Rhodobacter capsulatus*. *Journal of Bacteriology*, **168**, 655–9.

Biel, A. J. (1995). Genetic analysis and regulation of bacteriochlorophyll biosynthesis. In *Anoxygenic Photosynthetic Bacteria*, ed. R. E. Blankenship, M. L. T. Madigan & C. E. Bauer, pp. 1125–34. Dordrecht, The Netherlands: Kluwer Academic Publishers.

Biel, A. J. & Marrs, B. L. (1983). Transcriptional regulation of several genes for bacteriochlorophyll biosynthesis in *Rhodopseudomonas capsulata* in response to oxygen. *Journal of Bacteriology*, **156**, 686–94.

Biel, A. J. & Marrs, B. L. (1985). Oxygen does not directly regulate carotenoid biosynthesis in *Rhodopseudomonas capsulata*. *Journal of Bacteriology*, **162**, 1320–1.

Bollivar, D. W., Suzuki, J. Y., Beatty, J. T., Dobrowolski, J. M. & Bauer, C. E. (1994). Directed mutational analysis of bacteriochlorophyll *a* biosynthesis in *Rhodobacter capsulatus*. *Journal of Molecular Biology*, **237**, 622–40.

Buggy, J. J., Sganga, M. W. & Bauer, C. E. (1994a). Characterisation of a light-responding *trans*-activator responsible for differentially controlling reaction centre and light harvesting-I gene expression in *Rhodobacter capsulatus*. *Journal of Bacteriology*, **176**, 6936–43.

Buggy, J. J., Sganga, M. W. & Bauer, C. E. (1994b). Nucleotide sequence and characterisation of the *Rhodobacter capsulatus hvrB* gene: HvrB is an activator of S-adenosyl-L-homocysteine hydrolase expression and is a member of the LysR family. *Journal of Bacteriology*, **176**, 61–9.

Cheng, Y. L., Kalman, L. V. & Kaiser, D. (1994). The *dsg* gene of *Myxococcus xanthus* encodes a protein similar to translation initiation factor IF3. *Journal of Bacteriology*, **176**, 1427–33.

Cohen-Bazire, G., Sistrom, S. & Stanier, R. Y. (1957). Kinetic studies of pigment synthesis by non-sulphur purple bacteria. *Journal of Cellular and Comparative Physiology*, **49**, 25–68.

Coomber, S. A., Chaudri, M., Conner, A., Britton, G. & Hunter, C. N. (1990). Localised transposon Tn*5* mutagenesis of the photosynthetic gene cluster of *Rhodobacter sphaeroides*. *Molecular Microbiology*, **4**, 977–89.

Deisenhofer, J., Epp, P., Miki, K., Huber, R. & Michel, H. (1985). Structure of the protein subunits in the photosynthetic reaction centre of *Rhodopseudomonas viridis* at 3 Å resolution. *Nature*, **318**, 618–24.

Dierstein, R. (1983). Biosynthesis of pigment–protein complex polypeptides in bacteriochlorophyll-less mutant cells of *Rhodobacter capsulatus*. *FEBS Letters*, **160**, 281–6.

Drews, G. (1986). Adaptation of the bacterial photosynthetic apparatus to different light intensities. *Trends in Biochemical Sciences*, **11**, 255–7.

Drews, G. (1996). Forty-five years of developmental biology of photosynthetic bacteria. *Photosynthesis Research*, **48**, 325–52.

Drews, G. & Oelze, J. (1981). Organisation and differentiation of membranes of photosynthetic bacteria. *Advances in Microbial Physiology*, **22**, 1–92.

Eraso, J. M. & Kaplan, S. (1996). Complex regulatory activities associated with the histidine kinase PrrB in expression of photosynthesis genes in *Rhodobacter sphaeroides* 2.4.1. *Journal of Bacteriology*, **178**, 7037–46.

Fowler, G. J. S. & Hunter, C. N. (1996). The synthesis and assembly of functional high and low light LH2 antenna complexes from *Rhodopseudomonas palustris* in *Rhodobacter sphaeroides*. *Journal of Biological Chemistry*, **271**, 13356–61.

Gardiner, A. T., Cogdell, R. J. & Takaichi, S. (1993). The effect of growth conditions on the light-harvesting apparatus in *Rhodopseudomonas acidophila*. *Photosynthesis Research*, **38**, 159–67.

Gibson, L. C. D., McGlynn, P., Chaudri, M. & Hunter, C. N. (1992). A putative anaerobic coproporphyrinogen III oxidase in *Rhodobacter sphaeroides*. II. Analysis of a region of the genome encoding *hemF* and the *puc* operon. *Molecular Microbiology*, **6**, 3171–86.

Godefroy-Colburn, T., Wolfe, A. D., Dondon, J., Grunberg-Manago, M., Dessen, P. & Pantalini, D. (1975). Light scattering studies showing the effect of initiation factors on the reversible dissociation of *Escherichia coli* ribosomes. *Journal of Molecular Biology*, **94**, 461–78.

Gomelsky, M. & Kaplan, S. (1995*a*). *appA*, a novel gene encoding a *trans*-acting factor involved in the regulation of photosynthesis gene expression in *Rhodobacter sphaeroides* 2.4.1. *Journal of Bacteriology*, **177**, 4609–18.

Gomelsky, M. & Kaplan, S. (1995*b*). Genetic evidence that PpsR from *Rhodobacter sphaeroides* 2.4.1 functions as a repressor of *puc* and *bchF* expression. *Journal of Bacteriology*, **177**, 1634–7.

Gomelsky, M. & Kaplan, S. (1997). Molecular genetic analysis suggesting interactions between AppA and PpsR in regulation of photosynthesis gene expression in *Rhodobacter sphaeroides* 2.4.1. *Journal of Bacteriology*, **179**, 128–34.

Gong, L. & Kaplan, S. (1996). Translational control of *puf* operon expression in *Rhodobacter sphaeroides* 2.4.1. *Microbiology*, **142**, 2057–69.

Gong, L., Lee, J. K. & Kaplan, S. (1994). The Q-gene of *Rhodobacter sphaeroides*: its role in *puf* operon expression and spectral complex assembly. *Journal of Bacteriology*, **176**, 2946–61.

Gualerzi, C. O. & Pon, C. L. (1990). Initiation of mRNA translation in prokaryotes. *Biochemistry*, **29**, 5881–9.

Haggerty, T. J. & Lovett, S. T. (1993). Suppression of *recJ* mutations of *Escherichia coli* by mutations in translation initiation factor IF3. *Journal of Bacteriology*, **175**, 6118–25.

Hartz, D., McPheeters, D. S. & Gold, L. (1990). From polynucleotide to natural mRNA translation initiation: function of *Escherichia coli* initiation factors. In *The Ribosome: Structure, Function and Evolution*, ed. W. E. Hill, A. Dahlberg, R. A. Garrett, P. B. Moore, D. Schlessinger & J. R. Warner, pp. 275–80. American Society for Microbiology.

Hawthornthwaite, A. M. & Cogdell, R. J. (1991). Bacteriochlorophyll-binding proteins. In *The Chlorophylls* ed. H. Scheer, pp. 493–528. Boca Raton, FL: CRC Press.

Hoff, W. D., Sprenger, W. W., Postma, P. W., Meyer, T. E., Veenhuis, M., Leguijt, T. & Hellingwerf, K. J. (1994). The photoactive yellow protein from *Ectothiorhodospira halophila* as studied with a highly specific polyclonal antiserum: (intra)cellular localization, regulation of expression and taxonomic distribution of cross-reacting proteins. *Journal of Bacteriology*, **176**, 3920–7.

Hunter, C. N. (1995). Light-harvesting complex – rings of light. *Current Biology*, **5**, 826–8.

Hunter, C. N. & Coomber, S. A. (1988). Cloning and oxygen-regulated expression of the bacteriochlorophyll biosynthesis genes *bch E, B, A* and *C* of *Rhodobacter sphaeroides*. *Journal of General Microbiology*, **134**, 1491–7.

Hunter, C. N., Hundle, B. S., Hearst, J. E., Lang, H. P., Gardiner, A. T., Takaichi, S. & Cogdell, R. J. (1994). Introduction of new carotenoids into the bacterial photosynthetic apparatus by combining the carotenoid biosynthetic pathways of *Erwinia herbicola* and *Rhodobacter sphaeroides*. *Journal of Bacteriology*, **176**, 3692–7.

Hunter, C. N., McGlynn, P., Ashby, M. K., Burgess, J. G. & Olsen, J. D. (1991). DNA sequencing and complementation/deletion analysis of the *bchA-puf* operon region of *Rhodobacter sphaeroides*: *in vivo* mapping of the oxygen-regulated *puf* promoter. *Molecular Microbiology*, **5**, 2649–61.

Kalman, L. V., Cheng, Y. L. & Kaiser, D. (1994). The *Myxococcus xanthus dsg* gene product performs functions of translation initiation factor IF3 *in vivo*. *Journal of Bacteriology*, **176**, 1434–42.

Kiley, P. J. & Kaplan, S. (1988). Molecular genetics of photosynthetic membrane biosynthesis in *Rhodobacter sphaeroides*. *Microbiological Reviews*, **52**, 50–69.

Klug, G. (1993a). The role of mRNA degradation in the regulated expression of bacterial photosynthesis genes. *Molecular Microbiology*, **9**, 1–7.

Klug, G. (1993b). Regulation of expression of photosynthesis genes in anoxygenic photosynthetic bacteria. *Archives of Microbiology*, **159**, 397–404.

Klug, G. (1995). Post-transcriptional control of photosynthetic gene expression. In *Anoxygenic Photosynthetic Bacteria*, ed. R. E. Blankenship, M. L. T. Madigan & C. E. Bauer, pp. 1235–44. Dordrecht, The Netherlands: Kluwer Academic Publishers.

Klug, G., Gad'on, N., Jock, S. & Narro, M. L. (1991). Light and oxygen effects share a common regulatory DNA sequence in *Rhodobacter capsulatus*. *Molecular Microbiology*, **5**, 1235–9.

Klug, G., Kaufmann, N. & Drews, G. (1985). Gene expression of pigment-binding proteins of the bacterial photosynthetic apparatus: transcription and assembly in the membrane of *Rhodopseudomonas capsulata*. *Proceedings of the National Academy of Sciences, USA*, **82**, 6485–9.

Klug, G., Liebetanz, R. & Drews, G. (1986). The influence of bacteriochlorophyll biosynthesis on formation of pigment-binding proteins and assembly of pigment-protein complexes in *Rhodopseudomonas capsulata*. *Archives of Microbiology*, **146**, 284–91.

Kort, R., Hoff, W. D., van West, M., Kroon, A. R., Hoffer, S. M., Vlieg, K. H., Crielaard, W., van Beeumen, J. J. & Hellingwerf, K. J. (1996). The xanthopsins: a new family of eubacterial blue-light photoreceptors. *EMBO Journal*, **15**, 3209–18.

Lang, H. P., Cogdell, R. J., Gardiner, A. T. & Hunter, C. N. (1994). Early steps in carotenoid biosynthesis: sequences and transcriptional analysis of the *crtI* and *crtB* genes of *Rhodobacter sphaeroides* and overexpression and reactivation of *crtI* in *Escherichia coli* and *R. sphaeroides*. *Journal of Bacteriology*, **176**, 3859–69.

Lee, J. K. & Kaplan, S. (1992). *cis*-acting regulatory elements involved in oxygen and light control of *puc* operon transcription in *Rhodobacter sphaeroides*. *Journal of Bacteriology*, **174**, 1146–57.

Lee, J. K., Wang, S. Q., Eraso, J. M., Gardiner, J. & Kaplan, S. (1993). Transcriptional regulation of *puc* operon expression in *Rhodobacter sphaeroides* – involvement of an integration host factor binding sequence. *Journal of Biological Chemistry*, **268**, 24491–7.

McDermott, G., Prince, S. M., Freer, A. A., Hasthornthwaite-Lawless, A. M., Papiz, M. Z., Cogdell, R. J. & Isaacs, N. W. (1995). Crystal structure of an integral membrane light-harvesting complex from photosynthetic bacteria. *Nature*, **374**, 517–21.

McGlynn, P. & Hunter, C. N. (1992). Isolation and characterisation of a putative transcription factor involved in the regulation of the *Rhodobacter sphaeroides pucBA* operon. *Journal of Biological Chemistry*, **267**, 11098–103.

McGlynn, P. & Hunter, C. N. (1993). Genetic analysis of the *bchC* and *bchA* genes of *Rhodobacter sphaeroides*. *Molecular and General Genetics*, **236**, 227–34.

Meyer, T. E. (1985). Isolation and characterisation of soluble cytochromes, ferrodoxins and other chromophoric proteins from the halophilic phototrophic bacterium *Ectothiorhodospira halophila*. *Biochimica et Biophysica Acta*, **306**, 175–83.

Mizoguchi, H., Masuda, T., Nishimura, K., Shimada, H., Ohta, H., Shioi, Y. & Takamiya, K. (1997). Nucleotide sequence and transcriptional analysis of the flanking regions of the gene (*spb*) for the trans-acting factor that controls light-mediated expression of the *puf* operon in *Rhodobacter sphaeroides*. *Plant Cell Physiology*, **38**, 558–67.

Neidle, E. L. & Kaplan, S. (1993). 5-aminolevulinic acid availability and control of spectral complex formation in *hemA* and *hemT* mutants of *Rhodobacter sphaeroides*. *Journal of Bacteriology*, **175**, 2304–13.

O'Gara, J. P. & Kaplan, S. (1997). Evidence for the role of redox carriers in photosynthesis gene expression and carotenoid biosynthesis in *Rhodobacter sphaeroides* 2.4.1. *Journal of Bacteriology*, **179**, 1951–61.

Penfold, R. J. & Pemberton, J. M. (1994). Sequencing, chromosomal inactivation and functional expression in *Escherichia coli* or *ppsR*, a gene which represses carotenoid and bacteriochlorophyll synthesis in *Rhodobacter sphaeroides*. *Journal of Bacteriology*, **176**, 2869–76.

Phillips-Jones, M. K. & Hunter, C. N. (1994). Cloning and nucleotide sequence of *regA*, a putative response regulator gene of *Rhodobacter sphaeroides*. *FEMS Microbiology Letters*, **116**, 269–76.

Pollich, M., Jock, J. & Klug, G. (1993). Identification of a gene required for the oxygen-regulated formation of the photosynthetic apparatus of *Rhodobacter capsulatus*. *Molecular Microbiology*, **10**, 749–57.

Ponnampalam, S. N., Buggy, J. J. & Bauer, C. E. (1995). Characterisation of an aerobic repressor that coordinately regulates bacteriochlorophyll, carotenoid and light harvesting-II expression in *Rhodobacter capsulatus*. *Journal of Bacteriology*, **177**, 2990–7.

Sacerdot, C., Chiaruttini, C., Engst, K., Graffe, M., Milet, M., Dondon, J. & Springer, M. (1996). The role of the AUU initiation codon in the negative feedback regulation of the gene for translation initiation factor IF3 in *Escherichia coli*. *Molecular Microbiology*, **21**, 331–46.

Shimada, H., Iba, K. & Takamiya, K. (1992). Blue light irradiation reduces the expression of *puf* and *puc* operons of *Rhodobacter sphaeroides* under semi-aerobic conditions. *Plant and Cell Physiology*, **33**, 471–5.

Shimada, H., Ohta, H., Masuda, T., Shioi, Y. & Takamiya, K. (1993). A putative transcription factor binding to the upstream region of the *puf* operon in *Rhodobacter sphaeroides*. *FEBS Letters*, **328**, 41–4.

Shimada, H., Wada, T., Handa, H., Ohta, H., Mizoguchi, H., Nishimura, K., Masuda, T., Shioi, Y. & Takamiya, K. (1996). A transcription factor with a leucine-zipper motif involved in light-dependent inhibition of expression of the *puf* operon in the photosynthetic bacterium *Rhodobacter sphaeroides*. *Plant and Cell Physiology*, **37**, 515–22.

Schumacher, A. & Drews, G. (1979). The effects of light intensity on membrane differentiation in *Rhodopseudomonas capsulata*. *Biochimica et Biophysica Acta*, **547**, 417–28.

Sprenger, W. W., Hoff, W. D., Armitage, J. P. & Hellingwerf, K. J. (1993). The eubacterium *Ectothiorhodospira halophila* is negatively phototactic, with a wavelength dependence that fits the absorption spectrum of the photoactive yellow protein. *Journal of Bacteriology*, **175**, 3096–104.

Sussman, J. K., Simons, E. L. & Simons, R. W. (1996). *Escherichia coli* translation initiation factor 3 discriminates the initiation codon *in vivo*. *Molecular Microbiology*, **21**, 347–60.

Tadros, M. H. & Waterkamp, K. (1989). Multiple copies of the coding regions for the light-harvesting B800-850 alpha-polypeptide and beta-polypeptide are present in the *Rhodopseudomonas palustris* genome. *EMBO Journal*, **8**, 1303–8.

Tadros, M. H., Katsiou, E., Hoon, M. A., Yurkova, N. & Ramji, D. P. (1993). Cloning of a new antenna gene cluster and expression analysis of the antenna gene family of *Rhodopseudomonas palustris*. *European Journal of Biochemistry*, **217**, 867–75.

Varga, A. R. & Kaplan, S. (1993). Synthesis and stability of reaction center polypeptides and implications for reaction center assembly in *Rhodobacter sphaeroides*. *Journal of Biological Chemistry*, **268**, 19842–50.

Wong, D. K-H., Collins, W. J., Harmer, A., Lilburn, T. G. & Beatty, J. T. (1996). Directed mutagenesis of the *Rhodobacter capsulatus puhA* gene and *orf214*: pleiotropic effects on photosynthetic reaction centre and light-harvesting I complexes. *Journal of Bacteriology*, **178**, 2334–42.

Yeliseev, A. A., Eraso, J. M. & Kaplan, S. (1996). Differential carotenoid composition of the B875 and B800-850 photosynthetic antenna complexes in *Rhodobacter sphaeroides* 2.4.1: involvement of spheroidene and spheroidenone in adaptation to changes in light intensity and oxygen availability. *Journal of Bacteriology*, **178**, 5877–83.

Yeliseev, A. A., Krueger, K. E. & Kaplan, S. (1997). A mammalian mitochondrial drug receptor functions as a bacterial 'oxygen' sensor. *Proceedings of the National Academy of Sciences*, **94**, 5101–6.

Young, D. A., Bauer, C. E., Williams, J. C. & Marrs, B. L. (1989). Genetic evidence for superoperonal organisation of genes for photosynthetic pigments and pigment-binding proteins in *Rhodobacter capsulatus*. *Molecular and General Genetics*, **218**, 1–12.

Yurkova, N. & Beatty, J. T. (1996). Photosynthesis-independent regulation of bacteriochlorophyll synthesis by light intensity in *Rhodobacter capsulatus*. *FEMS Microbiology Letters*, **145**, 221–5.

Zeilstra-Ryalls, J. H. & Kaplan, S. (1995). Aerobic and anaerobic regulation in *Rhodobacter sphaeroides* 2.4.1: the role of the *fnrL* gene. *Journal of Bacteriology*, **177**, 6422–31.

Zeilstra-Ryalls, J. H. & Kaplan, S. (1996). Control of *hemA* expression in *Rhodobacter sphaeroides* 2.4.1: regulation through alterations in the cellular redox state. *Journal of Bacteriology*, **178**, 985–93.

Zhu, Y. S. & Hearst, J. E. (1986). Regulation of expression of genes for light harvesting antenna proteins LH-I and LH-II; reaction centre polypeptides RC-L, RC-M, and RC-H; and enzymes of bacteriochlorophyll and carotenoid biosynthesis in *Rhodobacter capsulatus* by light and oxygen. *Proceedings of the National Academy of Sciences, USA*, **83**, 7613–17.

Zhu, Y. S. & Kaplan, S. (1985). Effects of light, oxygen and substrates on steady-state levels of mRNA coding for ribulose-1,5-bisphosphate carboxylase and light-harvesting and reaction center polypeptides in *Rhodopseudomonas sphaeroides*. *Journal of Bacteriology*, **162**, 925–32.

Zucconi, A. P. & Beatty, J. T. (1988). Posttranscriptional regulation by light of the steady state levels of mature B800–850 light harvesting complexes in *Rhodobacter capsulatus*. *Journal of Bacteriology*, **170**, 877–82.

LIGHT REGULATION OF CAROTENOID SYNTHESIS IN *MYXOCOCCUS XANTHUS*

DAVID A. HODGSON AND ANDREW E. BERRY

Department of Biological Sciences, University of Warwick, Coventry CV4 7AL, UK

INTRODUCTION

Light and oxygen can be a lethal combination. Most cells contain photosensitizers and if light of the correct wavelength is absorbed by the molecule, a high energy species will be formed. This high energy state molecule can form specific adducts with and/or cleave proteins and lipids and so damage essential cell functions. This may not be life threatening because of the limited diffusion possibilities of many photosensitizers. However, it is also the property of some photosensitizers to interact with water and/or molecular oxygen to generate high energy, highly reactive species. These species include singlet oxygen ($^1O_2$), radicals like hydroxyl radical ($^{\cdot}OH$), radical ions like superoxide (O_2^-) and molecules like hydrogen peroxide. These species are very reactive and have the added disadvantage to the cell that they are readily diffusible and can be relatively long lived, for example $^1O_2$ can survive up to 100 μs in hydrophobic environments (see below).

The combination of light and oxygen is particularly acute for oxygenic phototrophs: that need to capture light energy to survive and generate oxygen in the process of acquiring reducing power by splitting water. They are replete with porphyrins, as are all phototrophs and aerobes (chlorophylls and haems of cytochromes), and these tetrapyrroles are particularly effective photosensitizers. One of the functions of carotenoids in phototrophs is to protect it against the consequences of light and oxygen. Another function is to act as accessory pigments for the capture of light energy not absorbed by chlorophylls.

In this chapter the mechanisms of light-induced damage in cells and the mechanisms by which carotenoids can protect the cell will be discussed. The occurrence of carotenoids in non-phototrophic bacteria and how carotenoid biosynthesis is controlled will then be discussed. The final section will concentrate on how the Gram-negative bacterium *Myxococcus xanthus* synthesizes carotenoids and the mechanism of blue-light induction of carotenogenesis.

PHOTOSENSITIVITY

Porphyrins as photosensitizers

Porphyrins are cyclic tetrapyrroles and are the backbone of haem, chlorophyll and vitamin B12, when iron, magnesium or cobalt, respectively, is chelated at the centre of the tetrapyrrole ring. They are well known for their property of photosensitization and it is this property that is specifically exploited in chlorophylls. They act as photosensitizers because they have the property of extensive electron delocalization within the tetrapyrrole ring. Light energy is absorbed and the high energy product is used to release electrons to generate reducing power. This property is present in all porphyrins and the energy captured from illumination is not always available to do useful work.

Light sensitivity is one of the symptoms of some forms of the genetic disease porphyria, where excess haem precursors are produced and excreted into the skin. The skin of sufferers exposed to light become reddened and, on prolonged exposure, blistering and scarring can result (Stine, 1989). Protoporphyrin IX (PPIX) is a precursor to both chlorophyll and haem and has been shown to be the cause of photosensitivity in *M. xanthus* (Burchard, Gordon & Dworkin, 1966; Burchard & Hendricks, 1969). All bacteria except some aerobes contain haem because they contain cytochromes, and so this potential photosensitivity lurks in most cells.

Light displays properties of both waves and particles, and although attributes such as wavelength and frequency are important, light must also be considered to be composed of a stream of discrete particles, called photons, each with a defined or 'quantized' amount of energy. The energy associated with a photon is inversely proportional to the wavelength of the light, that is, the shorter the wavelength of light, the higher the energy associated with it. The electrons within molecules also have wave-like properties and occupy orbitals which have distinct energies associated with them: they are also said to be quantized. Absorption of a photon by the molecule results in an electron moving to an orbital with higher potential energy. This elevation only occurs when the energy of the incident photon equals the energy required for the electronic transition. Therefore, light of specific wavelength will induce electrons to move between specific orbitals in specific molecules.

Electrons in bonding or sigma orbitals (σ) require more energy to undergo transition (σ to σ^*) than electrons occupying the non-bonding or pi (π) orbitals (π to π^*). Hence σ to σ^* transitions require shorter wavelength light (of the order of 200 nm) than do π to π^* transitions (in the visible range). In addition to its orbit, an electron has such properties as spin, vibrational energy and rotational energy. Electrons are charged and therefore their spin generates an associated magnetic field. This spin can only occur in certain limited or quantized directions; only two directions are possible. Two electrons occupying the same orbital cannot have the same direction of spin (the Pauli exclusion principle). During transition of one of a pair of electrons (π to π^* or σ to σ^*), if the electron retains its direction of spin, the molecule is said to have been excited to a singlet state. Alternatively, if during, or subsequent to, transition the electron has undergone a reversal in spin, the molecule is said to be in a triplet state.

Once an electron has been excited from its lowest energy level (ground state) to a singlet state, its higher potential energy can be released as emission of a photon (fluorescence) or as heat (non-radiative decay) as it returns to the ground state. Alternatively, the singlet state molecule can undergo a process called intersystem crossover, in which the direction of spin is flipped to form a triplet state. Decay to the ground state from the excited triplet state can occur by the emission of a photon in phosphorescence, or as heat. In both fluorescence and phosphorescence the wavelength of the emitted electron is longer than that of the photon absorbed, since some energy is lost in these processes. In most biomolecules of interest, including porphyrins, the triplet state has a considerably longer half-life than the singlet state and it is generally the triplet state of a photosensitizer which is the starting point for photochemical reactions. The extensive electron delocalization within the tetrapyrrole ring of porphyrins means there are plenty of π electrons capable of transition following absorption of blue-light. Porphyrins have a very characteristic absorption peak in the blue-light region (*c.* 410 nm) which is called the Soret band.

The primary event of photosensitization by a porphyrin is the absorption of a photon of appropriate wavelength to excite the tetrapyrrole to a short-lived singlet state with a half-life of 0.01 to 1 μs. This species then undergoes intersystem crossover resulting in the formation of the relatively stable triplet state. This metastable species has a half-life of 1 ms, which enables it to undergo chemical reactions. Triplet state porphyrins can cause damage to the cell via a number of processes. Recently, photosensitizing reactions have been divided into two classes (Foote, 1991). In the first class, triplet state porphyrins interact with substrate or solvent by hydrogen atom or electron transfer to form radicals, such as $\cdot$OH or radical ions, such as the O_2^- (Cox, Whitten & Gianotti, 1979). In the second class, the interaction is with oxygen to yield $^1O_2$ via transfer of excitation energy. A study in which the products of porphyrin photosensitizing reactions were trapped in reaction showed that products accumulated due to $^1O_2$ mechanisms were more prevalent than those arising from reaction with O_2^- (Cox & Whitten, 1982). Kranovsky (1979), observed that, during irradiation of PPIX in the presence of molecular oxygen, $^1O_2$ was generated more efficiently than the O_2^-. Further, the rate constant for reaction between ^{3}PPIX and the $^1O_2$ it generates are considerably lower than for chlorophyll and other porphyrin derivatives, with the result that $^1O_2$ is much more likely to escape into the cellular milieu.

High energy, reactive oxygen species

Oxygen is unusual among biological molecules in that its ground state is the triplet state ($^3O_2$). $^1O_2$ can be formed by the ground state collision of $^3O_2$ with either the triplet or the singlet state of porphyrin. This alleviates the spin restriction and makes $^1O_2$ considerably more reactive than $^3O_2$ and can be identified by its ultraweak chemiluminescence at 1270 nm. The species retains the physical characteristics of $^3O_2$, including its ready diffusibility. The cytotoxicity of $^1O_2$ has received much attention and is particularly damaging to membranes. This is because $^1O_2$ is a metastable species, having a life-time of 2–4 μs in hydrophilic environments and 25–100 μs in hydrophobic environments (Knox & Dodge, 1985). The species can react as an electrophile and as a nucleophile (Symons, Botte & Sybesma, 1985). Electron-rich functionalities are particularly prone to attack by $^1O_2$ and its reactions can be grouped into five categories: (i) ene reactions with olefins, (ii) 1,2 addition to electron-rich olefins, (iii) oxidation of sulphides to sulphoxides, (iv) photooxidation of phenols and (v) 1,4 addition to dienes and heterocycles (Foote, 1976). The first four of these reactions have biological significance.

(i) $^1O_2$ reacts with unsaturated fatty acids shifting the bond to the allylic position to produce the corresponding fatty acid hydroperoxide.

$$R—CH\!=\!CH—CH_2—R' + {}^1O_2 \rightarrow R—CH—CH\!=\!CH—R'$$
$$\underset{\displaystyle OOH}{|}$$

(ii) 1,2 addition of $^1O_2$ to electron-rich olefins lacking α-hydrogens or rigid olefins yields dioxetanes.

The dioxetane decays to a triplet state carbonyl which was thought to account for the weak chemiluminescence observed during lipid peroxidation. A recent study, however, has shown that this mechanism is not associated with lipid peroxidation-supported formation of electronically excited states (Di Mascio *et al.*, 1992).

(iii) The oxidation of methionine by $^1O_2$ to its corresponding sulphoxide is an important biological reaction.

$$CH_3—S—R + {}^1O_2 \rightarrow CH_3—S^+(O^-)—R$$

(iv) Tyrosine residues are susceptible to $^1O_2$ oxidation of the phenol side chain.

The reactions of $^1O_2$ with DNA, proteins and lipids have been well studied. *In vitro*, it has been shown that the amino acids tyrosine, tryptophan, cysteine, histidine, and methionine react most readily with $^1O_2$ (Matheson *et al.*, 1975). Similar processes occur *in vivo* and lead to enzyme inactivation. Therefore, there will be distinctive sites within the cell which are particularly prone to $^1O_2$ attack. Of particular relevance is the observation that the half-life of $^1O_2$ is prolonged by the hydrophobic interior of membranes (Suwa, Kimura & Schaap, 1977). In the presence of protoporphyrin, cell membrane components have been shown to be sensitive to $^1O_2$ (Anderson & Krinsky, 1973; Lamola, Yamane & Trozzolo, 1973). The $^1O_2$-specific product 3β-hydroxy-cholest-6-ene-5α-hydroperoxide was found in haematoporphyrin-sensitized liposomes (Suwa *et al.*, 1977). $^1O_2$ is capable of oxidizing sulphydryl groups, thereby disrupting the tertiary and/or quaternary structure of proteins with the potential for concomitant loss of functional activity. These sulphydryl groups have been shown to be important for the maintenance of membrane integrity (Edwards *et al.*, 1984). Damage to the membrane transport proteins and channel proteins will affect membrane permeability.

The general consensus is that the membrane is the major site of damage incurred due to photooxidative processes (Malik, Hanania & Nitzan, 1990). A useful review of this subject is given by Valenzeno (1987). Of particular relevance here is the phenomenon of lipid peroxidation. This is a free radical-initiated mechanism which proceeds via a carbon-centred radical. A single triggering event can result in the damage of thousands of polyunsaturated fatty acid molecules (McCord, 1985). This has severe consequences for membrane integrity and fluidity and is a form of oxidative stress which all aerobic organisms encounter.

The reaction can be represented as:

$$
\begin{array}{lll}
LH + {}^{\cdot}OH \rightarrow L^{\cdot} + H_2O & \text{Initiation} \\
L^{\cdot} + O_2 \rightarrow LO_2^{\cdot} & \text{Propagation} \\
LO_2^{\cdot} + LH \rightarrow L^{\cdot} + LOOH & \text{Propagation} \\
L^{\cdot} + L^{\cdot} \rightarrow L—L & \text{Termination} \\
LOO^{\cdot} + L^{\cdot} \rightarrow LOOL & \text{Termination} \\
LOO^{\cdot} + LOO^{\cdot} \rightarrow LOOL + O_2 & \text{Termination}
\end{array}
$$

where LH represents a polyunsaturated lipid (adapted from Ahmad, 1995).

The reaction is initiated by $\cdot$OH to form a carbon-centred lipid radical (L$\cdot$), which will react with oxygen to form the lipid peroxyl radical LO$_2\cdot$. This species will, in turn, react with lipids to generate the lipid radical and produce a lipid hydroperoxide (LOOH). $^1O_2$ attack on lipids results in the incorporation of oxygen into the molecule to generate the lipid hydroperoxide or a cyclic endoperoxide, as shown in reactions (i) and (ii), respectively (see above). Hydroperoxides can decompose to form malonaldehyde and hydroxynonals which are themselves highly reactive (Mannervik, 1985). Further, lipid hydroperoxides are catalytically decomposed by copper to free radicals such as LO$_2\cdot$ which contribute towards the propagation of the lipid preoxidation chain reaction (Schaich & Borg, 1985). A concise description of the process was given by Ahmad (1992):

> Oxidative stress is a chain process and a single initiating event caused by a prooxidant may cascade into a widespread chain reaction that produces many deleterious products in concentrations many times greater than the initiator.

CAROTENOIDS AS PROTECTORS AGAINST LIGHT DAMAGE

Interaction of molecular oxygen and water with triple state photosensitizers can generate $^1O_2$, O_2^-, $\cdot$OH and hydrogen peroxide (H_2O_2), with $^1O_2$ being the main product. Triplet state photosensitizers and their products can be dealt with by both enzymatic and non-enzymatic defences. Antioxidant enzymes are present in all aerobes and include superoxide dismutase, catalase and glutathione peroxidases. The small molecule antioxidants include ascorbic acid, α-tocopherol (vitamin E), glutathione and carotenoids. Carotenoids are an important part of the cell's antioxidant defence mechanism, and their hydrophobic nature makes them especially important in the protection of cell membrane components.

Carotenoids are a large family of yellow, orange and red pigments composed of at least 600 structurally distinct compounds. It is estimated that prokaryotes and eukaryotes synthesize 10^8 tonnes annually. They are products of the general isoprenoid biosynthetic pathway and can be split into two classes: the carotenes and the xanthophylls. Typically, carotenes consist of a C_{40} backbone built of eight, five-carbon, isoprenoid subunits, although C_{30}, C_{45}, C_{50} and C_{60} carotenes have been characterized. Xanthophylls are derivatives formed by incorporation of one or more oxygen functional groups. Carotenoids may be a-, mono- or bi-cyclic and the ring may be aromaticized. They may also be glycosylated and conjugated to a variety of molecules. The light-absorbing and photoprotective properties of the carotenoid molecule are determined by the number of conjugated double bonds and their isomerization state. Some carotenoids exist not only as all *cis* isomers but also as all *trans* isomers (Beyer, Mayer & Kleinig, 1989).

One of the first studies to obtain evidence of the photoprotective nature of carotenoids was the demonstration that a carotenoid-less mutant of

Micrococcus luteus was prone to killing by the effects of oxygen and light (Mathews & Sistrom, 1960). This protective effect was also in evidence in studies using liposome membranes (Anderson & Krinsky, 1973; Anderson *et al.*, 1974).

In vitro, carotenoids are extremely efficient quenchers of radicals and other excited species, but particularly of $^1O_2$. This property is conferred by their extensive conjugated bond system. It was demonstrated that at least seven or eight conjugated double bonds are required for quenching activity (Di Mascio, Kaiser & Sies, 1989). This was confirmed *in vivo* by the observation that a mutant of *Micrococcus luteus*, whose major carotenoid contained eight double bonds, was not resistant to light and oxygen (Mathews-Roth *et al.*, 1974). This suggested that the carotenoid molecule must have a triplet state whose energy is below that of the singlet state of oxygen (94 kJ mol^{-1}) in order for a singlet to triplet energy transfer to take place. The rates for these quenching reactions approach the upper limit set by diffusion; β-carotene has a rate constant for quenching of $^1O_2$ of 13×10^9–30×10^9 M s^{-1}, while lycopene, the most efficient *in vitro* quencher characterized thus far, has a rate of 31×10^9 M s^{-1} (Ahmad, 1992).

There are two types of quenching mechanisms; physical quenching and chemical quenching. In physical quenching, collision between the excited state or radical and the 'quenching molecule' results in the transfer of excitation energy to the quenching molecule, which is subsequently dissipated as heat or as reversible *cis–trans* isomerization of one or more double bonds. β-carotene is able to interact with the triplet state of many photosensitizers, including porphyrins and therefore, prevent the formation of excited states and radicals. Reaction with $^1O_2$ will yield the triplet state carotenoid and $^3O_2$. Since the carotenoid molecule is not destroyed in this interaction, one carotenoid molecule can be responsible for the quenching of many excited states or radicals. The rate constant for this process exceeds the reaction rate constants for reaction with the polyunsaturated fatty acid molecules and so, in theory, the presence of carotenoids in a relatively low concentration is effective in protecting the cell membrane. This process can be represented as shown below, where * denotes excitation energy.

$$^1O_2 + Car \rightarrow {}^3Car* + {}^3O_2$$
$$^3Car* \rightarrow Car + heat$$

Chemical quenching involves reaction between the excited state or radical with the antioxidant and often results in its incorporation into the molecule. Vitamin E (α-tocopherol) reacts readily with $^1O_2$. Under certain conditions carotenoids undergo reaction with peroxyl radicals which have been shown to add to the long chain of conjugated double bonds (Packer *et al.*, 1981). This function to eliminate peroxyl radicals may be an important weapon in the cell's armoury to counter the very damaging effects of lipid peroxidation, which has a devastating effect upon cell membranes. The peroxyl radical is a component of the chain reaction propagation step discussed above. The processes of chemical quenching by carotenoids has not been well characterized, as it is thought physical quenching predominates as the main mechanism of carotenoid light-protection.

In oxygenic phototrophs, carotenoids perform both light-protective and light harvesting roles. Upon absorption of light energy, a small proportion of chlorophyll molecules undergo intersystem crossover to form the triplet state. In this form the excitation energy could be transferred directly to oxygen (in oxygenic phototrophs) to form $^1O_2$. The chlorophyll molecule is extremely sensitive to auto-oxidation by reaction with this $^1O_2$ and this destruction of the principal location of energy capture in the cell would have dire consequences. Carotenoids, which are positioned very close to the chlorophyll molecule, are able to trap excitation energy by physically quenching the triplet state chlorophyll, thus preventing formation of $^1O_2$. The structure of the chloroplast reaction centre has been solved to 3.4 Angstrom resolution by X-ray crystallography. Three transmembrane helices are present and are held by ion pairs, which are also the ligands maintaining close contact between chlorophylls a and b at the centre of the complex. Two carotenoid molecules are held in close proximity to prevent the formation of $^1O_2$ (Kuhlbrandt, Wang & Fujiyoshi, 1994).

In anoxygenic phototrophs, carotenoids are also part of the light-harvesting complex, but in this case it is believed that their role is to act as accessory pigments. X-ray crystallography of the antenna complex (LH-II) of *Rhodopseudomonas acidophila* has revealed a quite remarkable structure with nonameric symmetry composed of two concentric cylinders constructed of helical protein subunits. Eighteen bacteriochlorophyll a molecules are positioned between the outer helices, with the bacteriochlorin rings perpendicular to the transmembrane helix axis. The carotenoid molecules are intertwined precisely with the phytol molecules which span the complex. Both carotenoids and bacteriochlorophyll molecules are bound non-covalently to the apoproteins, which are able to modulate their absorptive capacities (McDermott *et al.*, 1995; Freer *et al.*, 1996; Cogdell *et al.*, this volume).

The important role of carotenoids in photo-protection detailed above would imply that carotenoid-minus mutants of oxygenic phototrophs would be impossible to isolate. However, such mutants of maize have been isolated, but they are only able to survive in laboratory conditions of low light and low oxygen concentration (Siefermann-Harms, 1985). There is an interesting exception to this the rule: the phytopathogenic fungus *Cercopora nicotianae* produces cercosporin, a photosensitizer and effective $^1O_2$ generator. Carotenoid production was eliminated by targeted gene disruption, but this had no effect on either phytopathology nor sensitivity to light and cercosporin or other $^1O_2$ generators (Ehrenshaft, Jenns & Daub, 1995). This implies that *Cercopora nicotianae* has another mechanism of photoprotection beyond carotenoids.

OTHER PROTECTORS AGAINST SINGLET OXYGEN
IN BACTERIA

What other mechanisms are there for protection of the bacterial cell against $^1O_2$? It has been mentioned above that glutathione, or its equivalent, is present in the bacterial cytoplasm where it maintains a reducing environment. Could these molecules quench $^1O_2$ chemically? Mutant strains of *Salmonella typhimurium* and *Escherichia coli* devoid of glutathione (γ-L-glutamyl-L-cysteinyl-glycine) have been shown to exhibit no alteration in $^1O_2$ sensitivity, when exposed to the species in the absence of contact with a photosensitizer. Surface exposure to pure $^1O_2$ did cause killing but no mutagenic effects (Dahl, Midden & Hartman, 1988, 1989). This again implies that the main site of $^1O_2$ damage is the membrane not the cytoplasm. Expression of the *Erwinia herbicola* carotenoid biosynthetic genes protected *E. coli* from $^1O_2$ generated by irradiation of toluidine blue with visible light. These carotenoids also protected the host from near-UV (320–400 nm) but not far UV (200–300 nm) light (Tuveson, Larson & Kagan, 1988).

Dahl and his colleagues have shown that the lipopolysaccharide on the surface of the Gram-negative bacterium does, to some extent, protect the cell against exogenously generated $^1O_2$. Pure $^1O_2$ was generated by white-light irradiation of glass beads coated in photosensitizer and attached to a glass slide. The slide was held within one millimetre of a filter holding the bacteria. This ensured the bacteria did not come in contact with the photosensitizer, as the only thing that could diffuse across the 1 mm gap was $^1O_2$, which retains the diffusion characteristics of $^3O_2$. The surviving bacteria on the filter after different times of exposure were enumerated. Gram-negative species, such as *E. coli* and *S. typhimurium*, showed an initial slow rate of killing which increased during the time course. The initial rate for Gram-positive bacteria was faster and remained constant during the period of exposure. *S. typhimurium* TA1975, a deep rough mutant lacking nearly all of the outer lipopolysaccharide, was more susceptible to $^1O_2$ killing than a wild-type. Thus the lipopolysaccharide formed a physical barrier to $^1O_2$ penetration and initially protected the cell. However, upon prolonged exposure it became a source of secondary reaction products which penetrated the cytoplasmic membrane and caused damage. In Gram-positive bacteria, where the structure of the cell wall is fundamentally different, $^1O_2$ penetrates the peptidoglycan layer without producing any reactive secondary products and is directly incident upon the cytoplasmic membrane. Carotenoid-containing strains of *Micrococcus luteus* and *Staphylococcus aureus* were far more resistant to exogenous $^1O_2$ (Dahl, Midden & Hartman, 1987, 1988, 1989).

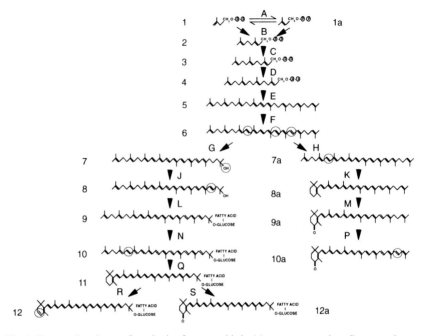

Fig. 1. Proposed pathway of synthesis of carotenoids in *Myxococcus xanthus*. Compounds are labelled with numbers and enzymatic actions with letters. The proposed product of a gene is given for each letter. The changes in the structures are highlighted by the circles. Data taken from Armstrong and Hearst (1996) and Botella *et al.* (1995). Key: 1, isopentenyl phosphate; 1a, dimethylallyl pyrophosphate; 2, geranyl pyrophosphate; 3, farnesyl pyrophosphate; 4, geranyl-geranyl pyrophosphate; 5, phytoene; 6, neurosporene; 7, hydroxyneurosporene; 7a, lycopene; 8, 7,8 dihydro, 3',4' dehydro-rhodopin; 8a, γ-carotene; 9, 7,8 dihydro, 3',4' dehydro-rhodopin glucoside ester; 9a, 4 keto-γ-carotene; 10, 3',4' dehydro-rhodopin glucoside ester; 10a, 4-keto-torulene; 11, 1',2' dihydro-1'-OH-torulene glucoside ester; 12, myxobactin ester; 12a, myxo-bacton ester; A, isomerization, ?; B, condensation, ?; C, condensation, ?; D, condensation, *crtE*; E, condensation, *crtB*; F, desaturation, *crtI*; G, hydroxylation, *crtC*; H, desaturation, *crtI*?; J, desaturation, *crtD*?; K, cyclization, *orf6*?; L, condensation, ?; M, ketonization, ?; N, desaturation, *orf2*? *crtI*?; P, desaturation, *crtD*? *orf2*?; Q, cyclization, *orf6*?; R, desaturation, *orf2*?; S, ketonization, ?.

CAROTENOID BIOSYNTHESIS

The biosynthetic pathways

The universal building block for carotenoids is the five-carbon isopentenyl pyrophosphate (IPP). An IPP molecule is isomerized to dimethylallyl pyrophosphate and condensed with IPP to form geranyl pyrophosphate. Sequential addition of two further IPP molecules leads to the 20-carbon molecule geranylgeranyl pyrophosphate via farnesyl pyrophosphate. Two geranylgeranyl pyrophosphate molecules are condensed to make the 40-carbon universal carotenoid precursor phytoene (Fig. 1). Anoxygenic photosynthetic bacteria, non-photosynthetic bacteria and fungi desaturate phytoene either three or four times to generate neurosporene or lycopene, respectively. The pathway is different in cyanobacteria and chloroplasts of algae and higher plants, where phytoene is converted to lycopene by two distinct sets of reactions, via the intermediate ζ-carotene. Lycopene and neurosporene are the starting blocks for many different reactions in many different organisms that lead to the wide diversity of carotenoid pigments mentioned above.

The genes encoding carotenoid biosynthesis have been isolated and characterized in a number of bacterial species. The reader is referred to Armstrong (1994), Armstrong and Hearst (1996) and Sandmann (1994) for recent reviews of the genetics of carotenoid biosynthesis in plants and micro-organisms. The bacteria that have received the most attention are *Rhodobacter sphaeroides* (Garí *et al.*, 1992), *R. capsulatus* (Armstrong *et al.*, 1989), *Erwinia herbicola* (Armstrong, Alberti & Hearst, 1990; To *et al.*, 1994), *Erw. uredovora* (Misawa *et al.*, 1990), *Synechococcus* sp. PCC7942 (Chamovitz, Pecker & Hirschberg, 1991), *Synechocystis* sp. PCC6803 (Martínez-Férez *et al.*, 1994), *Anabaena* sp. PCC7120 (Linden *et al.*, 1994), *Mycobacterium aurum* (Houssaini-Iraqui *et al.*, 1993), *Myc. marinum* (Ramakrishnan *et al.*, 1997), *Streptomyces griseus* (Schumann *et al.*, 1996), *Stm. setonii* (Kato *et al.*, 1995) and *Myxococcus xanthus* (Fontes, Ruiz-Vázquez & Murillo, 1993; Botella, Murillo & Ruiz-Vázquez, 1995). With the advent of total genome sequencing, many such structural genes for carotenoid biosynthesis should be turning up in the next few years. To discover more on the gene–enzyme relationships of those bacteria other than *M. xanthus*, the reader is recommended to the Armstrong and Sandmann reviews cited above.

Control of carotenoid biosynthesis

Carotenogenesis in bacteria is either constitutive or controlled by a number of environmental cues, depending on the biological role of the pigment. In the case of the oxygenic phototrophs, carotenoids are essential for protection and light capture and so are constitutive. In the case of anoxygenic phototrophs, the main role of carotenoids is photocapture, as photosynthesis only occurs in the absence of oxygen. The entire photosynthetic apparatus, including carotenoid production, is more dependent on the absence of oxygen than the presence of light (Armstrong, 1994; Phillips-Jones, this volume). In non-phototrophic bacteria, carotenogenesis is: constitutive, for example, *Erw. herbicola, Erw. uredovora Staphylococcus aureus* and *Micrococcus luteus*; light-induced, for example, *M. xanthus* and *Myc. marinum*; or cryptic, for example, *Stm. setonii* and *Stm. griseus*. Examples of both constitutive and light-inducible carotenogenesis can be found amongst *Mycobacterium* spp. There is a report of photoinduction of an orange pigment in five strains of the archaebacterial genus *Sulfolobus*. Preliminary characterization indicated these orange pigments to be hydroxylated analogues of β-carotene (Grogan, 1989). The rest of this chapter will address light-regulation of carotenogenesis, most specifically in *M. xanthus*.

CAROTENOGENESIS IN *MYXOCOCCUS XANTHUS*

Upon exposure to light, *M. fulvus* produces two major monocyclic, glycosy-lated carotenoid esters, myxobacton palmitate and myxobactin stearate (Fig. 1). They also produce a significant amount of 4-ketotorulene, an aglycosidic form of myxobacton that is probably a shunt product. The pathway of carotenogenesis in *M. fulvus* has been partially characterized (Kleinig, 1975) and appears to proceed from neurosporene rather than lycopene. The same major carotenoids and pathway of synthesis (Fig. 1) appears to be present in *M. xanthus* (Hodgson & Murillo, 1993; Botella *et al.*, 1995).

Light sensitivity in myxobacteria

Exposure of wild-type *M. xanthus* to light has one of two results: death and lysis (photolysis); or carotogenesis and survival. If the cells are in poor physiological condition, or if they are mutant and cannot synthesize carotenoids, photolysis is the only result of light exposure (Burchard & Dworkin, 1966). Therefore, carotenogenesis in *M. xanthus* is a light-induced, light-protective response. The action spectra of carotenogenesis and photo-lysis were obtained and found to be very similar with a major peak at *c*. 410 nm. A compound was isolated that had a similar absorption spectrum and production of the compound increased some 16-fold as the bacterial culture entered the stationary phase, that is, just as the cells were becoming most photosensitive. The compound was identified as protoporphyrin IX (PPIX) (Burchard *et al.*, 1966; Burchard & Hendricks, 1969). The work of Burchard and co-workers has been repeated in the author's laboratory and their results confirmed (Robson, 1992). From the foregoing discussion it is clear how light and PPIX cause photolysis of *M. xanthus*. Blue-light interacts with ground state PPIX to form ^{3}PPIX, which in turn interacts with $^3O_2$ to generate $^1O_2$ which then causes the membrane damage that results in lysis of the cell. Once carotenoids have been generated, they can quench ^{3}PPIX and $^1O_2$ and so protect the cell. A model is proposed below to explain how carotenogenesis is induced by blue-light with PPIX as the photosensitizer.

The genetics of carotenogenesis in myxobacteria

The work of Murillo and his colleagues has greatly helped to understand the genetic basis of carotenoid biosynthesis in *M. xanthus*. They identified

four genetically unlinked loci that are required for light-induced carotenogenesis; *carAB, carC, carD* and *carR*. Two classes of phenotype were
identified: those that could not produce carotenoids under any conditions
(Car⁻) and those that produced carotenoids constitutively (Car^c). There
were Car⁻ phenotypes associated with mutations at all four loci and Car^c
phenotypes associated with the *carR* and *carA* loci; however, there was a
difference in the Car^c associated with each locus. The *carR* mutant colonies
were red and the carotenoids produced were those typically produced
following light exposure of wild-type *M. xanthus*, only at much higher
concentration. The *carA* mutant colonies were pale orange and the
carotenoids produced in the light were predominantly phytoene and phytofluene. Exposure of *carA* cells to the light led to production of wild-type,
light-induced carotenoids. It was concluded that the *carA* mutants were
responsible for control of expression of only part of the carotenoid
biosynthesis pathway and that phytoene dehydrogenase was under different
control (Martínez-Laborda *et al.*, 1990). It was postulated that *carC*
encoded phytoene dehydrogenase and that the *carB* locus encoded
enzymes active before and after phytoene dehydrogenase (Martínez-
Laborda *et al.*, 1986, 1990; Balsalobre, Ruiz Vázquez & Murillo, 1987;
Martínez-Laborda & Murillo, 1989).

The *carBA* and *carC* loci were cloned and sequenced and it was shown that
carC encoded phytoene dehydrogenase (Fontes *et al.*, 1993) and that the
carBA locus contained eleven open reading frames, many with homology to
known carotenoid genes (Ruiz-Vázquez, Fontes & Murillo, 1993; Botella *et
al.*, 1995). The designation *car* came from the fungus *Phycomyces blakesleea-
nus* but has recently been changed to *crt* to comply with universal bacterial
nomenclature. Table 1 lists the *crt* genes of the *M. xanthus* carotenoid
regulon, Fig. 1 illustrates the proposed points of action in the biosynthetic
pathway and Fig. 2 illustrates the structure of the regulon.

There are two open reading frames (*orf7* and *orf8*) with no assigned
function, one (*orf2*) with an assigned function but an uncertain role and one
(*orf6*) with an assigned role but an uncertain function (Table 1). The
functions left to assign are: formation of the glucoside ester; condensation
of the glucoside ester with carotenoid; cyclization of 3′,4′ dehydro-rhodopin
glucoside ester and lycopene, tentatively assigned to *orf6* although there is no
obvious homology to known carotenoid cyclases (*crtL* and *crtY*); desaturation of hydroxyneurosporene, tentatively assigned to *crtD*; desaturation of
7,8 dihydro,3′,4′ dehydro-rhodopin glucoside ester; desaturation of 4 keto-γ-
carotene; and introduction of a ketone group into the ring structures. The
neurosporene to lycopene desaturation is presumed to be catalysed by the
phytoene dehydrogenase (*crtI*). For the two unassigned desaturation functions, the *crtI* and *crtD* products may be reused or there is the possibility that
orf2 product is a desaturase. There are also no genes for the steps prior to,
and including, farnesyl pyrophosphate synthesis.

Table 1. *The genes of the* Myxococcus xanthus *carotenoid regulon*

Original name	New name	Product size (M_r)	Proposed function
Structural genes			
carB	crtE	38641	Geraynlgeraynl diphosphate synthase
(orf1)			
orf3	crtB	37379	Phytoene synthase
carC	crtI	58427	Phytoene dehydrogenase
orf5	crtC	33970	Neurosporene hydratase
orf4	crtD	52432	Hydroxyneurosporene desaturase?
orf6	?	38664	Carotene cyclisation?
orf2	?	57790	Carotene desaturation?
orf7	?	14427	?
orf8	?	12665	?
Regulatory genes			
carA	?	25431	Repression of the crtEBDC promoter
(orf9)			
carA	?	32011	Repression of the crtEBDC promoter
(orf10)			(DNA binding domain?)
orf11	?	33088	Repression of the crtEBDC promoter? (DNA binding domain?)
carQ		19578	ECF family RNA polymerase sigma factor needed for control of crtI and carQRS promoters
carR		22997	Light-sensitive, membrane-bound anti-sigma factor
carS		12211	Activation of the crtEBDC promoter (anti-repressor?)
carD	?	34031	High mobility group I(Y)-like DNA-binding protein, required for expression of carQRS and genes involved in fruiting body formation

Regulation of carotenogenesis in myxobacteria

When the *carA1* allele was sequenced, two separate mutations were found, one in *orf9* and one in *orf10*. The hypothetical products of *orf10* and *orf11* appear to contain helix-turn-helix DNA-binding domains with homology to the MerR, mercury resistance regulatory protein family (Botella *et al.*, 1995). There is evidence that, in addition to the major *crtEBDC* promoter (Fig. 2), there is a low level constitutive promoter present for *orf9*, *orf10* and *orf11* (Fig. 2) that is induced to higher levels in the light (F. J. Murillo, pers. comm.). The simplest interpretation is that all, or a combination of some, of the products of *orf9*, *orf10* and *orf11* are involved in binding to the *crtEBDC* promoter and repressing its expression in the dark. The *carA1* mutation stops this from happening and so the operon becomes constitutive. As the *crtI* gene is not subject to repression by the 'CarA' complex constitutive expression of the *crtEBDC* operon does not lead to synthesis of 4-ketotorulene and the myxobactin and myxobacton esters, but rather to phytoene and phytofluene

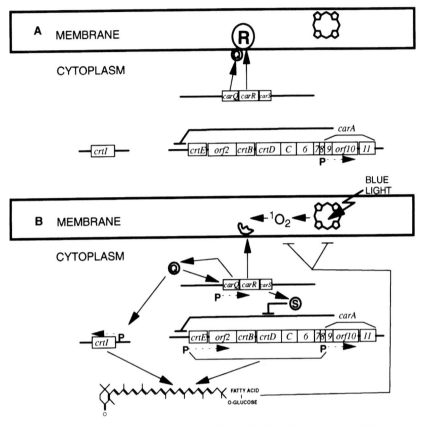

Fig. 2. Model of the proposed mechanism of light-induction of the carotenoid biosynthesis regulon of *Myxococcus xanthus*. A. In the dark: low-level expression from the *carQRS* promoter leads to translation of CarR, which is lodged in the inner membrane along with CarQ. There is no expression from the *crtI* gene beyond background expression. Some or all of the *carA* open reading frames, *orf9*, *orf10* and *orf11*, are expressed from a constitutive promoter. The products of these open reading frames repress transcription from the *crtEBDC* promoter. B. In the presence of light: blue-light is absorbed by protoporphyrin IX (porphyrin outline) and singlet oxygen ($^1O_2$) is generated. This species then leads to the destruction of CarR and hence release of CarQ. The sigma factor interacts with core RNA polymerase and initiates transcription from the *carQRS* and *crtI* promoters, provided carbon catabolite repression of *crtI* has been relieved. Transcription of the *carQRS* operon leads to production of more CarQ, CarR and CarS. The $^1O_2$ will activate destruction of CarR so there will be an exponential increase in CarQ and CarS. The CarS protein will inactivate CarA repression of the *crtEBDC* operon. The products of the *crtI* gene and the *crtEBDC* operon will lead to the synthesis of the myxobacton ester and related carotenoids which will quench the triplet state protoporphyrin IX and the $^1O_2$. Thus CarR will no longer be destroyed and all new CarQ synthesised will be sequestered to the inner membrane and hence will no longer be able to activate the *carQRS* operon and *crtI* gene. Thus there is product feedback regulation of the carotenoid regulon.

(Martínez-Laborda *et al.*, 1990). The phytofluene probably arises from the action of one of the carotene desaturases (*crtD* or *orf2*) encoded in the *crtEBDC* operon (Table 1).

The basis of the light regulation of the *M. xanthus* carotenoid regulon is the *carQRS* operon. The nomenclature *car* is retained for these regulatory genes as *crt* gene activity regulators. The role of the *carQRS* operon was elucidated because of the unexpected observation that a merodiploid containing a wild-type allele and a Tn5 insertion in the *carR* locus that caused a Car^c phenotype had the phenotype of the transposon mutation, that is, the mutant allele was dominant. One interpretation was that there was an activator of carotenogenesis and that the transposon had caused its constitutive expression. The corollary of this proposal was that the activator was driven from a blue-light-inducible promoter. Fusion of the region upstream from the proposed activator to a promoterless *E. coli lacZ* gene resulted in blue-light-inducible β-galactosidase activity. Deletion of the region downstream of the light-inducible promoter generated a Car$^-$ mutant, as predicted if an activator of carotenogenesis had been lost.

A number of red colony Car^c mutations had already been mapped to the *carR* locus (see above). When the light-inducible promoter *lacZ* fusion was introduced the Car^c *carR* mutants, the promoter was completely deregulated and there was as much a 5000-fold more β-galactosidase activity than in dark-grown wild-type cells containing the fusion. This demonstrated that the Car^c mutants had lost a negative regulator of the light-inducible promoter. Using cloned DNA from the locus it was possible to complement/gene convert the Car^c mutations to Car$^+$ phenotypes. The smallest piece of DNA that did this came from downstream of the light-inducible promoter. Therefore, there was a light-inducible promoter responsible for expression of the *crt* regulon and a negative regulator of the promoter (Hodgson, 1987, 1993).

Sequencing of the *carR* locus revealed three translationally coupled (the upstream stop codon overlaps the downstream start codon) open reading frames downstream of the light-inducible promoter (Fig. 2). Sequencing of the six Car^c *carR* mutations demonstrated that they all carried mutations in the second open reading frame for which the original designation of *carR* was retained. CarR had no known homologues within the databases but hydrophilicity plots strongly implied it was an integral inner membrane protein. A fusion protein consisting of CarR and staphylococcal protein A, when expressed in *E. coli*, was found to be localized to the inner membrane (Gorham *et al.*, 1996). Recently, using anti-CarR antiserum and Sarkosyl fractionation of the membranes, there is evidence that over-expressed CarR is exclusively found in the *M. xanthus* inner membrane (Browning, 1997).

Constitutive expression of the third open reading frame in the absence of the first open reading frame and *carR* led to a Car^c phenotype with orange colonies like those of the *carA* mutants and the original Tn5 insertion at *carR*. Further work revealed that the *crtEBDC* operon had been constitutively activated but not the *crtI* gene. Deletion of the third open reading frame abolished carotenoid biosynthesis and expression of the *crtEBDC* operon. The third open reading frame therefore encoded the carotenogenesis activator previously identified and named *carS*. CarS was necessary and sufficient for expression of the *crtEBDC* operon but has no homologue in the database, nor does it contain any known domains or motifs, based on sequence comparison (McGowan, Gorham & Hodgson, 1993).

The first open reading frame was shown to be required for carotenogenesis by generation of an in-frame insertion mutation that gave a Car⁻ phenotype. The insertion did not disrupt *carR* or *carS* translation which was particularly important as it had been shown that these two genes were essential for properly regulated carotenogenesis. The open reading frame was named *carQ* and was shown to be essential for expression of the *carQRS* and *crtI* promoters. If the *lacZ carQRS* promoter fusion was introduced into strain lacking the *carQRS* operon, no expression was observed. If a wild-type *carQ* gene was added back, promoter expression was recovered, although it was constitutive. Carotenogenesis was not recovered because *carS* was not present. The lack of *carR* explained the constitutive expression of the promoter (McGowan *et al.*, 1993).

While both the *carQRS* operon and the *crtI* gene are CarQ-dependent and 'CarA'-independent, it was observed that the expression of *crtI* only became maximal as the culture entered stationary phase (Fontes *et al.*, 1993). The *carQRS* promoter was expressed maximally, independently of culture phase (Robson, 1992). It was demonstrated that the *crtI* culture phase dependency was due to carbon source repression of the promoter and that as carbon source was depleted in the stationary phase of the culture, derepression of *crtI* became possible (Fontes *et al.*, 1993). It has always been found possible to induce carotenogenesis at any phase of a *M. xanthus* culture (Robson, 1992). This implies that either low expression of *crtI* is necessary for carotenogenesis or that the carotenoids formed in exponential phase are not the final carotenoids (myxobacton ester, etc.) but rather phytoene and phytofluene, the products of sole expression of the *crtEBDC* operon.

Sequence analysis of the *carQ* product revealed it belonged to a new family of RNA polymerase sigma factors. This family was unusual in that they were small (M_r c. 20 000) and involved in control of functions within the cell membrane, cell wall or extracellular milieu. This latter property led to the naming of the family as the **extra-cytoplasmic function** sigma factors. The small size reflects the almost total loss of the N-terminal region 1 and truncation of the regions 3A and 3B compared to those seen in the σ^{70} family. However, the DNA binding regions (regions 2 and 4) were intact (Lonetto *et al.*, 1994). A sigma factor activity explains the requirement of CarQ for *carQRS* and *crtI* promoter activities. Sigma factor activity of CarQ *in vitro* has recently been demonstrated using renatured CarQ, pure *E. coli*

RNA polymerase core enzyme and the *carQRS* promoter region (Browning, 1997).

In summary, three gene products are responsible for the light-inducible expression of the *M. xanthus* carotenoid regulon. CarQ is a sigma factor, CarR is a negative regulator of the *carQRS* promoter and has the appearance of an integral inner membrane protein and CarS is a small (M_r 12 211) specific activator of the *crtEBDC* operon. As CarS has no obvious DNA binding domain, it has been suggested that it is an anti-repressor that disrupts the 'CarA' complex repression of the *crtEBDC* promoter.

It has been proposed that CarR is an inner membrane-bound anti-sigma factor that sequesters CarQ to the inner membrane in the dark and that light leads to release and hence expression of the regulon (Gorham *et al.*, 1996). The evidence for this proposal is circumstantial. Epistasis experiments reveal *carQ carR* double mutants have the phenotype of *carQ* mutants, that is, Car$^-$. Therefore, CarQ acts downstream of CarR in the regulatory cascade. Addition of a single extra copy of a wild-type *carQ* gene led to a Car^c phenotype whereas addition of an extra copy of a wild-type *carR* gene had no effect on the carotenoid phenotype. The *carQ* and *carR* genes are coupled translationally. This is a common prokaryote mechanism to ensure a precise stoichiometry of expression of the coupled genes. If translational coupling was uncoupled by separating the intact *carQ* and *carR* genes, then a Car^c phenotype was observed. If the uncoupled genes were kept within 20 kb of each other, the Car^c phenotype was mild, that is Car$^{+/c}$, but if the genes were separated by a third of the *M. xanthus* chromosome, an extreme Car^c phenotype was seen. It was proposed that CarQ and CarR were associated in a one-to-one stoichiometry at the membrane and that proximal translation of the two allowed efficient transfer of CarQ to the membrane and so avoided complexing with core RNA polymerase. If there was an extra copy of the *carQ* gene, then CarQ proteins would be free to activate the *carQRS* and *crtI* promoters constitutively. Expression of the *carQRS* operon leads to expression of CarS and hence expression of *crtEBDC* operon.

The *carD* gene has been identified as a regulatory gene of the carotenoid regulon (Nicolás, Ruiz Vázquez & Murillo, 1994). This statement contains two misnomers: (i) the *carD* mutant was blocked in both fruiting and carotenogenesis, therefore, it is not a *car* or *crt* gene; and (ii) sequencing of the gene revealed it to encode a transcription factor with high mobility group I(Y)-like DNA binding domains (Nicolás *et al.*, 1996). This protein is made constitutively, although it does decrease during fruiting body formation, and is not light modulated, therefore it is no more a regulatory protein than are the proteins of the RNA polymerase complex. It is an interesting question why this protein is necessary for transcription of *carQRS* and genes involved in fruiting body formation. There is the potential for serine phosphorylation of this protein and serine/threonine protein kinases have been identified in

M. xanthus (Zhang *et al.*, 1992). The effect of loss of these protein kinases on carotenoid biosynthesis has not been reported.

The central proposal of this model is the light-induced activation of CarQ by release from CarR and the inner membrane. It is proposed that light-dependent destruction of CarR leads to CarQ release and activation. This proposed mechanism arose from the observation that, when a CarR:LacZ fusion was overexpressed in *M. xanthus*, the protein could not be found if the cells were grown in the light, but could if the cells were grown in the dark. There was an extra copy of *carQ* so it is known that the gene encoding the fusion protein was transcribed at very high level in the light; therefore, either translation of the fusion protein was being inhibited or the protein itself was light sensitive (Gorham *et al.*, 1996). Recently it has been shown, using anti-CarR antiserum, that the native protein is also light sensitive and that it is most light sensitive in the late exponential/early stationary phase of growth (Browning, 1997). It is not clear what the basis of light sensitivity is in this protein. Presuming that it is the protein itself that is light sensitive and not its translation, it could be that there is a 'light-inducible' CarR protease or that CarR is marked in someway by light so that it becomes more susceptible to proteolysis.

Singlet oxygen as a second messenger

There are several lines of evidence that $^1O_2$ is the direct cause of CarR disappearance in the cell and hence induction of the carotenoid regulon. When *M. xanthus* bearing *lacZ* fused to the *carQRS* promoter was placed in a fermenter and pure oxygen was bubbled through the culture in the dark, both carotenogenesis and the promoter were induced. If the cells were cultured such that oxygen was the growth-limiting substrate, it proved impossible to activate the promoter or induce carotenogenesis in the light. It was concluded that an active oxygen species was an important intermediate in the caroteno-genesis induction pathway. Addition of H_2O_2 or O_2^- (generated from methyl viologen) failed to induce carotenogenesis. However, red light plus 100 pM toluidine blue did induce carotenogenesis and *carQRS* promoter expression, whereas red light alone had no effect. Toluidine blue is known to generate $^1O_2$ in red light. The implication was that $^1O_2$ was the second messenger and that it was generated *in situ* by blue light and PPIX (Robson, 1992).

In confirmation of this proposal, it was found that carotenoid quenched the light-generated signal. The activity of the *carQRS* promoter (assayed as a *lacZ* fusion) initially increased in the presence of light. However, even if cells were subjected to continuous illumination, the promoter was inactivated with time. The peak of promoter activity coincided with the synthesis of carotenoids (Robson, 1992). If the Car$^-$ mutants were subject to the same light regime, the *carQRS* promoter activity continued to increase for as long as the cells were subjected to light. Thus it appeared that carotenoids protected the promoter against light as well as the cells. In further confirmation of the carotenoid quenching effect, it was found to be difficult to induce the *carQRS* promoter in a *carA1* mutant background where there was constitutive production of phytoene and phytofluene (Hodgson, 1993). It could be argued that this lack of response of the *carQRS* promoter in a *carA* background was because one or more of the 'CarA' open reading frames was needed for *carQRS* promoter activity. However, in a Car$^-$ *crtI carA* double mutant the *carQRS* promoter behaves as in a Car$^-$ *crtI carA$^+$* mutant (D. Whitworth, pers. comm.).

It could be argued that the quenching effect of carotenoids could be due to their quenching of ^{3}PPIX (as detailed above). However, it was found that increasing concentrations of 1,4-diazobicyclo[2,2,2,]octane, a specific $^1O_2$ quenching agent, decreased the cell's ability to activate the *carQRS* promoter in the light (Robson, 1992).

Lang-Feulner and Rau (1975) showed that red-light and a variety of dyes, including toluidine blue, would induce carotenogenesis in the fungus *Fusarium aqueductuum*, whereas normally the fungus needed light of wavelength shorter than 520 nm. Again, this implies that $^1O_2$ is also an intermediate in fungal light-induced carotenogenesis. It has recently been reported that $^1O_2$ is the inducer of carotenogenesis in the yeast *Phaffia rhodozyma* (Schroeder & Johnson, 1995). Early work with *Mycobacterium marinum* implicated a porphyrin as a photosensitizer for light-induced carotenogenesis as action spectra revealed a peak at the Soret band (410 nm) (Mathews, 1963; Batra & Rilling, 1964). Again, there may be a role for $^1O_2$ as an intermediate in the activation cascade in mycobacteria. Work with an unidentified mycobacterium species implicated the role for a flavin as a photosensitizer (Batra & Rilling, 1964).

It is proposed that $^1O_2$ induces the destruction of CarR and so CarQ is released to activate the carotenoid regulon. This, in turn, leads to the synthesis of the myxobacton and myxobactin esters, which quench the ^{3}PPIX and $^1O_2$ and so leads to the feedback reduction of carotenoid biosynthesis. A similar role of $^1O_2$-induced proteolysis of a protein has been proposed to occur during photoinhibition of photosystem II in chloroplasts. If plants are exposed to high light, photobleaching occurs, whereby the photosystem II core proteins D1 and, to some extent, D2 are degraded. This reduces the chlorophyll content of the cell and so it is less susceptible to photodamage. If isolated photosystem II protein cores, including D1, D2, a 43-kDa protein and two other proteins were exposed to light, degradation of D1 occurred. Binding of the serine protease inhibitor diisopropyl fluorophosphate revealed that the 43 kDa protein was the protease, even though there was no sequence homology between this protein and any other proteases and it had previously been characterized as a chlorophyll a binding protein. There was evidence that the protease was active in the dark but that degradation of D1 was activated by light-induced damage to the substrate (Salter *et al.*, 1992). The same laboratory has produced evidence that $^1O_2$, produced from triplet state chlorophyll, was involved in the process of D1 damage (Vass *et al.*, 1992).

A MODEL FOR LIGHT INDUCTION OF CAROTENOGENESIS IN *MYXOCOCCUS XANTHUS*

Figure 2 illustrates our proposed model for the blue-light-induction of carotenoid biosynthesis in *M. xanthus*. In the dark there will be low-level expression from the *carQRS* promoter and so CarQ will be formed. However, the presence of CarR will lead to sequestration of the sigma factor to the inner membrane. Hence there will be little transcription from the *carQRS* and *crtI* promoters. Some or all of the *carA* open reading frames, *orf9*, *orf10* and *orf11*, are expressed from a constitutive promoter and the products of these open reading frames will repress transcription from the *crtEBDC* promoter.

Blue-light will be absorbed by PPIX, which due to its hydrophobic nature, is expected to be lodged in the inner membrane. The ^{3}PPIX will generate $^1O_2$ which can survive in the membrane 25 to 100 μs. This species then leads to release of CarQ from CarR. The method of release of CarQ could be any of a number of methods: (i) Interaction of $^1O_2$ with CarR could lead to an allosteric change that leads to CarQ release: (ii) $^1O_2$ stops the translation of CarR but not that of CarQ; (iii) CarR is destroyed by a $^1O_2$-activated specific CarR protease; and (iv) CarR is destroyed because interaction of CarR amino acid residues with $^1O_2$ leads to a damaged protein that is recognised as substrate by non-specific proteases in the cell. The observation of loss of over-expressed CarR in *M. xanthus* grown in the light implies that one of the last three mechanisms is more likely than the first. The released sigma factor will interact with core RNA polymerase and initiate transcription from the *carQRS* and *crtI* promoters, provided carbon catabolite repression of *crtI* has been relieved. Transcription of the *carQRS* operon will lead to production of more CarQ, CarR and CarS. Provided the blue-light is still present, $^1O_2$ will continue to inactivate CarR so there will be an exponential increase in CarQ and CarS. The CarS protein will inactivate repression of the *crtEBDC* operon by the product of the 'CarA' open reading frames.

The products of the *crtI* gene and the *crtEBDC* operon will lead to the synthesis of the myxobacton ester and related carotenoids which will quench the triplet state PPIX and $^1O_2$. Thus CarR will no longer be destroyed and all new CarQ synthesized will be sequestered to the inner membrane. This will mean the sigma factor will no longer be able to activate the *carQRS* operon and *crtI* gene. Thus there is product feedback regulation of the carotenoid regulon and carotenoid synthesis will cease once enough have been produced to suppress membrane damaging species produced in the light.

CONCLUSIONS

A detailed model has been evolved for the mechanism of light-induction of carotenogenesis in *M. xanthus*. The model arose from a number of genetic and physiological observations and reveals a number of testable hypotheses. At present the authors' laboratory is testing these hypotheses using more biochemical approaches. The areas still to be tested include: proof that CarQ is sequestered to the membrane by CarR; the nature of the interaction between CarQ and CarR; the mechanism of disappearance of CarR; the open reading frame products involved in repression of the *crtEBDC* promoter; the mechanism of repression of the *crtEBDC* promoter; the mechanism of inhibition of 'CarA' repression of the *crtEBDC* promoter by CarS; the mechanism of the carbon catabolite repression of *crtI*; and the role of CarD in expression of *carQRS*; the role, if any, of phosphorylation of CarD.

The dissection of *carQRS* promoter has begun and it has been found to be very large, requiring, as it does 145 base pairs of DNA upstream of the transcription start site. There is also evidence of bending of DNA in this promoter and the *crtEBDC* promoter but not of the *crtI* promoter (Berry, 1997). Having resolubilized active CarQ (Browning, 1997) it is now possible to begin *in vitro* analysis of the promoters dependent upon it.

The proposal that $^1O_2$ may be an active participant in light-induction cascade in *M. xanthus* carotenogenesis is novel. However, the observations that: $^1O_2$ could activate carotenogenesis in the yeast *Phaffia rhodozyma* (Schroeder & Johnson, 1995); that dyes in the presence of light that would generate $^1O_2$ activated carotenogenesis in the fungus *Fusarium aqueductuum* (Lang-Feulner & Rau, 1975); and that a porphyrin was a photosensitizer for light-induced carotenogenesis in *Mycobacterium marinum* (Mathews, 1963; Batra & Rilling, 1964), might imply this mechanism is more widely distributed in nature.

REFERENCES

Ahmad, S. (1992). Biochemical defence of pro-oxidant plant allelochemicals by herbivorous insects. *Biochemical Systematics and Ecology*, **20**, 269–96.

Ahmad, S. (1995). Antioxidant mechanisms of enzymes and proteins. In *Oxidative Stress and Antioxidant Defences in Biology*, ed. S. Ahmad, pp. 238–72. New York: Chapman & Hall.

Anderson, S. M. & Krinsky, N. I. (1973). Protective action of carotenoids against photodynamic damage to liposomes. *Photochemistry and Photobiology*, **18**, 403–8.

Anderson, S. M., Krinsky, N. I., Stone, M. J. & Clagett, D. C. (1973). Effect of singlet oxygen quenchers on oxidative damage to liposomes initiated by photosensitization or by radiofrequency discharge. *Photochemistry and Photobiology*, **20**, 65–9.

Armstrong, G. A. (1994). Eubacteria show their true colors: genetics of carotenoid pigment biosynthesis from microbes to plants. *Journal of Bacteriology*, **176**, 4795–802.

Armstrong, G. A. & Hearst, J. E. (1996). Genetics and molecular biology of carotenoid pigment biosynthesis. *FASEB Journal*, **10**, 228–37.

Armstrong, G. A., Alberti, M. & Hearst, J. E. (1990). Conserved enzymes mediate the early reactions of carotenoid biosynthesis in nonphotosynthetic and photosynthetic prokaryotes. *Proceedings of the National Academy of Sciences, USA*, **87**, 9975–9.

Armstrong, G. A., Alberti, M., Leach, F. & Hearst, J. E. (1989). Nucleotide sequence, organization, and nature of the protein products of carotenoid biosynthesis gene cluster of *Rhodobacter capsulatus*. *Molecular and General Genetics*, **216**, 254–68.

Balsalobre, J. M., Ruiz-Vázquez, R. M. & Murillo, F. J. (1987). Light induction of gene expression in *Myxococcus xanthus*. *Proceedings of the National Academy of Sciences, USA*, **84**, 2359–62.

Batra, P. P. & Rilling, H. C. (1964). On the mechanism of photoinduced carotenoid synthesis: aspects of the photoinductive reaction. *Archives of Biochemistry and Biophysics*, **107**, 485–92.

Berry, A. E. (1997). Towards a molecular mechanism for light-induction of gene transcription in *Myxococcus xanthus*. PhD thesis, University of Warwick, Coventry, UK.

Beyer, P., Mayer, M. & Kleinig, K. (1989). Molecular oxygen and the state of geometric isomerism of intermediates are essential in the carotene desaturation and cyclization reactions in daffodil chromoplasts. *European Journal of Biochemistry*, **184**, 141–50.

Botella, J. A., Murillo, F. J. & Ruiz-Vázquez, R. M. (1995). A cluster of the structural and regulatory genes for light-induced carotenogenesis in *Myxococcus xanthus*. *European Journal of Biochemistry*, **233**, 238–48.

Browning, D. F. (1997). A molecular investigation of light-induced carotenoid synthesis in *Myxococcus xanthus*. PhD thesis, University of Warwick, Coventry, UK.

Burchard, R. P. & Dworkin, M. (1966). Light-induced lysis and carotenogenesis in *Myxococcus xanthus*. *Journal of Bacteriology*, **91**, 535–41.

Burchard, R. P. & Hendricks, S. B. (1969). Action spectrum for carotenogenesis in *Myxococcus xanthus*. *Journal of Bacteriology*, **97**, 1165–8.

Burchard, B. P., Gordon, S. A. & Dworkin, M. (1966). Action spectrum for the photolysis of *Myxococcus xanthus*. *Journal of Bacteriology*, **91**, 896–7.

Chamovitz, D. I., Pecker, I. & Hirschberg, J. (1991). The molecular basis of resistance to the herbicide norflurazon. *Plant Molecular Biology*, **16**, 967–74.

Cox, G. S. & Whitten, D. G. (1982). Mechanisms for the photooxidation of protoporphyrin IX in solution. *Journal of the American Chemical Society*, **37**, 516–21.

Cox, G. S., Whitten, D. G. & Gianotti, C. (1979). Interaction of porphyrin and metalloporphyrin excited states with molecular oxygen. Energy transfer versus electron transfer quenching mechanisms in photooxidation. *Chemical Physics Letters*, **67**, 511–5.

Dahl, T. A., Midden, W. R. & Hartman, P. E. (1987). Pure singlet oxygen cytotoxicity for bacteria. *Photochemistry and Photobiology*, **46**, 345–52.

Dahl, T. A., Midden, W. R. & Hartman, P. E. (1988). Pure exogenous singlet oxygen: nonmutagenicity in bacteria. *Mutation Research*, **201**, 127–36.

Dahl, T. A., Midden, W. R. & Hartman, P. E. (1989). Comparison of killing of Gram-negative and Gram-positive bacteria by pure singlet oxygen. *Journal of Bacteriology*, **171**, 2188–94.

Di Mascio, P., Kaiser, S. & Sies, H. (1989). Lycopene as the most efficient biological carotenoid singlet oxygen quencher. *Archives of Biochemistry and Biophysics*, **274**, 532–8.

Di Mascio, P., Sunquist, A. R., Devasagayam, T. P. A. & Sies, H. (1992). Assay of lycopene and other carotenoids as quenchers of singlet oxygen. *Methods in Enzymology*, **213**, 429–38.

Edwards, J. C., Chapman, D., Cramp, W. A. & Yatvin, M. B. (1984). The effects of ionising radiation on biomembrane structure and function. *Progress in Biophysics and Molecular Biology*, **43**, 71–93.

Ehrenshaft, M., Jenns, A. E. & Daub, M. E. (1995). Targeted gene disruption of carotenoid biosynthesis in *Cercospora nicotianae* reveals no role for carotenoids in photosensitizer resistance. *Plant–Microbe Interactions*, **8**, 569–75.

Fontes, M., Ruiz-Vázquez, R. M. & Murillo, F. J. (1993). Growth phase dependence of the activation of a bacterial gene for carotenoid synthesis by blue light. *The EMBO Journal*, **12**, 1265–75.

Foote, C. S. (1976). Photosensitised oxidation and singlet oxygen: consequences in biological systems. In *Free Radicals and Biology*, volume 2, ed. W. A. Pryor, pp. 85–113. New York: Academic Press.

Foote, C. S. (1991). Definition of type I and type II photosensitized oxidation. *Photochemistry and Photobiology*, **54**, 659.

Freer, A., Prince, S., Sauer, K., Papiz, M., Hawthornthwaite-Lawless, A., McDermott, G., Cogdell, R. & Isaacs, N. W. (1996). Pigment–pigment interactions and energy-transfer in the antenna complex of the photosynthetic bacterium *Rhodopseudomonas acidophila*. *Structure*, **4**, 449–62.

Garí, E., Toledo, J. C., Gilbert, I. & Barbé, J. (1992). Nucleotide sequence of the methoxyneurosporene dehydrogenase gene from *Rhodobacter sphaeroides*: comparison with other bacterial carotenoid dehydrogenases. *FEMS Microbiology Letters*, **93**, 103–8.

Gorham, H. C., McGowan, S. J., Robson, P. R. H. & Hodgson, D. A. (1996). Light-induced carotenogenesis in *Myxococcus xanthus*: light-dependent membrane sequestration of ECF sigma factor CarQ by anti-sigma factor CarR. *Molecular Microbiology*, **19**, 171–86.

Grogan, D. W. (1989). Phenotypic characterization of the archaebacterial genus *Sulfolobus*: comparison of five wild-type strains. *Journal of Bacteriology*, **171**, 6710–19.

Hodgson, D. A. (1987). Light-inducible promoter. International patent application No. PCT/GB87/0040. New British patent application (1986) No. 8615263.

Hodgson, D. A. (1993). Light induced carotenogenesis in *Myxococcus xanthus*: genetic analysis of the *carR* region. *Molecular Microbiology*, **7**, 471–88.

Hodgson, D. A. & Murillo, F. J. (1993). Genetics of regulation and pathway of synthesis of carotenoids. In *Myxobacteria II*, ed. M. Dworkin & D. Kaiser, pp. 157–81. Washington, CD: ASM Press.

Houssaini-Iraqui, M., David, H. L., Clavel-Sérès, S., Hilali, F. & Rastogi, N. (1993). Characterisation of *car α*, *car Lep*, and *Crt I* genes controlling the biosynthesis of carotenes in *Mycobacterium aurum*. *Current Microbiology*, **27**, 317–22.

Kato, F., Hino, T., Nakaji, A., Tanaka, M. & Koyama, Y. (1995). Carotenoid synthesis in *Streptomyces setonii* ISP5395 is induced by the gene *crtS*, whose product is similar to a sigma factor. *Molecular and General Genetics*, **247**, 387–90.

Kleinig, H. (1975). On the utilization *in vivo* of lycopene and phytoene as precursors for the formation of carotenoid glucoside ester and on the regulation of carotenoid biosynthesis in *Myxococcus fulvus*. *European Journal of Biochemistry*, **57**, 301–8.

Knox, J. P. & Dodge, A. D. (1985). Singlet oxygen and plants. *Phytochemistry*, **24**, 889–96.

Krasnovsky, A. A. Jr. (1979). Photoluminescence of singlet oxygen in pigment solutions. *Photochemistry and Photobiology*, **29**, 29–36.

Kuhlbrandt, W., Wang, D. N. & Fujiyoshi, Y. (1994). Atomic model of plant harvesting complex by electron crystallography. *Nature*, **367**, 614–21.

Lamola, A. A., Yamane, T. & Trozzolo, A. M. (1973). Cholesterol hyperoxide formation in red cell membranes and photohemolysis in erythropoietic protoporphyria. *Science*, **170**, 1131–3.

Lang-Feulner, J. & Rau, W. (1975). Redox dyes as artificial photoreceptors in light-dependent carotenoid synthesis. *Photochemistry and Photobiology*, **21**, 179–83.

Linden, H., Misawa, N., Saito, T. & Sandmann, G. (1994). A novel carotenoid biosynthesis gene coding for ζ-carotene desaturase: functional expression, sequence and phylogenetic origin. *Plant Molecular Biology*, **24**, 369–79.

Lonetto, M. A., Brown, K. L., Rudd, K. E. & Buttner, M. J. (1994). Analysis of the *Streptomyces coelicolor sigE* gene reveals the existence of a subfamily of eubacterial RNA polymerase σ factors involved in the regulation of extracytoplasmic functions. *Proceedings of the National Academy of Sciences, USA*, **91**, 7573–7.

Malik, Z., Hanania, J. & Nitzan, Y. (1990). New trends in photobiology. Bactericidal effects of photoactivated porphyrins – an alternative approach to anti-microbial drugs. *Journal of Photochemistry and Photobiology*, Series B, **5**, 281–93.

McCord, J. M. (1985). Oxygen-derived free radicals in postischemic injury. *New England Journal of Medicine*, **312**, 159–63.

McDermott, G., Prince, S. M., Freer, A. A., Isaacs, N. W., Papiz, M. Z., Hawthornthwaite-Lawless, A. M. & Cogdell, R. J. (1995). The three-dimensional structure of the light-harvesting antenna complex (LH-II) from *Rhodopseudomonas acidophila*, strain-10050, at 2.5 Angstrom resolution. *Protein Engineering*, **8**, 43.

McGowan, S. J., Gorham, H. C. & Hodgson, D. A. (1993). Light induced carotenogenesis in *Myxococcus xanthus*: DNA sequence analysis of the *carR* region. *Molecular Microbiology*, **10**, 713–35.

Mannervik, B. (1985). Glutathione peroxidase. *Methods in Enzymology*, **113**, 490–5.

Martínez-Férez, I. M., Fernández-González, B., Sandmann, G. & Vioque, A. (1994). Cloning and expression in *Escherichia coli* of the gene coding for phytoene synthase from the cyanobacterium *Synechocystis* sp. PCC6803. *Biochimica et Biophysica Acta*, **1218**, 145–52.

Martínez-Laborda, A. & Murillo, F. J. (1989). Genic and allelic interactions in the carotenogenic response of *Myxococcus xanthus* to blue light. *Genetics*, **122**, 481–90.

Martínez-Laborda, A., Elais, M., Ruiz-Vázquez, R. & Murillo, F. J. (1986). Insertions of Tn*5* linked to mutations affecting carotenoid synthesis in *Myxococcus xanthus*. *Molecular and General Genetics*, **205**, 107–14.

Martínez-Laborda, A., Balsalobre, J. M., Fontes, M. & Murillo, F. J. (1990). Accumulation of carotenoids in structural and regulatory mutants of the bacterium *Myxococcus xanthus*. *Molecular and General Genetics*, **223**, 205–10.

Matheson, I. B. C., Etheridge, R. D., Kratowich, N. R. & Lee, J. (1975). The quenching of singlet oxygen by amino acids and proteins. *Photochemistry and Photobiology*, **21**, 165–71.

Mathews, M. M. (1963). Studies on the localization, function, and formation of the carotenoid pigments of a strain of *Mycobacterium marinum*. *Photochemistry and Photobiology*, **2**, 1–8.

Mathews, M. M. & Sistrom, W. R. (1960). The function of carotenoid pigments in *Sarcina lutea*. *Archiv für Microbiologie*, **35**, 139–46.

Mathews-Roth, M. M., Wilson, T., Fujimori, E. & Krinsky, N. I. (1974). Carotenoid chromaphore length and protection against photosensitisation. *Photochemistry and Photobiology*, **19**, 217–22.

Misawa, N., Nakagawa, M., Kobayashi, K., Yamano, S., Izawa, Y., Nakamura, K. & Harashima, K. (1990). Elucidation of the *Erwinia uredovora* carotenoid biosynthetic pathway by functional analysis of the gene products expressed in *Escherichia coli. Journal of Bacteriology*, **172**, 6704–12.

Nicolás, F. J., Ruiz-Vázquez, R. M. & Murillo, F. J. (1994). A genetic link between light response and multicellular development in the bacterium *Myxococcus xanthus. Genes and Development*, **8**, 2375–87.

Nicolás, F. J., Cayuela, M. L., Martínez-Argudo, I. M., Ruiz-Vázquez, R. M. & Murillo, F. J. (1996). High mobility group I(Y)-like DNA binding domains on a bacterial transcription factor. *Proceedings of the National Academy of Sciences, USA*, **93**, 6881–5.

Packer, J. E., Mahood, J. S., Mora-Arellano, V. O., Slater, T. F., Willson, R. L. & Wolfenden, B. S. (1981). Free-radicals and singlet oxygen scavengers – reaction of a peroxyradical with β-carotene, diphenylfuran and 1,4 Diazobicyclo(2,2,2)-octane. *Biochemical and Biophysical Research Communications*, **98**, 901–6.

Ramakrishnan, L., Tran, H. T., Federspiel, N. A. & Falkow, S. (1997). A *crtB* homologue essential for photochromogenicity in *Mycobacterium marinum*: isolation, characterization and gene disruption via homologous recombination. *Journal of Bacteriology*, **179**, 5862–8.

Robson, P. R. H. (1992). Towards the mechanism of carotenogenesis in *Myxococcus xanthus*. PhD thesis, University of Warwick, Coventry, UK.

Ruiz-Vázquez, R. M., Fontes, M. & Murillo, F. J. (1993). Clustering and co-ordinated activation of carotenoid genes in *Myxococcus xanthus* by blue light. *Molecular Microbiology*, **10**, 25–34.

Salter, A. H., Virgin, I., Hagman, A. & Andersson, B. (1992). On the molecular mechanism of light-reduced D1 protein degradation in photosystem II core particles. *Biochemistry*, **31**, 3990–8.

Sandmann, G. (1994). Carotenoid biosynthesis in microorganisms and plants. *European Journal of Biochemistry*, **223**, 7–24.

Schaich, K. M. & Borg, D. C. (1985). Peroxidising lipids as propagators of free-radical damage-reactions with amino acids and nucleic-acid bases. Abstracts of papers of the American Chemical Society, **189**, No. APR-, 14-AGFD.

Schroeder, W. A. & Johnson, E. A. (1995). Singlet oxygen and peroxyl radicals regulate carotenoid biosynthesis in *Phaffia rhodozyma. Journal of Biological Chemistry*, **270**, 18374–9.

Schumann, G., Nürnberger, H., Sandmann, G. & Krügel, H. (1996). Activation and analysis of cryptic *crt* genes for carotenoid biosynthesis from *Streptomyces griseus. Molecular and General Genetics*, **252**, 658–66.

Siefermann-Harms, D. (1985). Carotenoids in photosynthesis. 1. Location in photosynthetic membranes and light-harvesting function. *Biochimica et Biophysica Acta*, **811**, 325–55.

Stine, G. J. (1989). *The New Human Genetics*. Dubuque, Iowa: W. C. B. Press.

Suwa, K., Kimura, T. & Schaap, A. P. (1977). Reactivity of singlet molecular oxygen with cholesterol in a phospholipid matrix model for oxidative damage of membranes. *Biochemical and Biophysical Research Communications*, **75**, 785.

Symons, M., Botte, P. & Sybesma, C. (1985). The surface and membrane-potentials of chromatophores of photosynthetic bacteria as studied by carotenoid absorbance changes. *Biochimica et Biophysica Acta*, **806**, 161–7.

To, K-Y., Lai, E-M., Lee, L-Y., Lin, T-P., Hung, C-H., Chen, C-H., Chang, Y-S. & Liu, S-T. (1994). Analysis of the gene cluster encoding carotenoid biosynthesis in *Erwinia herbicola* Eho 13. *Microbiology*, **140**, 331–9.

Tuveson, R. W., Larson, R. A. & Kagan, J. (1988). Role of cloned carotenoid genes expressed in *Escherichia coli* in protecting against inactivation by near-UV light and specific phototoxic molecules. *Journal of Bacteriology*, **170**, 4675–80.

Valenzeno, D. P. (1987). Photomodification of biological membranes with emphasis on singlet oxygen mechanisms. *Photochemistry and Photobiology*, **46**, 147–60.

Vass, I., Styring, S., Hundal, T., Koivuniemi, A., Aro, E-M. & Andersson, B. (1992). Reversible and irreversible intermediates during photoinhibition of photosystem II: stable reduced Q_A species promote chlorophyll triplet formation. *Proceedings of the National Academy of Sciences, USA*, **89**, 1408–12.

Zhang, W. D., Muñoz-Dorado, M., Inouye, M. & Inouye, S. (1992). Identification of a putative eukaryotic-like protein kinase family in the developmental bacterium *Myxococcus xanthus*. *Journal of Bacteriology*, **174**, 5450–3.

RESPONSES TO BLUE-LIGHT IN *NEUROSPORA CRASSA*

G. MACINO, G. ARPAIA, H. LINDEN AND P. BALLARIO[1]

Istituto Pasteur, Fondazione Cenci Bolognetti, Dipartimento di Biotecnologie Cellulari, Sezione di Genetica Molecolare, Università di Roma 'La Sapienza', Viale Regina Elena, 324, 00161 Roma, Italy
[1]*Dipartimento di Genetica e Biologia Molecolare, Centro di Studio per gli Acidi Nucleici, Piazzale A. Moro 5, Università di Roma 'La Sapienza', 00185 Roma, Italy*

INTRODUCTION

Organisms are able to sense and respond to fluctuations in environmental light signals via complex mechanisms in which the stimuli are transformed into signals that are transmitted through biochemical changes in the cell. The process of blue-light perception, the transduction of the light signal and the blue-light responses have been studied extensively in plants, algae, fungi and bacteria (Senger, 1980). In fungi the elaborated behaviour of Phycomcyes in response to blue-light has been reviewed (Corrochano & Cerda Olmedo, 1992).

The ascomycete *Neurospora crassa* has been proven to be a paradigm for photobiological, biochemical and genetic studies of the enigmatic process of blue-light regulation. *N. crassa* is able to sense light only in the blue region of the spectrum, and this triggers diverse developmental, morphological and physiological responses (for review see Degli Innocenti & Russo, 1984). During conidiation, blue-light stimulates the formation of the conidial structures and influences the number of conidia produced; during the sexual cycle, it influences development of protoperithecia and the number of perithecia formed and it stimulates positive phototropism of the perithecial beaks. During vegetative growth, light induces the production of carotenoids and acts to entrain a circadian rhythm (Table 1).

All of these light-induced phenomena are abolished in two *N. crassa* mutants that are 'blind' to light: *white collar-1* (*wc-1*) and *white collar-2* (*wc-2*) (Perkins *et al.*, 1982; Nelson *et al.*, 1989). No epistatic relationships have been described between *wc-1* and *wc-2*. The pleiotropic phenotypes of these white collar mutants have suggested a role for the white collar gene products in the light signal transduction pathway.

Genes under the control of blue-light, such as the carotenogenic albino genes *al-1*, *al-2* and *al-3*, the circadian clock gene *frq*, the blue-light-induced

Table 1. *Light-regulated processes in* Neuropsora crassa

Light-regulated process	Refs.
During asexual cycle	
Mycelial carotenoid biosynthesis	Harding & Turner (1981)
Formation of conidia	Klemm & Ninneman (1978)
	Lauter *et al.* (1996)
Conidiophore phototropism	Siegel, Matsuyama & Urey (1968)
	Faull (1930)
Suppression and phase shifts of the circadian rhythm of conidiation	Sargent & Briggs (1967)
Change of membrane potential and input resistance	Kritsky *et al.* (1984)
	Potapova *et al.* (1984)
During sexual cycle	
Formation of protoperithecia	Degli-Innocenti *et al.* (1983)
Carotenoid biosynthesis in perithecial walls	Perkins (1988)
Phototropism of perithecial beaks	Harding & Melles (1983)

genes *bli-3* and *bli-4*, the conidiation genes *con-5* and *con-10*, the clock-controlled genes *ccg-4* and *ccg-6* and *wc-1*, have been shown to lose their light-inducible transcriptional activation in the *white collar* mutant backgrounds (Nelson *et al.*, 1989; Schmidhauser *et al.*, 1990; Arpaia *et al.*, 1993, 1995; Dunlap, 1996; Lauter & Russo, 1991).

There has recently been a consolidated effort to elucidate the *N. crassa* blue-light signal transduction pathway. Biochemical analysis of light signal transduction has already been carried out by several groups to find putative transduction components and second messengers involved in a variety of blue-light responses, but results have often been contradictory. Although the possibility that cAMP participates in blue light signal transduction has also been explored (Shaw & Harding, 1987; Kallies, Gebauer & Rensing, 1996), different studies have led to conflicting results, so the involvement of cAMP remains questionable. Both the *Neurospora* adenylate cyclase (Kore-eda, Murayama & Uno, 1991) and a member of the cAMP-dependent protein kinase family have been cloned (Yarden *et al.*, 1992) but no information suggesting a possible role of these enzymes in blue light signalling has been published. A possible function of inositol phosphates in light signalling has also been examined. However, no changes in the levels of inositol phosphates have been detected following light induction (Lakin-Thomas, 1993; Kallies *et al.*, 1996).

Recently, a selection system was developed in order to isolate mutants affected in blue light signal transduction (Carattoli *et al.*, 1995). This system was applied in a first attempt to isolate the BLR (blue-light regulator) mutants showing down-regulation of photoregulated genes. The same screening system was subsequently used for the isolation of mutants which showed a constitutive carotenoid biosynthesis and the mutants seemed to be

```
WC2        EYVCTDCGTLDSPEWRKGPSGPKTLCNACGLR
WC1        ...C..C.T..TPEWR.GPSG...LCN.CGLR
NTL1       ...C..C....TPEWR.GP.....LCNACGL.
SRD1       ...C..C....T..WR.GP.....LC..CGL.
GATA-1     ...C..C.T..TPLWRR.P.G_..LCNACGL.
GATA-2     ...C..CG...TPLWRR...G_..LCNACGL.
GATA-6     ...C..CG...TPLWRR...G_..LCNACGL.
AREA       ...CT.C.T..TPLWRR.P.G_..LCNACGL.
NIT2       ...CT.C....TPLWRR.P.G_..LCNACGL.
                        *
```

Fig. 1. Multiple alignment of GATA zinc fingers: *N. crassa* WC2 (Y09119), *N. crassa* WC1 (X94300), *Nicotiana tabacum* NTL1 (S46418) and *S. cerevisiae* SRD1 (X06322) are zinc fingers with a X_{18} loop. Human GATA-1 (P15976), GATA-2 (P23769), GATA-6 (U51335), *A. nidulans* AREA (X52491) and *N. crassa* NIT2 (M33956) are zinc fingers with a X_{17} loop. The asterisk indicates the conserved leucine of canonical GATA proteins (X17), points correspond to non-conserved residues, - indicates the gap necessary for the alignment.

involved in the transcriptional repression of light-regulated genes (Linden, Rodriguez-Franco & Macino, 1997*a*). The isolation of *white collar* (*wc*) mutants were also possible using this screening system. In spite of a comprehensive search, only two different *wc* mutants have been described, *wc-1* and *wc-2* (Linden, Ballario & Macino, 1997*b*).

WC-1 AND WC-2: BLUE-LIGHT SPECIFIC TRANSCRIPTION FACTORS

wc-1 and *wc-2* genes have been cloned recently by two different strategies: *wc-1* by chromosome walking and mutant complementation (Ballario *et al.*, 1996) and *wc-2* by insertional mutagenesis using a selection system developed specifically for the identification of blue-light regulatory mutants (Linden & Macino, 1997; Carattoli *et al.*, 1995).

The deduced amino acid sequence of WC1 and WC2 suggests a role for these proteins as transcription factors involved in light regulation. Structural characteristics of WC1 and WC2 include the presence of a zinc finger domain similar to that previously described for vertebrate GATA factors and also for general regulators of nitrogen metabolism in fungi like the *A. nidulans areA* gene, the *N. crassa nit-2* gene, and the *S. cerevisiae* GLN3 and DAL80 genes (Crawford & Arst, 1993; Cunningham & Cooper, 1991). The vertebrate GATA factors have two zinc fingers, while the single zinc finger present in the fungal proteins is similar to the carboxy-proximal zinc finger of vertebrates. A comparison of the zinc finger domain of WC1 and WC2 with those of other members of the class-four zinc fingers is shown in Fig. 1. All proteins with this class-four zinc finger motif (17 aa finger loop), have been demonstrated to be transcriptional regulators capable of binding to promoter GATA sequences. Among the class-four zinc finger proteins, WC1 and WC2 have the unusual characteristic of possessing an 18 aa loop in the zinc finger and lack a leucine residue in the loop that is normally conserved (indicated by the asterisk in Fig. 1). These unusual features of WC1 and WC2 are shared with other recently isolated putative transcription factors such as the tobacco putative nitrogen-regulatory protein NTL1 (Daniel-Vedele & Caboche, 1993) and with the *S. cerevisiae* protein SRD1 (Hesse, Stanford & Hopper, 1994). The leucine residue conserved in the majority of class-four zinc finger proteins has been shown to be involved in the binding of the chicken GATA-1 protein to the GATA site of its cognate promoter (Omichinski *et al.*, 1993) and has also been proposed to be responsible for GATA-binding of the AREA protein (Kudla *et al.*, 1990).

The amino-terminal region of WC1 contains a stretch of 28 glutamine residues, embedded in a region rich in glutamine and histidine residues, while proline-rich and acidic regions were identified in WC2. These domains have been described for many other transcription factors and have been implied in transcriptional activation (Mermod *et al.*, 1989; Hope, Mahadevan & Struhl, 1988). Whether or not one of these regions of WC1 or WC2 actually plays a role in transcriptional activation is not known.

The idea that WC1 and WC2 are transcription factors is reinforced by the band-shift experiments which demonstrated that WC1 and WC2 are able to bind to a specific *Neurospora* promoter sequence. The *al-3* promoter had been characterized by deletion and site-directed mutagenesis (Carattoli *et al.*, 1994). These results indicate that the region of the *al-3* promoter composed of nucleotides -226 and -55, contains all the necessary information for blue-light photoinduction. This region contains two directly repeated GATA sequences separated by 17 nt. A 41-mer oligonucleotide containing the *al-3* GATA sequence region has been shown to be retarded in band shift experiments. In order to investigate the importance of the GATA sequences for specific binding of WC1 and WC2, competition experiments were performed with unlabelled oligonucleotides within which the GATA sequence had been mutated. As expected, these mutated oligonucleotides did not displace the native *al-3* GATA oligonucleotide at low and high molar ratios. However, at very high molar ratios (1:100) the mutant GATA begins to compete, suggesting the existence of other bases involved in the WC1 and WC2 DNA-binding. Further experiments on WC1 and WC2 DNA-binding and the characterization of the binding properties of the other 18 amino acid loop proteins will demonstrate whether or not these proteins belong to a distinct class of GATA-like proteins.

PROTEIN KINASE C: A NOVEL COMPONENT OF BLUE-LIGHT TRANSDUCTION PATHWAY

It has been shown both genetically and biochemically that WC1 and WC2 are transcription factors that bind to, and affect, light regulation of the *al-3* gene but what is the mechanism by which light activates the rapid responses via WC1 and WC2? WC1 is characterized by several putative phosphorylation sites, that are non-randomly distributed within the polypeptide. This pattern is reminiscent of the organization of putative phosphorylation sites in other transcription factors (Crawford & Arst, 1993). The presence of potential post-translational modification sites in WC1 suggests the possibility that WC1 becomes activated by specific light-induced phosphorylation or dephosphorylation.

In higher plants, a biochemical approach has proven to be very successful for elucidating the mechanisms involved in phytochrome signalling (Bowler *et al.*, 1994; Neuhaus *et al.*, 1993). Similarly, in *Neurospora*, pharmacological experiments have been designed to interfere with light signal transduction by utilizing agonists and antagonists to either induce or block signal transduction. These data have revealed a role for protein kinase C in the blue light transduction pathway in *N. crassa* (G. Arpaia *et al.*, unpublished observations). It has been shown that protein kinase C is able to mediate light induction of conidiation-specific *albino-3* transcripts, representing a light-specific positively acting element. Gene expression and promoter studies suggest that different stimuli address distinct regulatory *cis* elements in promoters (Bell-Pedersen, Dunlap & Loros, 1996; Arpaia *et al.*, 1993, 1995) but the overall response due to the interplay of the different stimuli is strictly coordinated and finely tuned. Light induction of almost all *Neurospora* light-induced genes is dependent on the products of the regulatory genes *wc-1* and *wc-2*, which are probably involved in transcriptional activation. The hypothesis that components binding at light regulatory *cis* elements are regulated by the action of protein kinase C action is strengthened by the finding that the WC1 protein is a substrate for protein kinase C, as demonstrated by the fact that a portion of an *E. coli* expressed WC1 protein, corresponding to the zinc finger region and bearing several putative protein kinase C phosphorylation sites, is phosphorylated *in vitro*.

PAS DOMAINS AS PUTATIVE TRANSDUCTION ELEMENTS

An obvious question is whether WC1 and WC2, both of which are necessary for light induction, interact either directly or indirectly. A putative dimerization domain called PAS has been identified in each WC protein. PAS is an acronym derived for a dimerization domain conserved in the proteins Per, Arnt and Sim (Reisz-Porszasz *et al.*, 1994; Huang, Edery & Rosbach, 1993). The ability of the PAS domains in the WC proteins to form dimers has been proposed (Linden & Macino, 1997) and was recently demonstrated *in vitro* (Ballario & Macino, unpublished observations). The ability of the WC proteins to form hetero- or homodimers could enable them to recognize several different binding sites in light-regulated promoters. Moreover, it is possible that light may induce the formation of WC dimers, thus activating their function as transcription factors. This working model is purely hypothetical at present, as there is no direct evidence for, or against, the

```
NifL       32 EHAPIASI TDLKANILYANRAFRTITGYGSEEVLGKNESILS
Bat1      132 PIGITISDATDPEEPIIYINDSFEDITGYSPEDVVGANHRFLQ
WC1       390 SCAFVVCDVTLNDCPIIYVSDNFQNLTGYSRHEIVGRNCRFLQ
               T      I Y    F   TGY    E   G N

NifL       NGTTPRLVYQ...........ALWGRLAQKKPWSGVLVNRRKD
Bat1       GPKTNEEPRG...........GFWTAITEDHDTQVVLRNYRKD
WC1        APDGNVEAAGTKREFNENNAVYTLKKTIAEGQEIQQSLINYRKG
                                          L N RK

NifL       KTLYLAELTVAPVLNEAGETIYYLGMHRD
Bat1       GSLFWNQVDISPIYDEDGTVSHYVGFQMD
WC1        GKPFLNLLTMIPIPWDTEEIRYFIGFQID
               P    E       G    D
```

Fig. 2. Putative sensory domains. Multiple alignment of WC1 with NifL from *A. vinelandii* (S19883) and Bat1 from *H. halobium* (A43650). The residues conserved in two proteins are in bold. The residues conserved in all the proteins considered are indicated below and underlined.

```
WC1       LSLKG  LFLYLSPACK  KVLEYDASDL  VGT.SLSSIC  HPSDIVPVTR
WC2       LDANG  RIKHVSPSVE  PLTGYKPPEI  IDL.FLRDLI  HPDDVGVFTA
PER       HTATG  IISHVDSAAV  SALGYLPQDL  MGR.SIMDLY  HHDDLPVIKE
ARNT      HNIEG  IFTFVDHRCV  ATVGYQPQEL  LGK.NIVEFC  HPEDQQLLRD
PYP       LDGDG  NILQYNAAEG  DITGRDPKEV  IGKNFFKDVA  PCTDSPEFYG
MESPHY    SDEYG  CCTENNPAME  KLTGVRREDV  IGR.MLM...  ..GDFGSALR
PHYA      VDSDG  LVNGWNTKIA  ELTGLSVDEA  IGKHFLTLVE  DSSVEIVKRM
```

Fig. 3. PAS domains. Multiple alignment of WC1 and WC2 with *Drosophila melanogaster* PER (P12348), *Homo sapiens* ARNT (P27540), *E. halophila* PYP (P16113), *Mesotaenium caldariorum* MESPHY (U31283), *Arabidopsis thaliana* PHYA (P14712).

ability of light to initiate the formation of WC dimers. The first of the two PAS repeats of the WC1 protein (amino acid 399 to 504) (see Fig. 2) contains striking identity (35%) to Bat, a transcription factor of *Halobacterium halobium* required for the oxygen-mediated expression of the bacteriopsin gene. Bat is also proposed to be the oxygen-sensing element in this signal transduction cascade (Gropp & Betlach, 1994). The BAT-like region in the WC1 protein also contains identity (29%) with NifL, a protein that regulates *nif* gene transcription in response to changes in environmental oxygen concentrations in *Klebsiella pneumoniae* and *Azotobacter vinelandii* (Blanco *et al.*, 1993). It has been shown that NifL is a flavoprotein, with FAD as prosthetic group. NifL does not sense molecular oxygen directly, but is responsive to the oxidation state of the chromophore. It therefore represents an example of a redox-sensitive regulator (Hill *et al.*, 1996).

An additional homology in the second PAS repeat of WC1 and the unique PAS of WC2 has been identified in other photoreceptor genes. In these cases this region is thought to be involved in signal transduction (see Fig. 3). The gene *mesphyl* cloned from the green alga *Mesotaenium caldariorum* has been proposed to encode a member of the lower plant phytochrome subfamily (Lagarias, Shu-Hsing & Lagarias, 1995). The deduced *mesphyl* protein contains an internal repeat with partial sequence identity to NifL, Bat, PYP, and to the PAS domains of ARNT, PER, WC1 and WC2. Mutations in the corresponding regions of phytochrome A or phytochrome B of angiosperms blocks their biological activity, despite the fact that their ability to absorb light and dimerize is unaffected (Quail *et al.*, 1996). It has therefore been concluded that this region is not involved in light perception but instead is involved in signal transduction. By analogy, a model for light signal transduction in *Neurospora* may involve a similar role for the PAS domains in signal transduction of the WC proteins in addition to their dimerization function discussed above.

WHAT IS THE NATURE OF THE CRYPTOCHROME IN *NEUROSPORA*?

Despite extensive research over the last decades and much published data, no photoreceptor responsible for blue-light/UV perception has been identified unequivocally. This is true not only for *Neurospora* and other fungi but also for higher plants. The first blue-light photoreceptor candidate was cloned from higher plants only recently (Ahmad & Cashmore, 1993). The CRY1 protein reveals close homology to bacterial DNA photolyase and was shown to bind flavin (Lin *et al.*, 1995). In *N. crassa* a photolyase gene has been cloned (Yajima *et al.*, 1991). Mutants in the photolyase gene created by RIPing (a process that mutates DNA duplications prior to meiosis in *Neurospora*) revealed a normal light response (Linden, Ballario & Macino, 1997b) indicating that the *Neurospora* photolyase does not function as blue-light photoreceptor in carotenogenesis and light-regulated transcription. Action spectra for different *Neurospora* blue light responses were recorded and indicated the involvement of flavin- or carotene-type photoreceptors (Zalokar, 1955; Sargent & Briggs, 1967; De Fabo, Harding & Shropshire, 1976). Another approach for the identification of blue light/UV photoreceptors was the investigation of light-induced optical absorbance changes (LIAC). Such absorbance changes have been reported for many organisms (Galland & Senger, 1988) and have also been described for *Neurospora* (Munoz, Brody & Butler, 1974). The LIAC in *Neurospora* were correlated to the photoreduction of a *b*-type cytochrome and an action spectrum for this photoreduction suggested again a flavin-type photoreceptor (Munoz & Butler, 1975).

Genetic evidence for the participation of flavins and cytochromes in blue-light perception came from the study of different mutants. The effect of flavin deficiency on blue-light perception was studied in the *rib-1* and *rib-2* mutants (Paietta & Sargent, 1981). The flavin deficiency mainly affected the photo-suppression of conidial banding suggesting a flavin-type photoreceptor for this blue-light effect. The authors put forward a hypothesis where several flavin-type photoreceptors are responsible for the blue-light perception in *Neurospora*. Paietta & Sargent (1983) used the flavin analogues, 1-deazariboflavin and roseoflavin, to supplement the flavin deficient mutants, *rib-1* and *rib-2*. They found that both compounds can act as photoreceptors for photosuppression and phase shifting of the circadian conidiation.

At present, no definite answer can be given about the nature of the *Neurospora* blue-light photoreceptor. However, the experimental data points toward a flavin-type photoreceptor. The similar action spectra for the various blue-light responses may indicate the presence of only one UV/blue-light photoreceptor in *Neurospora*.

CONCLUSIONS

The *Neurospora* blind mutants, *wc-1* and *wc-2*, have always been considered to be the best candidates for the isolation of the blue-light photoreceptor gene. The molecular data reviewed herein, confirms a central role for these proteins in blue-light signal transduction in *Neurospora*. Based on the data in hand, a working model can be postulated for the function of the WC proteins in blue light signal transduction. In this model the two transcription factors form a complex via their PAS domains present in both proteins. This heterodimerization may be influenced by the binding of a cofactor involved either in light signal transduction or light perception. This simple mechanism, which does not exclude the involvement of second messengers or post-translational changes (i.e. protein phosphorylation), could be sufficient to explain rapid light responses, such as light-induced carotenogenesis. The findings reported open up an extremely fascinating field of study relating to the regulation of WC1 and WC2 with respect to their role as effectors of the blue-light response both singularly and in combination.

ACKNOWLEDGEMENTS

The authors thank academic colleagues for stimulating discussion. This work was supported in part by grants from Istituto Pasteur Fondazione Cenci Bolognetti and from Ministero delle Risorse Agricole, Alimentari e Forestali, Piano Nazionale Biotecnologie Vegetali.

REFERENCES

Ahmad, M. & Cashmore, A. R. (1993). The HY4 gene of *A. thaliana* encodes a protein with characteristics of a blue-light photoreceptor. *Nature*, **366**, 162–6.
Arpaia, G., Loros, J. J., Dunlap, J. C., Morelli, G. & Macino, G. (1993). The interplay of light and circadian clock. Independent dual regulation of clock-controlled gene *ccg-2* (*eas*). *Plant Physiology*, **102**, 1299–305.

Arpaia, G., Loros, J. J., Dunlap, J. C., Morelli, G. & Macino, G. (1995). Light induction of the clock-controlled gene *ccg-1* is not transduced through the circadian clock in *Neurospora crassa*. *Molecular and General Genetics*, **247**, 157–63.

Ballario, P., Vittorioso, P., Magrelli, A., Talora, C., Cabibbo, A. & Macino, G. (1996). White collar-1, a central regulator of blue light responses in *Neurospora*, is a zinc finger protein. *EMBO Journal*, **15**, 1650–7.

Bell-Pedersen, D., Dunlap, J. C. & Loros, J. J. (1996). Distinct *cis*-acting elements mediate clock, light, and developmental regulation of the *Neurospora crassa eas* (*ccg-2*) gene. *Molecular and Cellular Biology*, **16**, 513–21.

Blanco, G., Drummond, M., Woodley, P. & Kennedy, C. (1993). Sequence and molecular analysis of the *nifL* gene of *Azotobacter vinelandii*. *Molecular Microbiology*, **9**, 869–79.

Bowler, C., Yamagata, H., Neuhaus, G. & Chua, N. H. (1994). Phytochrome signal transduction pathways are regulated by reciprocal control mechanisms. *Genes and Development*, **8**, 2188–202.

Carattoli, A., Cogoni, C., Morelli, G. & Macino, G. (1994). Molecular characterization of upstream regulatory sequences controlling the photoinduced expression of the albino-3 gene of *Neurospora crassa*. *Molecular Microbiology*, **13**, 787–95.

Carattoli, A., Kato, E., Rodriguez-Franco, M., Stuart, W. D. & Macino, G. (1995). A chimeric light-regulated amino acid transport system allows the isolation of blue light regulator (*blr*) mutants of *Neurospora crassa*. *Proceedings of National Academy of Sciences, USA*, **92**, 6612–16.

Corrochano, L. M. & Cerda-Olmedo, E. (1992). Sex, light and carotenes: the development of Phycomyces. *Trends in Genetics*, **8**, 268–74.

Crawford, N. M. & Arst, H. N. (1993). The molecular genetics of nitrate assimilation in fungi and plants. *Annual Review of Genetics*, **27**, 115–46.

Cunningham, T. S. & Cooper, T. G. (1991). Expression of the DAL80 gene, whose product is homologous to the GATA factors and is a negative regulator of multiple nitrogen catabolic genes in *S. cerevisiae* is sensitive to nitrogen catabolite repression. *Molecular and Cellular Biology*, **12**, 6205–15.

Daniel-Vedele, F. & Caboche, M. (1993). A tobacco cDNA clone encoding a GATA-1 Zinc-finger protein homologous to regulators of nitrogen metabolism in fungi. *Molecular and General Genetics*, **240**, 365–73.

De Fabo, E. C., Harding, R. W. & Shropshire, W. (1976). Action spectrum between 260 and 800 nanometers for the photoinduction of carotenoid biosynthesis in *Neurospora crassa*. *Plant Physiology*, **57**, 440–5.

Degli-Innocenti, F. & Russo, V. E. A. (1984). Isolation of new white collar mutants of *Neurospora crassa* and studies on their behaviour in the blue light-induced formation of protoperithecia. *Journal of Bacteriology*, **159**, 757–61.

Degli-Innocenti, F., Pohl, U. & Russo, V. E. A. (1983). Photoinduction of protoperithecia in *Neurospora crassa* by blue light. *Photochemistry and Photobiology*, **37**, 49–51.

Dunlap, J. C. (1996). Genetic and molecular analysis of circadian rhythms. *Annual Review of Genetics*, **30**, 579–601.

Faull, A. F. (1930). On the resistance of *Neurospora crassa*. *Mycologia*, **22**, 288–303.

Galland, P. & Senger, H. (1988). The role of pterins in the photoreception and metabolism of plants. *Photochemistry and Photobiology*, **48**, 811–20.

Gropp, F. & Betlach, M. C. (1994). The bat gene of *Halobacterium halobium* encodes a transacting oxygen inducibility factor. *Proceedings of National Academy of Sciences, USA*, **91**, 5475–9.

Harding, R. W. & Melles, S. (1983). Genetic analysis of the phototrophism of *Neurospora crassa* perithecial beaks using white collar and albino mutants. *Plant Physiology*, **72**, 996–1000.

Harding, R. W. & Turner, R. V. (1981). Photoregulation of the carotenoid biosynthetic pathway in albino and white collar mutants of *Neurospora crassa*. *Plant Physiology*, **68**, 745–9.

Hesse, S. M., Stanford, D. R. & Hopper, A. K. (1994). SRD1, an *S. cerevisiae* gene affecting pre-rRNA processing contains a C2/C2 zing-finger motif. *Nucleic Acids Research*, **7**, 1265–71.

Hill, S., Austin, S., Eydmann, T., Jones, T. & Dixon, R. (1996). *Azotobacter vinelandii* NIFL is a flavoprotein that modulates transcriptional activation of nitrogen-fixation genes via a redox-sensitive switch. *Proceedings of National Academy of Sciences, USA*, **93**, 2143–8.

Hope, I. A., Mahadevan, S. & Struhl, K. (1988). Structural and functional characterization of the short acidic transcriptional activation region of yeast GCN4 protein. *Nature*, **333**, 635–40.

Huang, Z. J., Edery, I. & Rosbach, M. (1993). PAS is a dimerization domain common to *Drosophila* Period and several transcription factors. *Nature*, **364**, 259–62.

Kallies, A., Gebauer, G. & Rensing, L. (1996). Light effects on cyclic nucleotide levels and phase shifting of the circadian clock in *Neurospora crassa*. *Photochemistry and Photobiology*, **63**, 336–43.

Klemm, E. & Ninnemann, H. (1978). Correlation between absorbance changes and a physiological response induced by blue light in *Neurospora crassa*. *Photochemistry and Photobiology*, **28**, 227–30.

Kore-eda, S., Murayama, T. & Uno, I. (1991). Isolation and characterization of the adenylate cyclase structural gene of *Neurospora crassa*. *Japanese Journal of Genetics*, **66**, 317–34.

Kritsky, M. S., Sokolovsky, V. Y., Belozerskaya, T. A. & Chernysheva, E. K. (1982). Relationship between cyclic AMP levels and accumulation of carotinoid pigments in *Neurospora crassa*. *Archives of Microbiology*, **133**, 206–8.

Kritsky, M. S., Afanasieva, T. P., Belozerskaya, T. A., Chailakhian, L. M., Chernysheva, E. K., Filippovich, S. Y., Levina, N. N., Potapova, T. V. & Sokolovsky, V. Y. (1984). Photoreceptor mechanism of Neurospora crassa: control over the electrophysiological properties of cell membrane and over the level of nucleotide regulators. In *Blue Light Effects in Biological Systems*, ed. H. Senger, pp. 207–10. Berlin: Springer-Verlag.

Kudla, B., Caddick, M. X., Langdon, T., Martinez-Rossi, N. M., Bennet, C. F., Sisley, S., Davies, R. W. & Arst, H. N. (1990). The regulatory gene *areA* mediating nitrogen metabolite repression in *Aspergillus nidulans*. Mutations affecting specificity of gene activation alter a loop residue of a putative Zn-finger. *EMBO Journal*, **9**, 1355–64.

Lagarias, D. M., Wu, S-H. & Lagarias, J. C. (1995). Atypical phytochrome gene structure in the green alga *Mesotaenium caldariorum*. *Plant Molecular Biology*, **29**, 1127–42.

Lakin-Thomas, P. L. (1993). Evidence against a direct role for inositol phosphate metabolism in the circadian oscillator and the blue-light signal transduction pathway in *Neurospora*. *Biochemical Journal*, **292**, 813–18.

Lauter, F. R. & Russo, V. E. A. (1991). Blue light induction of conidiation-specific genes in *Neurospora crassa*. *Nucleic Acids Research*, **19**, 6883–6.

Lin, C., Robertson, D. E., Ahmad, M., Raibekas, A. A., Shuman-Jorns, M., Dutton, P. L. & Cashmore, A. R. (1995). Association of flavin adenine dinucleotide with the *Arabidopsis* blue-light receptor CRY1. *Science*, **269**, 968–9.

Linden, H. & Macino, G. (1997). White collar 2, a partner in blue light signal transduction, controlling expression of light-regulated genes in *Neurospora crassa*. *EMBO Journal*, **16**, 98–109.

Linden, H., Rodriguez-Franco, M. & Macino, G. (1997*a*). Regulatory mutants of *Neurospora crassa* in blue light perception. *Molecular and General Genetics*, **254**, 111–18.

Linden, H., Ballario, P. & Macino, G. (1997*b*). Blue light regulation in *Neurospora crassa*. *Fungal Genetics and Biology*, in press.

Mermod, N., O'Neill, E. A., Kelly, T. J. & Tjan, R. (1989). The proline-rich transcriptional activator of CTF/NF-1 is distinct from the replication and DNA binding domain. *Cell*, **598**, 741–53.

Munoz, V. & Butler, W. L. (1975). Photoreceptor pigment for blue light in *Neurospora crassa*. *Plant Physiology*, **55**, 421–6.

Munoz, V., Brody, S. & Butler, W. L. (1974). Photoreceptor pigment for blue light responses in *Neurospora crassa*. *Biochemistry and Biophysics Research Communications*, **58**, 322–7.

Nelson, M. A., Morelli, G., Carattoli, A., Romano, N. & Macino, G. (1989). Molecular cloning of *Neurospora crassa* carotenoid biosynthetic gene (albino-3) regulated by blue light and the products of the white collar genes. *Molecular and Cellular Biology*, **9**, 1271–6.

Neuhaus, G., Bowler, C., Kern, R. & Chua, N. H. (1993). Calcium/calmodulin-dependent and -independent phythochrome signal transduction pathways. *Cell*, **73**, 937–52.

Omichinski, J. G., Clore, G. M., Schaad, O., Fensenfeld, G., Trainor, C., Appella, E., Stahl, S. J. and Gronenborn, A. M. (1993). NMR structure of a specific DNA complex of Zn-containing DNA binding domain of GATA-1. *Nature*, **261**, 438–46.

Paietta, J. & Sargent, M. L. (1981). Photoreception in *Neurospora crassa*: correlation of reduced light sensitivity with flavin deficiency. *Proceedings of the National Academy of Sciences, USA*, **78**, 5573–7.

Paietta, J. & Sargent, M. L. (1983). Modification of blue light photoresponses by riboflavin analogs in *Neurospora crassa*. *Plant Physiology*, **72**, 764–6.

Perkins, D. D. (1988). Photoinduced carotenoid synthesis in perithecial wall tissue of *Neurospora crassa*. *Fungal Genetics Newsletter*, **35**, 38–9.

Perkins, D. D., Radford, A., Newmeyer, D. & Bjorkmann, M. (1982). Chromosomal loci of *Neurospora crassa*. *Microbiological Review*, **46**, 426–570.

Potapova, T. V., Levina, N. N., Belozerskaya, T. A., Kritsky, M. S. & Chailakhian, L. M. (1984). Investigation of electrophysiological responses of *Neurospora crassa* to blue light. *Archives of Microbiology*, **137**, 262–5.

Quail, P. H., Boylan, M. T., Parks, B. M., Short, T. W., Xu, Y. & Wagner, D. (1996). Phytochromes: photosensory perception and signal transduction. *Science*, **268**, 675–80.

Reisz-Porszasz, S., Probst, M. R., Fukunaga, B. N. & Hankinson, O. (1994). Identification of functional domains of the aryl hydrocarbon receptor nuclear translocator protein (ARNT). *Molecular and Cellular Biology*, **14**, 6075–86.

Sargent, M. L. & Briggs, W. R. (1967). The effects of light on a circadian rhythm of conidiation in *Neurospora*. *Plant Physiology*, **42**, 1504–10.

Schmidhauser, T. J., Lauter, F. R., Russo, V. E. A. & Yanofsky, C. (1990). Cloning, sequence, and photoregulation of *al-1*, a carotenoid biosynthetic gene of *Neurospora crassa*. *Molecular and Cellular Biology*, **10**, 5064–70.

Senger, H. (1980). The Blue Light Syndrome. Berlin: Springer–Verlag.

Shaw, N. M. & Harding, R. W. (1987). Intracellular and extracellular cyclic nucleotides in wild-type and white collar mutant strains of *Neurospora crassa*. *Plant Physiology*, **83**, 377–83.

Siegel, R. W., Matsuyama, S. S. & Urey, J. C. (1968). Induced macroconidia formation in *Neurospora crassa*. *Experientia*, **24**, 1179–81.

Sommer, T., Chambers, J. A. A., Eberle, J., Lauter, F. R. & Russo, V. E. A. (1989). Fast light-regulated genes of *Neurospora crassa*. *Nucleic Acids Research*, **14**, 5713–23.

Yajima, H., Inoue, H., Oikawa, A. & Yasui, A. (1991). Cloning and functional characterization of a eucaryotic DNA photolyase gene from *Neurospora crassa*. *Nucleic Acids Research*, **19**, 5359–62.

Yarden, O., Plamann, M., Ebbole, D. J. & Yanofsky, C. (1992). *cot-1*, a gene required for hyphal elongation in *Neurospora crassa*, encodes a protein kinase. *EMBO Journal*, **11**, 2159–66.

Zalokar, M. (1955). Biosynthesis of carotenoids in *Neurospora*. Action spectrum of photoactivation. *Archives of Biochemistry and Biophysics*, **56**, 318–25.

CIRCADIAN RHYTHMS IN CYANOBACTERIA

SUSAN S. GOLDEN[1], MASAHIRO ISHIURA[2], CARL HIRSCHIE JOHNSON[3] AND TAKAO KONDO[2]

[1]*Department of Biology, Texas A&M University, College Station, TX 77843-3258, USA*
[2]*Division of Biological Science, Graduate School of Science, Nagoya University, Chikusa, Nagoya, 464-01, Japan*
[3]*Department of Biology, Vanderbilt University, Nashville, TN 37235, USA*

INTRODUCTION

Organisms at all levels of cellular complexity manifest daily rhythms of biological activity that are controlled by an endogenous biochemical oscillator. These periodic fluctuations, called circadian rhythms, are variations that peak and trough with a recurring period of about (*circa*) a day (*diem*). Circadian rhythms allow diverse organisms to adapt to environmental cycles of such relevant conditions as light, temperature, and humidity (Aschoff, 1989; Sweeney, 1987). In many cases, an ecological significance can be presumed: for example, the anticipation of dawn for optimizing pupal eclosion or photosynthesis, or the anticipation of dusk for minimizing predation while foraging (Daan, 1981; DeCoursey, 1986; Pittendrigh, 1981; Sweeney, 1987). The potential advantage of other daily rhythms is more cryptic, such as that of leaf movements in plants (Sweeney, 1987). In addition to its obvious role in controlling daily patterns, the circadian oscillator appears to be a key component in the timing mechanism that controls photoperiodic behaviour (Bünning, 1960: Reppert, 1989; Sweeney, 1987). In humans, circadian rhythmicity has clinical significance (Moore-Ede, Czeisler & Richardson, 1983), as it is involved in such maladies as insomnia (Strogatz, Kronauer & Czeisler, 1987), sudden cardiac death (Muller & Tofler, 1991; Willich *et al.*, 1987), jet-lag (Tajima *et al.*, 1991), shift-work maladjustment (Czeisler *et al.*, 1990) and depression (Healy, 1987; Lewy *et al*, 1987), as well as in other aspects of mental health (Satlin *et al.*, 1991; Van Cauter *et al.*, 1991) and ageing (Brock, 1991).

It is not yet known whether the circadian clocks of diverse organisms share a biochemical mechanism (Dunlap, 1996), but several universal features of the biological clock can be functionally defined. Rhythms that are controlled by a circadian oscillator persist in constant conditions (that is, in the absence of environmental cues) with a period close to 24 h, reset to a new phase in response to light/dark signals and maintain very nearly the same period over a range of temperatures (Pittendrigh, 1993).

The circadian circuitry of an organism can be thought of as comprising three subsystems: an oscillator that keeps time, input pathways that synchronize the clock with the external environment and output pathways that transfer the timekeeping information from the oscillator to biological processes (Johnson & Hastings, 1986). None of these has been described in biochemical detail for any organism, although the field is progressing rapidly towards the goal of uncovering the oscillator in a number of eukaryotes, particularly *Drosophila* and *Neurospora* (Dunlap, 1996; Dunlap *et al.*, this volume; Kyriacou *et al.*, this volume). The sequences of genes whose products are likely to contribute to the central timekeeping mechanism have been reported for *Drosophila* (Bargiello, Jackson & Young, 1984; Myers *et al.*, 1995; Reddy *et al.*, 1984), *Neurospora* (Crosthwaite, Dunlap & Loros, 1997; McClung *et al.*, 1989) and most recently, mouse (King *et al.*, 1997). A shared domain called PAS provides a link among these genes that may hint at their origins (Crosthwaite *et al.*, 1997; King *et al.*, 1997) but a more comprehensive view of the clock components in a single organism is needed to derive a molecular model that explains the function of the circadian oscillator. An opportunity to rapidly achieve a complete genetic profile of the circadian oscillator has arisen with the discovery that at least some prokaryotes – namely, the cyanobacteria – possess a clock that has the same properties as have been described for eukaryotic organisms.

CYANOBACTERIA: CIRCADIAN RHYTHMS IN PROKARYOTES

Temporal separation of nitrogen fixation and photosynthesis

A number of laboratories reported throughout the 1980s that some diazotrophic cyanobacteria grown in a cycle of alternating light and dark (LD) temporally separate the processes of nitrogen fixation and photosynthesis. Nitrogen-fixing cyanobacteria are unique among diazotrophs in that their photosynthetic metabolism produces molecular oxygen, which is a poison for the nitrogenase enzyme (Gallon, 1992). Thus, temporal separation provided an explanation for how unicellular diazotrophic cyanobacteria, unable to segregate nitrogen fixation to special cell types as the heterocystous strains do, could cope with these apparently incompatible processes.

The first report of temporal separation of nitrogen fixation and photosynthesis was related to culture age in *Gloeocapsa* grown in continuous light but no daily periodicity was evident (Gallon, LaRue & Kurz, 1974). A few years later it was demonstrated that *Gloeocapsa* fixes nitrogen daily, specifically during the dark periods, when grown in a diurnal cycle of 12 h light–12 darkness (LD 12:12) (Millineaux, Gallon & Chaplin, 1981). In contrast, *Anabaena cylindrica*, which segregates nitrogenase to specialized heterocysts and carries out oxygen-evolving photosynthesis in vegetative cells, fixes nitrogen preferentially in the light, when photosynthetically derived ATP and reductant for the nitrogen fixation reaction would be most abundant.

Diurnal alterations of nitrogen fixation and photosynthesis do not inherently suggest circadian control, as they could reflect direct regulation by environmental cues. However, additional experiments with several species suggested an endogenous rhythm that continues when environmental conditions change. For example, when *Gloeocapsa* grown in LD 12:12 was exposed to LD 16:8, nitrogenase activity continued to peak 8 h after the onset of the darkness; however, in the new light cycle the nitrogenase peak fell at the onset of the light period, rather than during the dark as it did in LD 12:12. The first strong evidence for an endogenous oscillation came from experiments with a non-heterocystous *Oscillatoria* species (filamentous), which not only exhibited nitrogenase activity exclusively during the dark period of an LD 16:8 cycle but continued to cycle when the cells were transferred to continuous light (LL) (Stal & Krumbein, 1985). Another characteristic of circadian rhythms can be seen in these data, in that the nitrogenase activity began to rise before the onset of darkness in the LD cycle, indicating the cells' ability to anticipate the environmental change.

Mitsui *et al.* (1986) reported that two strains of unicellular marine *Synechococcus* (spp. Miami BG 43511 and 43522) separate the processes of nitrogen fixation and photosynthesis into different phases of the cell division cycle. Each activity showed one peak per day, with photosynthetic oxygen evolution, carbohydrate synthesis, and nitrogen fixation each in a different phase. These patterns of metabolic activity continued when the cultures were transferred to LL. In these experiments the *Synechococcus* spp. doubled approximately once per 24 h, and the authors concluded that the metabolic oscillations were tied to specific phases of the cell cycle. More recent data from Mitsui and Suda (1995) showed that both nitrogenase activity and photosynthesis peak and trough daily in non-growing cultures, indicating that the cell cycle alone cannot account for the rhythms.

The first systematic test of the possibility that cyanobacterial oscillations of nitrogen fixation and photosynthesis might be timed by a circadian clock, like that described in eukaryotes, was conducted with a strain of freshwater *Synechococcus* called RF-1 (Grobbelaar *et al.*, 1986; Huang & Chow, 1986). This strain, like *Gloeocapsa*, restricted nitrogen fixation to the dark portion of an LD cycle (Huang & Chow, 1986). Transfer to LL after an adaptive period in LD resulted in persistence of the pattern with a period of about 24 h. Furthermore, the rhythms of both nitrogen fixation and amino acid uptake rates were temperature compensated, with approximate 24 h periods under constant temperatures in the range of 21°C–37 °C (Chen *et al.*, 1991: Huang & Chow, 1990). The phases of the rhythms could be reset by temperature or LD cycles, fulfilling the last of the three criteria for circadian phenomena (Pittendrigh, 1993).

The circadian alternation of nitrogen fixation and photosynthesis has since been documented ultrastructurally in *Cyanothece* sp. strain ATCC 51142 (Schneegurt *et al.*, 1994). This strain accumulates carbohydrate storage granules during photosynthetically active times which are depleted during nitrogen-fixing phases. The rhythmic appearance and disappearance of these subcellular granules, like the metabolic processes of photosynthesis and nitrogen fixation, persist independent of an LD cycle.

The planktonic cyanobacterium *Trichodesmium*, which contributes significantly to primary production in tropical oligotrophic waters, shows an endogenous rhythm of nitrogen fixation which is likely to be circadian-clock controlled (for review see Capone *et al.*, 1997). A temperature-compensated 24 h recurrence of cell division was reported in a marine strain of *Synechococcus*, WH7803, demonstrating that cellular functions other than nitrogen fixation and photosynthesis are under the control of the clock in cyanobacteria (Sweeney & Borgese, 1989).

An artificial 'hand of the clock'

Metabolic processes like nitrogen fixation and photosynthesis are tedious to assay for a circadian phenotype, which requires periodic sampling over several days. Because searching for mutants requires the examination of thousands of clones, a simpler circadian phenotype in cyanobacteria, whose measurement could be automated, was desirable for genetic studies. Kondo *et al.* (1993) developed a reporter strain (AMC149) of the transformable cyanobacterium, *Synechococcus* sp. strain PCC 7942, in which bioluminescence served as an observable circadian behaviour or 'hand of the clock'. The *luxAB* genes encoding the luciferase enzyme from *Vibrio harveyi* were fused to the *psbAI* gene, which encodes a photosystem II protein and is expressed strongly in the cyanobacterium (Li & Golden, 1993). When this reporter gene was integrated into the *Synechococcus* chromosome, the cells produced light in the presence of the luciferase substrate decanal. Bioluminescence from cells entrained to an LD cycle oscillated in LL with peaks (and troughs) at 24 h intervals; furthermore, the phase of the rhythm could be reset by dark pulses and was temperature compensated (Kondo *et al.*, 1993). A similar bioluminescent reporter strain of *Synechocystis* sp. strain PCC 6803 also showed rhythmic light production (Aoki, Kondo & Ishiura, 1995). Thus, bioluminescence provided a robust circadian phenotype for analysis of circadian rhythms in cyanobacteria, as had been demonstrated previously in eukaryotes for the naturally bioluminescent dinoflagellate *Gonyaulax* (Krasnow *et al.*, 1980; see Roenneberg & Rehman, this volume) and transgenic *Arabidopsis* plants that carry the firefly luciferase gene, *luc*, (Millar *et al.*, 1992). This opened the door for circadian research using model unicellular strains in which genetic manipulation is straightforward, which was not the case for the diazotrophs in which the clock was first recognized.

Identification of 'clock genes'

Huang and colleagues succeeded in isolating mutants affected in the nitrogenase activity and amino acid uptake rhythms of *Synechococcus* RF-1 (Huang, Wang & Grobbelaar, 1993). Two mutants were identified in which neither rhythm was evident, and two others in which uptake of an amino acid remained rhythmic but the nitrogenase rhythm was lost. The difficulty of transforming this strain has impeded the analysis of the genetic defects responsible for these phenotypes.

Kondo and Ishiura (1994) demonstrated that the bioluminescence rhythms of AMC149 could be observed from isolated colonies on agar plates. A screening device capable of monitoring bioluminescence repeatedly from a dozen plates, each of which can carry approximately 1000 colonies, was developed to search for circadian rhythm mutants (Kondo *et al.*, 1994). Mutants with periods ranging from 16 h to 60 h, and arhythmic clones, were identified following chemical mutagenesis of the reporter strain (Kondo *et al.*, 1994). Complementation of several mutants with DNA from wild-type *Synechococcus* identified a segment of the genome that carries three open reading frames (ORFs); sequence analysis of this region from the mutant strains revealed that a lesion in any of the three ORFs can cause an altered circadian phenotype (M. Ishiura, S. Aoki, S. Kutsuna, C. R. Andersson, H. Iwasaki, C. H. Johnson, S. S. Golden & T. Kondo, unpublished observations). It is likely that these ORFs encode proteins that are central to the circadian oscillator mechanism of *Synechococcus*.

Global circadian control of gene expression

In order to determine what proportion of the genome is controlled by the circadian clock in *Synechococcus* sp. strain PCC 7942, a promoterless *luxAB* gene set was fused randomly throughout the chromosome (Liu *et al.*, 1995). All resulting transformants that were bioluminescent, indicating fusion of the luciferase genes downstream of a promoter, showed a circadian rhythm of light production. Although most of the clones (approximately 80%) exhibited rhythms in the same phase and with the same waveform as AMC149, some showed bioluminescence patterns that peaked at different times or that rose and fell with different rates. One clone whose bioluminescence peaked in opposite phase to that of AMC149 was AMC287, in which *luxAB* was fused to the *purF* gene, which encodes an enzyme of the purine biosynthesis pathway (Liu *et al.*, 1995, 1996). The *purF* product glutamine PRPP aminotransferase, like nitrogenase, is sensitive to oxygen, suggesting that temporal separation from the peak of oxygen-evolving photosynthesis may be beneficial to the cell.

The ubiquity of rhythmicity among *luxAB* fusion clones suggested that the circadian clock globally controls transcription in *Synechococcus*, and that there may be no special *cis* information required for circadian expression of genes (Liu *et al.*, 1995). Consistent with this proposal, rhythmic bioluminescence could be observed from strains that carried *luxAB* fused to an *E. coli* consensus promoter (N. V. Lebedeva & S. S. Golden, unpublished data). However, the occurrence of some clones that show rhythms peaking in other phases indicates that there is some gene-specific control by the clock. The cyanobacterial circadian clock may impose a 'generic' rhythm on transcription, creating cycles of gene expression in a default phase unless additional information is present to modify the rhythms (Liu *et al.*, 1995). Genes expressed in unusual phases may carry *cis* information, recognized by specific factors, that alter the default rhythm.

An output pathway

Amplitude of the oscillation of gene expression is another feature of a circadian rhythm that can be influenced by specific factors. A mutant of AMC149 named M16, which shows a very low amplitude of the bioluminescence rhythm, was identified following mutagenesis by random integration of foreign DNA into the cyanobacterial chromosome (Tsinoremas *et al.*, 1996). The insertion was shown to lie within the *rpoD2* gene, which encodes a group 2 sigma 70-type transcription factor, and inactivation of this gene was demonstrated to be responsible for the mutant phenotype. Specifically, the sigma factor appeared to be involved in generating the trough of the cycle, as peak values of bioluminescence were similar to (or higher than) in AMC149.

When *rpoD2* was inactivated in reporter strains in which other promoters were fused to *luxAB*, most were not affected, and showed normal patterns of bioluminescence (Tsinoremas *et al.*, 1996). However, one other gene fusion, in which the *ndhD* gene drives expression of *luxAB*, also showed a low-amplitude phenotype when *rpoD2* was silenced. Thus, RpoD2 appears to be part of an output pathway that carried information for generating the trough to a subset of *Synechococcus* genes.

Input pathways

Light, and its cycle with darkness, is the dominant pathway for setting the circadian clock in diverse organisms (Dunlap, 1996; Pittendrigh, 1981). In plants, genes under control of the red-sensing photoreceptor phytochrome are responsive to light only during certain phases of the circadian cycle

(Millar & Kay, 1996); conversely, phytochrome and other plant photoreceptors influence the clock, suggesting that the signal transduction pathways for light-regulated gene expression and input to the circadian clock are closely intertwined (Anderson & Kay, 1996). A cyanobacterial phytochrome has been identified in *Synechocystis* sp. strain PCC 6803 which shows the red/far-red photoreversible absorption spectra exhibited by the plant photoreceptor (Hughes *et al.*, 1997), and a gene whose product is similar to phytochrome was shown to be involved in chromatic adaptation of *Fremyella diplosiphon* (Kehoe & Grossman, 1996). Some photosynthesis genes are regulated by light in *Synechococcus* sp. strain PCC 7942, including the *psbAI* gene that drives *luxAB* in the bioluminescent reporter strain AMC149 (Golden, 1995). These genes respond to a change in either the intensity or quality of incident light, showing a response to low-fluence blue light that is similar to the effect of high-fluence white light; red light can attenuate the response to blue light (Tsinoremas, Schaefer & Golden, 1994). The possible role of phytochrome or other photoreceptors in setting the circadian clock has not yet been addressed but these data suggest several avenues of investigation.

THE CLOCK AND THE CELL DIVISION CYCLE

Initial reports of circadian rhythms in cyanobacteria were based on cultures growing with a variety of doubling times, including once per 24 h (Mitsui *et al.*, 1986), more slowly than once a day (Sweeney & Borgese, 1989) or in the absence of cell division (Huang & Pen, 1994; Kondo *et al.*, 1994; Mitsui & Suda, 1995). Research in eukaryotic organisms had suggested that a circadian clock could not operate under conditions in which cells double faster than once per day (Ehret & Wille, 1970; Pittendrigh, 1993). Because cyanobacteria can grow at a rate of several doublings per day, these organisms provided an opportunity to question experimentally whether the cell can keep track of the circadian timing circuit when several generations arise during one cycle. Kondo *et al.* (1996) showed that AMC149 growing exponentially with a 12 h doubling time produced an increase in oscillating bioluminescence consistent with a circadian rhythm in every cell and exponential growth of the culture. Transcript abundance from the *psbAII* gene of the parent strain, *Synechococcus* sp. strain PCC 7942, oscillated with a 24 h rhythm under high-light conditions, in which the doubling time was once per 5–6 h.

These data showed that circadian rhythms can persist regardless of the timing of the cell division. However, other experiments indicated that the cell division cycle is influenced by the clock (Mori, Binder & Johnson, 1996). Certain phases of the circadian cycle appear to forbid cell division in continuously growing cultures. Although the persistence of the circadian clock in rapidly dividing cyanobacterial cells differs from what was predicted from eukaryotic organisms, the influence of the clock over the timing of the cell division cycle has been demonstrated in both (Edmunds, 1988; Mori *et al.*, 1996).

CONCLUDING REMARKS

Several recent advances came together to make this an exciting time for studying circadian rhythms in cyanobacteria: the demonstration that rhythms can be monitored easily in genetically malleable strains (Aoki *et al.*, 1995; Kondo *et al.*, 1994), the completion of the sequence of an entire cyanobacterial genome (Kaneko *et al.*, 1996) and the identification of several new clock genes from eukaryotes (Crosthwaite *et al.*, 1997; Gekakis *et al.*, 1995; King *et al.*, 1997; Myers *et al.*, 1995). The products of these genes share a motif called PAS (Kay, 1997*a*,*b*), which is not encoded by the genes that complement the *Synechococcus* circadian mutants (M. Ishiura, S. Aoki, S. Kutsuna, C. R. Andersson, H. Iwasaki, C. H. Johnson, S. S. Golden & T. Kondo, unpublished observations). This raises the possibility that the cyanobacterial clock is not homologous to the circadian oscillator of animals. The range of clock components available for comparison will be expanded soon, when plant genes related to the circadian mechanism are characterized (Millar *et al.*, 1995). Given the special relationship between cyanobacteria and plants (Woese, 1987), the comparison of clock genes between these two groups may be particularly enlightening in discovering the origin(s) of the circadian biological clock.

ACKNOWLEDGEMENTS

The authors' research has been supported by grants from the NIMH (MH43836 and MH01179 to CHJ), NIH (GM37040 to SSG), and NSF (MCB-9219880 and MCB 9633267 to CHJ; MCB-9311352 and MCB-9513367 to SSG); and from the Ministry of Education, Science and Culture (07670097, 07554045, 07253228, 07558103, 08404053, and 08454244 to TK and MI). NSF and JSPS supported a joint USA–Japan Co-operative Programme grant for travel among the four collaborators (NSF INT9218744 and JSPS BSAR382). The Human Frontier Science Program funds collaborative research of the four co-authors.

REFERENCES

Anderson, S. L. & Kay, S. A. (1996). Illuminating the mechanism of the circadian clock in plants. *Trends in Plant Science*, **1**, 51–7.
Aoki, S., Kondo, T. & Ishiura, M. (1995). Circadian expression of the *dnaK* gene in the cyanobacterium *Synechocystis* sp. strain PCC 6803. *Journal of Bacteriology*, **177**, 5605–11.
Aschoff, J. (1989). Temporal orientation: circadian clocks in animals and humans. *Animal Behaviour*, **37**, 881–96.

Bargiello, T. A., Jackson, F. R. & Young, M. W. (1984). Restoration of circadian behavioural rhythms by gene transfer in *Drosophila*. *Nature*, **312**, 752–4.

Brock, M. A. (1991). Chronobiology and aging. *Geriatric Bioscience*, **39**, 74–91.

Bünning, E. (1960). Circadian rhythms and the time measurement in photoperiodism. *Cold Spring Harbor Symposium on Quantitative Biology*, **25**, 249–56.

Capone, D. G., Zehr, J. P., Paerl, H. W., Bergman, B. & Carpenter, E. J. (1997). *Trichodesmium*, a globally significant marine cyanobacterium. *Science*, **276**, 1221–9.

Chen, T-H., Chen, T-L., Hung, L-M. & Huang, T-C. (1991). Circadian rhythm in amino acid uptake by *Synechococcus* RF-1. *Plant Physiology*, **97**, 55–9.

Crosthwaite, S. K., Dunlap, J. C. & Loros, J. J. (1987). *Neurospora wc-1* and *wc-2*: transcription, photoresponses, and the origins of circadian rhythmicity. *Science*, **276**, 763–9.

Czeisler, C. A., Johnson, M. P., Duffy, J. F., Brown, E. N., Ronda, J. M. & Kronquer, R. E. (1990). Exposure to bright light and darkness to treat physiologic maladaptation to night work. *New England Journal of Medicine*, **322**, 1253–9.

Daan, S. (1981). Adaptive daily strategies in behavior. In *Biological Rhythms*, ed. J. Aschoff, pp. 275–98. New York: Plenum.

DeCoursey, P. J. (1986). Light-sampling behavior in photoentrainment of a rodent circadian rhythm. *Journal of Comparative Physiology*, **159**, 161–9.

Dunlap, J. C. (1996). Genetic and molecular analysis of circadian rhythms. *Annual Review of Genetics*, **30**, 579–601.

Edmunds, L. N. (1988). *Cellular and Molecular Bases of Biological Clocks*. New York: Springer-Verlag.

Ehret, C. F. & Wille, J. J. (1970). The photobiology of circadian rhythms in protozoa and other eukaryotic microorganisms. In *Photobiology of Microorganisms*, ed. P. Halldal, pp. 369–416. New York: Wiley.

Gallon, J. R. (1992). Tansley Review No. 44. Reconciling the incompatible: N_2 fixation and O_2. *New Phytology*, **122**, 571–609.

Gallon, J. R., LaRue, T. A. & Kurz, W. G. W. (1974). Photosynthesis and nitrogenase activity in the blue-green alga *Gloeocapsa*. *Canadian Journal of Microbiology*, **20**, 1633–7.

Gekakis, N., Saez, L., Delahaye-Brown, A-M., Myers, M. P., Sehgal, A., Young, M. W. & Weitz, C. J. (1995). Isolation of *timeless* by PER protein interaction: defective interaction between *timeless* protein and long-period mutant PER^L. *Science*, **270**, 811–15.

Golden, S. S. (1995). Light-responsive gene expression in cyanobacteria. *Journal of Bacteriology*, **177**, 1651–4.

Grobbelaar, N., Huang, T-C., Lin, H-Y. & Chow, T. J. (1986). Dinitrogen-fixing endogenous rhythm in *Synechococcus* RF-1. *FEMS Microbiology Letters*, **37**, 173–7.

Healy, D. (1987). Rhythm and blues: neurochemical, neuropharmacological and neuropsychological implications of a hypothesis of a circadian rhythm dysfunction in the affective disorders. *Psychopharmacology*, **93**, 271–85.

Huang, T-C. & Chow, T-J. (1986). New type of N_2-fixing unicellular cyanobacterium (blue-green alga). *FEMS Microbiology Letters*, **36**, 109–10.

Huang, T-C. & Chow, T-J. (1990). Characterization of the rhythmic nitrogen-fixing activity of *Synechococcus* sp. RF-1 at the transcription level. *Current Microbiology*, **20**, 23–6.

Huang, T-C. & Pen, S-Y. (1994). Induction of a circadian rhythm in *Synechococcus* RF-1 while the cells are in a 'suspended state'. *Planta*, **194**, 436–8.

Huang, T-C., Wang, S-T. & Grobbelaar, N. (1993). Circadian rhythm mutants of the prokaryotic *Synechococcus* RF-1. *Current Microbiology*, **27**, 249–54.

Hughes, J., Lamparter, T., Mittmann, F., Hartmann, E., Gärtner, W., Wilde, A. & Börner, T. (1997). A prokaryotic phytochrome. *Nature*, **386**, 663.

Johnson, C. H. & Hastings, J. W. (1986). The elusive mechanism of the circadian clock. *American Scientist*, **74**, 29–36.

Kaneko, T., Sato, S., Kotani, H., Tanaka, A., Asamizu, E., Nakamura, Y., Miyajima, N., Hirosawa, M., Sugiura, M., Sasamoto, S., Kimura, T., Hosouchi, T., Matsuno, A., Muraki, A., Nakazaki, N., Naruo, K., Okumura, S., Shimpo, S., Takeuchi, C., Wada, T., Watanabe, A., Yamada, M., Yasuda, M. & Tabata, S. (1996). Sequence analysis of the genome of the unicellular cyanobacterium *Synechocystis* sp. strain PCC 6803. II. Sequence determination of the entire genome and assignment of potential protein-coding regions. *DNA Research*, **3**, 185–209.

Kay, S. A. (1997*a*). As time PASses: the first mammalian clock gene. *Science*, **276**, 1093.

Kay, S. A. (1997*b*). PAS, present, and future: clues to the origins of circadian clocks. *Science*, **276**, 753.

Kehoe, D. & Grossman, A. R. (1996). Similarity of a chromatic adaptation sensor to phytochrome and ethylene receptors. *Science*, **273**, 1409–12.

King, D. P., Zhao, Y., Sangoram, A. M., Wilsbacher, L. D., Tanaka, M., Antoch, M. P., Steeves, T. D. L., Vitaterna, M. H., Kornhauser, J. M., Lowrey, P. L., Turek, F. W. & Takahashi, J. S. (1997). Positional cloning of the mouse circadian *clock* gene. *Cell*, **89**, 641–53.

Kondo, T. & Ishiura, M. (1994). Circadian rhythms of cyanobacteria: monitoring the biological clocks of individual colonies of bioluminescence. *Journal of Bacteriology*, **176**, 1881–5.

Kondo, T., Strayer, C. A., Kulkarni, R. D., Taylor, W., Ishiura, M., Golden, S. S. & Johnson, C. H. (1993). Circadian rhythms in prokaryotes: luciferase as a reporter of circadian gene expression in cyanobacteria. *Proceedings of the National Academy of Sciences, USA*, **90**, 5672–6.

Kondo, T., Tsinoremas, N. F., Golden, S. S., Johnson, C. H., Kutsuna, S. & Ishiura, M. (1994). Circadian clock mutants of cyanobacteria. *Science*, **266**, 1233–6.

Kondo, T., Mori, T., Lebedeva, N. V., Aoki, S., Ishiura, M. & Golden, S. S. (1996). Circadian rhythms in rapidly dividing cyanobacteria. *Science*, **275**, 224–7.

Krasnow, R., Dunlap, J. C., Taylor, W., Hastings, J. W., Vetterling, W. & Gooch, V. (1980). Circadian spontaneous bioluminescent glow and flashing of *Gonyaulax polyedra*. *Journal of Comparative Physiology*, **138**, 19–26.

Lewy, A. J., Sack, R. L., Miller, L. S. & Hoban, T. M. (1987). Antidepressant and circadian phase-shifting effects of light. *Science*, **235**, 352–4.

Li, R. & Golden, S. S. (1993). Enhancer activity of light-responsive regulatory elements in the untranslated leader regions of cyanobacterial *psbA* genes. *Proceedings of the National Academy of Sciences, USA*, **90**, 11678–82.

Liu, Y., Tsinoremas, N. F., Johnson, C. H., Lebedeva, N. V., Golden, S. S., Ishiura, M. & Kondo, T. (1995). Circadian orchestration of gene expression in cyanobacteria. *Genes and Development*, **9**, 1469–78.

Liu, Y., Tsinoremas, N. F., Golden, S. S., Kondo, T. & Johnson, C. H. (1996). Circadian expression of genes involved in the purine biosynthetic pathway of the cyanobacterium *Synechococcus* sp. strain PCC 7942. *Molecular Microbiology*, **20**, 1071–81.

McClung, C. R., Fox, B. A., Dunlap, J. C., Jackson, F. R. & Young, M. W. (1989). The *Neurospora* clock gene *frequency* shares a sequence element with the *Drosophila* clock gene *period*. *Nature*, **339**, 558–62.

Millar, A. J. & Kay, S. A. (1996). Integration of circadian and phototransduction pathways in the network controlling *CAB* gene transcription in *Arabidopsis*. *Proceedings of the National Academy of Sciences, USA*, **93**, 15491–6.

Millar, A. J., Short, S. R., Chua, N-H. & Kay, S. A. (1992). A novel circadian phenotype based on firefly luciferase expression in transgenic plants. *Plant Cell*, **4**, 1075–87.

Millar, A. J., Carré, I. A., Strayer, C. A., Chua, N-H. & Kay, S. A. (1995). Circadian clock mutants in *Arabidopsis* identified by Luciferase imaging. *Science*, **267**, 1161–6.

Millineaux, P. M., Gallon, J. R. & Chaplin, A. E. (1981). Acetylene reduction (nitrogen fixation) by cyanobacteria grown under alternating light–dark cycles. *FEMS Microbiology Letters*, **10**, 245–7.

Mitsui, A. & Suda, S. (1995). Alternative and cyclic appearance of H_2 and O_2 photoproduction activities under non-growing conditions in an aerobic nitrogen-fixing unicellular cyanobacterium *Synechococcus* sp. *Current Microbiology*, **30**, 1–6.

Mitsui, A., Kumazawa, S., Takahashi, A., Ikemoto, H., Cao, S. & Arai, T. (1986). Strategy by which nitrogen-fixing unicellular cyanobacteria grow photoautotrophically. *Nature*, **323**, 720–2.

Moore-Ede, M., Czeisler, C. & Richardson, G. S. (1983). Circadian timekeeping in health and disease. *New England Journal of Medicine*, **309**, 469–76 and 530–6.

Mori, T., Binder, B. & Johnson, C. H. (1996). Circadian gating of cell division in cyanobacteria growing with average doubling times of less than 24 hours. *Proceedings of the National Academy of Sciences, USA*, **93**, 10183–8.

Muller, J. E. & Tofler, G. H. (1991). Circadian variation and cardiovascular disease. *New England Journal of Medicine*, **325**, 1038–9.

Myers, M. P., Wager-Smith, K., Wesley, C. S., Young, M. W. & Sehgal, A. (1995). Positional cloning and sequence analysis of the *Drosophila* clock gene, *timeless*. *Science*, **270**, 805–8.

Pittendrigh, C. S. (1981). Circadian systems: general perspective and entrainment. In *Handbook of Behavioral Neurobiology: Biological Rhythms*, ed. J. Aschoff, pp. 57–80 and 95–124. New York: Plenum.

Pittendrigh, C. S. (1993). Temporal organization: reflections of a Darwinian clock-watcher. *Annual Review of Physiology*, **55**, 17–54.

Reddy, P., Zehring, W. A., Wheeler, D. A., Pirrotta, V., Hadfield, C., Hall, J. C. & Rosbash, M. (1984). Molecular analysis of the period locus in *Drosophila melanogaster* and identification of a transcript involved in biological rhythms. *Cell*, **44**, 21–32.

Reppert, S. M. (1989). *Development of Circadian Rhythmicity and Photoperiodism in Mammals*. Perinatology Press.

Satlin, A., Teicher, M. H., Lieberman, H. R., Baldessavini, R. J., Volicer, L. & Rheaume, Y. (1991). Circadian locomotor activity rhythms in Alzheimer's Disease. *Neuropsychopharmacology*, **5**, 115–26.

Schneegurt, M. A., Sherman, D. M., Nayar, S. & Sherman, L. A. (1994). Oscillating behavior of carbohydrate granule formation and dinitrogen fixation in the cyanobacterium *Cyanothece* sp. strain ATCC 51142. *Journal of Bacteriology*, **176**, 1586–97.

Stal, L. J. & Krumbein, W. E. (1985). Nitrogenase activity in the non-heterocystous cyanobacterium *Oscillatoria* sp. grown under alternating light–dark cycles. *Archives of Microbiology*, **143**, 67–71.

Strogatz, S. H., Kronauer, R. E. & Czeisler, C. A. (1987). Circadian pacemaker interferes with sleep onset at specific times of each day: role in insomnia. *American Journal of Physiology*, **253**, R172–8.

Sweeney, B. M. (1987). *Rhythmic Phenomena in Plants*. San Diego: Academic Press.

Sweeney, B. M. & Borgese, M. B. (1989). A circadian rhythm in cell division in a prokaryote, the cyanobacterium *Synechococcus* WH7803. *Journal of Phycology*, **25**, 183–6.

Tajima, N., Uematsu, M., Asukata, I., Yamamoto, K., Sasaki, M. & Hokary, M. (1991). Recovery of circadian rhythm of plasma cortisol levels after a 3-day trip between Tokyo and San Francisco. *Aviation Space and Environmental Medicine*, **62**, 325–7.

Tsinoremas, N. F., Ishiura, M., Kondo, T., Andersson, C. R., Tanaka, K., Takahashi, H., Johnson, C. H. & Golden, S. S. (1996). A sigma factor that modifies the circadian expression of a subset of genes in cyanobacteria. *EMBO Journal*, **15**, 2488–95.

Tsinoremas, N. F., Schaefer, M. R. & Golden, S. S. (1994). Blue and red light reversibly control *psbA* expression in the cyanobacterium *Synechococcus* sp. strain PCC 7942. *Journal of Biological Chemistry*, **269**, 16143–7.

Van Cauter, E., Linkowski, P., Kerhofs, M., Hubain, P., L-Hermite-Baleriaux, M., Leelerq, R., Brasseur, M., Copinschi, G. & Mendlewicz, J. (1991). Circadian and sleep-related endocrine rhythms in schizophrenia. *Archives of General Psychiatry*, **48**, 348–56.

Willich, S. N., Levy, D., Rocco, M. B., Tofler, G. H., Stone, P. H. & Muller, J. E. (1987). Circadian variation in the incidence of sudden cardiac death in the Framingham Heart Study Population. *American Journal of Cardiology*, **60**, 801–6.

Woese, C. R. (1987). Bacterial evolution. *Microbiological Reviews*, **51**, 221–71.

SURVIVAL IN A TEMPORAL WORLD – THE CIRCADIAN PROGRAMME OF THE MARINE UNICELL *GONYAULAX*

TILL ROENNEBERG AND JALEES REHMAN

Institute for Medical Psychology, Chronobiology Group, Ludwig-Maximilians University, Goethestr. 31, D-80336 München 2, Germany

INTRODUCTION

Ecological research tries to understand the complex interactions between organisms as well as their interaction with the environment but it predominantly investigates the world's spatial aspects, its biotopes and niches. However, evolution, adaptation and interactions between organisms operate at the temporal level. The sun and the moon create a temporal environment on our planet which is as predictable in its qualities as the spatial world. The four cyclic structures: tides, day, lunar month and year, offer niches (chronotopes) to which organisms can adapt and in which they can specialize, and could thus also be described as temporal spaces (Roenneberg, 1992). The main difference between the ecology of the spatial and temporal structures lies in the fact that organisms can fully specialize to their specific spatial niche and may never leave it, while they always have to live through all the temporal niches of the respective temporal structure (e.g. dawn, day, dusk, and night). A prerequisite for the biological strategies which have developed to cope with environment appears to be an endogenous representation of the specific external structure. On the temporal level, this task is accomplished by biological clocks or temporal programmes (Pittendrigh, 1993). These are endogenous mechanisms which oscillate, even under constant experimental conditions, with a period length close to the one of the cyclic temporal space.

Our insights into biological clocks are especially advanced concerning the endogenous representation of the 24 h day. Circadian (about one day) systems enable organisms to schedule functions at all organismic levels, such as activity and sleep, feeding and drinking, gene expression and enzyme activity, into the appropriate temporal niches of the solar day. A minimal concept of the mechanisms behind the circadian system involves (i) an endogenous biochemical oscillator which is capable of producing an oscillation in the 24 h range, (ii) an input pathway by which the oscillator can be entrained with the outside world (responding to environmental signals called *zeitgeber*), and (iii) output pathways by which the oscillator controls the various recordable circadian rhythms.

The capability to produce a circadian oscillation appears to be implemented at the cellular level, even in higher animals (Takahashi, 1987; Cahill & Besharse, 1993; Michel *et al.*, 1993; Pickard & Tang, 1993, 1994; Welsh *et al.*, 1995). The unicellular marine dinoflagellate *Gonyaulax polyedra* provides an excellent model system for circadian research both at the organismal and at the cellular level. To fully understand the mechanisms of circadian programmes, it is helpful to investigate as many different circadian rhythms as possible. In *Gonyaulax*, many physiological functions oscillate on a circadian basis (Hastings, Rusak & Boulos, 1991; Roenneberg & Morse, 1994; Roenneberg & Mittag, 1996) and several of these can be recorded for many weeks in constant conditions (Roenneberg, Merrow & Eisensamer, 1997). The list of rhythmic outputs of the *Gonyaulax* circadian system is growing rapidly, especially concerning the activity and abundance of enzymes and other proteins; the concentration of many substrates and metabolites is also found to be rhythmic. The most important rhythms are summarized in Fig. 1.

The information gathered in more than 40 years of research into the circadian system of *Gonyaulax* has shown that this temporal programme is controlled by highly complex mechanisms involving several oscillators and feedback loops at the system's input and output side (Roenneberg, 1996; Roenneberg & Mittag, 1996).

THE TEMPORAL ENVIRONMENT OF *GONYAULAX*

To understand the circadian programme of an organism, it helps to know the specific temporal framework of its environment (and its resources) to which it has adapted. Like other phytoplankton, *Gonyaulax* cells migrate vertically over a great distance. Some dinoflagellates sink and rise more than 33 m between the surface and the bottom of the ocean every day (J. Brenner, personal communication) which corresponds to a distance two million times their diameter. The onsets of both rise and decline anticipate rather than are triggered by the environmental light changes. During the day, *Gonyaulax* cells form dense swarms (aggregations) at the ocean's surface, where they photosynthesize, while they take advantage of the higher nutrient concentrations at greater depth during the night (Riley & Skirrow, 1965; Harrison, 1979). Due to the differential light absorption of seawater (Lythgoe, 1979), the cells migrate through gradients of both intensity and spectral quality. In addition to these photic gradients, they encounter gradients of temperature and nutrients. All three of these resource gradients – light, temperature, and nutrients – can act as *zeitgeber* (i.e. are relevant input signals for the circadian programme).

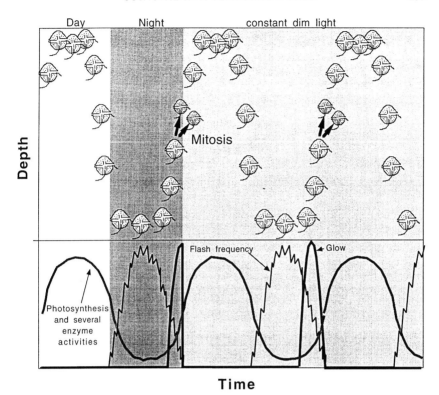

Fig. 1. Circadian rhythms in the marine unicell *Gonyaulax polyedra*. The timing of the daily vertical rise and decline and the phase of mitosis is shown schematically in the top part of the figure, first under light:dark conditions (on the left) and then in constant dim light. The bottom of the graph indicates, under the same light conditions, the rhythms of photosynthesis and several enzyme activities as well as the two forms of bioluminescence, flashing and glowing. Over the course of the night, bioluminescence can be recorded in two forms: the frequency of brief, action-potential-like flashes reaches a maximum around midnight, while the level of a sustained glow reaches its maximum at the end of the night (redrawn after Roenneberg & Morse, 1994).

ENTRAINMENT BY LIGHT

Light is by far the best understood *zeitgeber* for all circadian systems (Roenneberg & Foster, 1997). It can affect both the period and the phase of the circadian oscillator. Different intensities of constant light lead to different period lengths (Aschoff, 1979) and single light pulses can shift the oscillation (Pittendrigh, 1981). Direction and amount of these phase shifts are a function of the time of day when the light pulses are given (phase response curve, PRC). Experiments with *Gonyaulax* have shown that light can reach the oscillator via at least two separate input pathways, one of which is itself under circadian control. An action spectrum for phase shifting by light pulses has shown that the *Gonyaulax* circadian system responds both to short and to long wavelength light (with maxima at 475 and 650 nm; Hastings & Sweeney, 1960). In photosynthesizing organisms, the combined action of both red and blue light is generally interpreted as an involvement of either a chlorophyll or a phytochrome. The action of phytochromes has not been shown in *Gonyaulax* but experiments with the photosynthesis inhibitor Diuron indicate that a chlorophyll may be involved in circadian light responses (Johnson & Hastings, 1989; Roenneberg, 1994).

More detailed experiments revealed the involvement of separate light input pathways into the *Gonyaulax* circadian system. A first indication for more than one photoreceptor was found when the period's dependence on the light's fluence rate was measured under different spectral light conditions (Roenneberg & Hastings, 1988). Increasing fluence rates of short wavelength light shorten the period, while an increase of red light lengthens it. The different period dependencies for red and blue light also predict different phase responses to light pulses in the appropriate wavelengths. When red or blue light pulses are given during the subjective day (the phase of the circadian cycle which corresponds to the light phase under entrained conditions) and the first few hours of the subjective night, both light pulse types elicit indistinguishable phase responses. However, when the two light pulse types are given during the remaining part of the subjective night, blue pulses greatly advance the rhythm (i.e. all light pulses given during that part of the cycle reset the system to subjective dawn) while red pulses still lead to considerable delays (Roenneberg & Deng, 1997). The light input pathway which largely advances in the subjective night is mainly blue sensitive, the other is both red and blue sensitive, delaying the rhythm throughout most of the cycle. Even when high fluences are applied ($> 100 \mu$mol m^{-2}s^{-1}), the relatively small responses to red light can not be changed into the strong advances of the blue sensitive system (leading to phase shifts of more than 12 h already at 5μmol m^{-2}s^{-1}). These physiological experiments not only prove the existence of two separate light inputs but also show that the blue-sensitive light input involves feedback mechanisms, i.e. that it is itself under circadian control.

The two light input pathways can also be distinguished pharmacologically. The mammalian phosphagen, creatine, shortens the circadian period in constant white or blue but not in red light (Roenneberg, Nakamura & Hastings, 1988). This shortening in constant light is due to the fact that creatine increases the sensitivity of the blue sensitive phase shifting mechanism (Roenneberg & Taylor, 1994). Allopurinol, an inhibitor of xanthine oxidase, is another chemical that can help to dissect the light input of the *Gonyaulax* circadian system. It specifically inhibits the effects of white and blue light pulses given in the subjective night and does not change the effect of red or blue pulses during the subjective day (Deng & Roenneberg, 1997). Again, allopurinol does not only affect the responses to single light pulses but shows also effects in constant light, lengthening the period in blue or white light but not in constant red light.

The molecular mechanisms of light effects on the *Gonyaulax* circadian system are still not known, but they apparently involve protein phosphorylation because the kinase inhibitor 6-dimethylaminopurine (DMAP) also blocks the effect of light pulses. But it does so in any circadian phase and is independent of the spectral composition of the light pulse (Comolli, Taylor & Hastings, 1994). Thus, DMAP appears to act more downstream in the light transduction than creatine and allopurinol. We still do not know how creatine acts biochemically. In mammals, it is known as a transporter for high energy phosphate groups but its role or even its existence in plants (including *Gonyaulax*) has not been shown (Roenneberg *et al.*, 1991). In contrast, xanthine oxidase, the enzyme which is inhibited by allopurinol (both *in vivo* and *in vitro*) appears to be part of the *Gonyaulax* biochemistry. Its activity can actually be measured in cell extracts and it is also under circadian control, showing a 15-fold difference over the course of one cycle (T-S. Deng and T. Roenneberg, unpublished results).

ENTRAINMENT BY NITRATE

Although light is by far the most investigated *zeitgeber*, circadian systems also respond to 'non-photic' signals. The temporal programmes of poikilotherms and plants, respond to temperature; so does the *Gonyaulax* circadian system (Sweeney & Hastings, 1960). Acoustic signals (Gwinner, 1966*a*; Wever, 1979), restricted feeding (Aschoff, 1987; Jilge, 1991; Kennedy, Coleman & Armstrong, 1991; Chabot & Menaker, 1992; Hau & Gwinner, 1992; Stephan 1992*a*, *b*) or even scheduled activity (Mrosovsky *et al.*, 1992; Van Reeth *et al.*, 1994; Fleissner & Fleissner, 1996) can also entrain circadian systems. It has recently been shown that the *Gonyaulax* circadian system is not only affected by light and temperature but also by a nutrient. In testing the components of natural seawater and of the cells' laboratory culture medium (Guillard & Ryther, 1962), it was found that only nitrate has *zeitgeber* properties (Roenneberg & Rehman, 1996). Its addition to the medium of nitrate-depleted cells shifts the phase and changes period and amplitude. The components of the transduction pathway mediating these effects appear to use, at least partially, the known steps of nitrogen assimilation (Kleinhofs & Warner, 1990).

These findings are extremely useful for the dissection of the *Gonyaulax* circadian system because, unlike the case of light, the nitrate transduction pathway is experimentally far more accessible. Experiments show that, as in the case of the light, elements of the nitrate transduction cascade are themselves under circadian control. Both nitrate uptake (T. Roenneberg & M. Merrow, unpublished observations) and the activity of the transduction's first enzyme, nitrate reductase (Ramalho, Hastings & Colepicolo, 1995), vary considerably over the course of a circadian cycle. The subsequent step in nitrate assimilation, the reduction of nitrite to ammonium, catalyzed by nitrite reductase, appears to be also under circadian control (B. Eisensamer & T. Roenneberg, unpublished observations). In contrast, the activity of glutamine synthetase, a subsequent enzyme in the pathway, does not vary over the circadian cycle (B. Eisensamer & T. Roenneberg, unpublished observations). Current experiments indicate that transduction of the 'nitrate' signal to the circadian system also involves the photorespiratory system, specifically the peroxisomal biochemistry (T. Roenneberg & Merrow, unpublished observations).

Biochemical and physiological experiments on the effects of light and nitrate show that both *zeitgeber* have strong mutual interactions. The amount of phase shifting by light depends on the available nitrate concentration and, inversely, the degree by which nitrate affects the circadian oscillator depends on the quantity and the quality of the constant light (T. Roenneberg, K. Lindgren & J. Rehman, unpublished observations). The circadian effects of several substrates and inhibitors of the nitrate transduction pathway are similarly light dependent (T. Roenneberg & Merrow, unpublished observations).

THE INVOLVEMENT OF MORE THAN ONE OSCILLATOR

One of the important questions in circadian biology is how many oscillators orchestrate the temporal daily programme. An indication for more than one came from the finding that different output rhythms can run with more or less independent periodicities under constant conditions (internal desynchronization). The most prominent example for internal desynchronization is the uncoupling of the temperature rhythm and the sleep/wake cycle in humans (Aschoff, 1965; Aschoff, Gerecke & Wever, 1967). Other examples have been found in animals (Hoffman, 1971; Boulos & Rusak, 1982; Honma, Honma & Hiroshige, 1986; Schardt, Wilhelm & Erkert, 1989; Meijer *et al.*, 1990; Honma, Honma & Hiroshige, 1991; Ebihara & Gwinner, 1992; Tosini & Menaker, 1995, 1996). Recent results show that even single cells of the circadian pacemaker in mammals, the suprachiasmatic nucleus (SCN) can have independent periods in tissue culture (Shinohara *et al.*, 1995; Welsh *et al.*, 1995). It was thought, however, that the circadian programme of single cells was driven by only a single oscillator (McMurry & Hastings, 1972).

One of the great advantages of the *Gonyaulax* model system, the possibility to record several different circadian rhythms simultaneously (Roenneberg *et al.*, 1997), has led to the discovery of at least two circadian oscillators within one cell (Roenneberg & Morse, 1993). When *Gonyaulax* cultures are kept in constant red light (Roenneberg & Morse, 1993) or when nitrate is added to cells in bright white light (Roenneberg & Rehman, 1996; Roenneberg *et al.*, 1997), the rhythms of bioluminescence and aggregation can desynchronize, showing the existence of two circadian oscillators within one cell controlling aggregation (A-oscillator) and bioluminescence (B-oscillator). This uncoupling is due to different sensitivities of the underlying oscillators to light and nutrients (Morse, Hastings and Roenneberg, 1994; Roenneberg & Rehman, 1996). The B-oscillator is mainly sensitive to blue light and strongly reacts to nitrate steps or pulses, while the A-oscillator also responds to changes of long wavelength light, but shows only small responses to nitrate.

Detailed experiments investigating different aspects of the *Gonyaulax* behaviour, the rhythms of vertical migration, swarm-formation (aggregations), and self-selection of light intensities, showed that this complex behaviour is also controlled by different circadian oscillators (Eisensamer & T. Roenneberg, unpublished observations). When cultures are kept in horizontal light gradients, the daytime aggregations accumulate at locations of preferred intensities (Roenneberg & Hastings, 1992); the preferred light intensity changes systematically over the course of a circadian day, as already shown for birds (Gwinner, 1966*b*). Only this phototactic aspect of the aggregation phenomenon appears to be controlled by the A-oscillator, while the rhythm of vertical migration, together with bioluminescence, is driven by the B-oscillator (Roenneberg *et al.*, 1997).

CONTROL OF OUTPUT RHYTHMS

Theoretically, the circadian pacemaker can use all the available forms of cellular regulation – transcription, translation, and post-translational modifications – to control any of its output variables ('hands of the clock'). In *Neurospora* (Loros & Dunlap, 1991; Dunlap *et al.*, this volume), cyanobacteria (Kondo *et al.*, 1993; Liu *et al.*, 1995; Golden *et al.*, this volume), some algae (Jacobshagen & Johnson, 1994), and higher plants (Piechulla & Gruissem, 1987; Meyer, Thienel & Piechulla, 1989; Millar & Kay, 1991; Oberschmidt, Hücking & Piechulla, 1995) the various characterized outputs are all transcriptionally controlled. In contrast, the outputs in *Gonyaulax* appear to be controlled on the translational level. These differences may, however, be due to the respective methods applied and not so much to a species-specific specialization. The experiments in *Neurospora* specifically looked for clock-controlled genes (*ccg*) before even knowing their biological function (Loros & Dunlap, 1991; Dunlap *et al.*, this volume), while the bioluminescence rhythm in *Gonyaulax*, one of the best understood circadian output pathways, was investigated because one wanted to understand the rhythmic control of a known biological function. All the components of the bioluminescence system are under circadian control (Nicolas *et al.*, 1987; Fritz, Morse & Hastings, 1990; Morse, Fritz & Hastings, 1990): the enzyme luciferase, its substrate luciferin, and a luciferin-binding protein (LBP) (Fogel & Hastings, 1971; Morse *et al.*, 1989*a*; Morse, Pappenheimer & Hastings, 1989*c*; Nicolas *et al.*, 1991; Desjardins & Morse, 1993). While the binding protein is synthesized at the beginning of the night and degraded before dawn, its mRNA is constitutively expressed (Dunlap & Hastings, 1981; Johnson, Roeber & Hastings, 1984; Morse *et al.*, 1989*b*). Thus, the circadian clock controls this rhythm at the translational level. Recent experiments have shown that this translational control acts on a short sequence in the 3′ untranslated region of the *lbp* mRNA. An inhibitory protein-factor, circadian controlled translational regulator (CCTR) (Mittag, Lee & Hastings, 1994, 1995; Mittag & Hastings, 1996), binds to this *cis*-element more strongly during the day, preventing translation of the existing mRNA. Its affinity decreases at the beginning of the night and LBP is synthesized (Mittag *et al.*, 1994, 1995; Mittag & Hastings, 1996).

CONCLUSIONS

The current state of knowledge and hypotheses about the circadian system of the marine unicellular dinoflagellate *Gonyaulax polyedra* are summarized in Fig. 2. The effects of light and nitrate show the complexity of circadian input pathways. Several functions and enzymes along the transduction pathway of nitrate to the circadian oscillator (e.g. uptake, nitrate reductase, nitrite reductase) are themselves under circadian control and/or are influenced by light. Thus, entrainment is not only ensured by the impact of environmental signals (*zeitgeber* = time giver) but also by the endogenously controlled 'readiness' or responsiveness of several elements (*zeitnehmer* = time taker) along their transduction. Another example for a *zeitnehmer* in *Gonyaulax* is the blue light input which also receives feedback from the oscillator. The circadian strategies, which have been discovered in *Gonyaulax*, apparently can be generalized, since other examples have since been characterized in different organisms. Light effects in the mammalian SCN are mediated by the two immediate early genes, *c-fos* and *junB*. As part of the light transduction mechanisms, these genes give rise to transcription factors which then act on other genes. Both *c-fos* and *junB* oscillate themselves with a circadian period, the former on the protein level (Franziska Wollnik, personal communication), the latter on the mRNA level (Guido, Rusak & Robertson, 1996), and thus also represent input feedback loops.

Work has shown that circadian systems containing more than one oscillator are not restricted to multicellular animals and plants but can also exist in unicellular organisms. The two oscillators in *Gonyaulax* respond differently to environmental stimuli and control different output rhythms. The complexity of the *Gonyaulax* circadian system, as summarized in Fig. 2, adds several new aspects to the function of biological temporal programmes. Basic biological functions dealing with important environmental resources such as photosynthesis or nitrate metabolism can be timed independently to reach an optimal rate at certain times of the day. However, the actual external availability of resources feeds back on to the circadian system,

Environmental
inputs
'zeitgeber'

Phenotypic
outputs

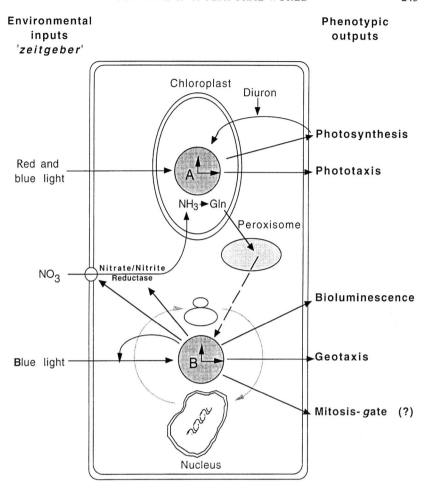

Fig. 2. Summary and working hypothesis of the complex mechanism of the *Gonyaulax* circadian system, showing the different environmental inputs (*zeitgeber*), light and nitrate, on the left and some of the prominent phenotypic outputs on the right. One of the two oscillators (B) is presumed to be cytosolic and is hypothetically drawn in analogy to the molecular oscillators in *Drosophila* (Rosbash, 1995; Kyriacou *et al.*, this volume) and *Neurospora* (Dunlap, 1996; Dunlap *et al.*, this volume) although there is no information, as yet, as to whether these exist in *Gonyaulax*. The other oscillator (A) is drawn in the chloroplast because results indicate its close relationship to photosynthesis. The discovery that photosynthesizing cyanobacteria control their metabolism with a circadian system (Johnson *et al.*, 1996), taken together with the endosymbiont theory (Margulis, 1981), increase the likelihood that chloroplasts may possess their own circadian oscillator. Feedback loops exist in the blue-sensitive light input and the transduction of nitrate effects. The results of current experiments indicate that this transduction involves the biochemistry of the chloroplast and the peroxisome but that its final target is the cytosolic B-oscillator.

thereby optimizing the temporal programme. As a result of this strategy, organisms cannot only anticipate the predictable presence of a resource but can also acutely adapt to the actual resource availability. This plasticity allows organisms to optimize cellular functions in a temporal environment which is largely predictable but can also change from day to day and within the course of the year.

ACKNOWLEDGEMENTS

Our work was supported by grants from the Deutsche Forschungsgemeinschaft, the Friedrich-Baur-Stiftung, and the Meyer-Struckmann-Stiftung.

REFERENCES

Aschoff, J. (1965). Circadian rhythms in man. *Science*, **148**, 1428–32.

Aschoff, J. (1979). Circadian rhythms: influences of internal and external factors on the period measured under constant conditions. *Zeitschrift für Tierpsychologie*, **49**, 225–49.

Aschoff, J. (1987). Effects of periodic food availability on circadian rhythms. In *Comparative Aspects of Circadian Clocks*, ed. T. Hiroshige & K. Honma, pp. 19–40. Sapporo: Hokkaido University Press.

Aschoff, J., Gerecke, U. & Wever, R. (1967). Desynchronization of human circadian rhythms. *The Japanese Journal of Physiology*, **17**, 450–7.

Boulos, Z. & Rusak, B. (1982). Circadian phase response curves for dark pulses in the hamster. *Journal of Comparative Physiology*, **A146**, 411–17.

Cahill, G. M. & Besharse, J. C. (1993). Circadian clock functions localized in *Xenopus* retinal photoreceptors. *Neuron*, **10**, 573–7.

Chabot, C. C. & Menaker, M. (1992). Circadian feeding and locomotor rhythms in pigeons and house sparrows. *Journal of Biological Rhythms*, **7**, 287–300.

Comolli, J., Taylor, W. & Hastings, J. W. (1994). An inhibitor of protein phosphorylation stops the circadian oscillator and blocks light-induced phase shifting in *Gonyaulax polyedra*. *Journal of Biological Rhythms*, **9**, 13–26.

Deng, T-S. & Roenneberg, T. (1997). Photobiology of the *Gonyaulax* circadian system II: allopurinol inhibits blue light effects. *Planta*, **202**, 502–9.

Desjardins, M. & Morse, D. (1993). The polypeptide components of scintillons, the bioluminescence organelles of the dinoflagellate *Gonyaulax polyedra*. *Biochemistry and Cell Biology*, **71**, 176–82.

Dunlap, J. C. (1996). Genetics and molecular analysis of circadian rhythms. *Annual Reviews in Genetics*, **30**, 579–601.

Dunlap, J. C. & Hastings, J. W. (1981). The biological clock in *Gonyaulax* controls luciferase activity by regulation turnover. *Journal of Biological Chemistry*, **256**, 10509–18.

Ebihara, S. & Gwinner, E. (1992). Different circadian pacemakers control feeding and locomotor activity in European starlings. *Journal of Comparative Physiology*, **A171**, 63–7.

Fleissner, G. & Fleissner, G. (1997). The scorpion's clock as a model for feedback mechanisms in circadian systems. In *Scorpion Biology and Research*, ed. P. Brownell & G. Polis. Oxford: Oxford University Press, in press.

Fogel, M. & Hastings, J. W. (1971). A substrate binding protein in the *Gonyaulax* luminescence reaction. *Archives in Biochemistry and Biophysics*, **142**, 310–21.

Fritz, L., Morse, D. & Hastings, J. W. (1990). The circadian bioluminescence rhythm of *Gonyaulax* is related to daily variations in the number of light-emitting organelles. *Journal of Cell Science*, **95**, 321–8.

Guido, M. E., Rusak, B. & Robertson, H. A. (1996). Spontaneous circadian and light-induced expression of junB mRNA in the hamster suprachiasmatic nucleus. *Brain Research*, **732**, 215–22.

Guillard, R. R. L. & Ryther, J. H. (1962). Studies on marine planktonic diatoms: *Cyclotella nana* Hufstedt and *Denotula confervacea* (Cleve) Gran. *Canadian Journal of Microbiology*, **8**, 229–39.

Gwinner, E. (1966a). Entrainment of a circadian rhythm in birds by species-specific song cycles (Aves, Fringillidae; *Carduelis spinus, Serinus serinus*). *Experientia*, **22**, 765.

Gwinner, E. (1966b). Tagesperiodische Schwankungen der Vorzugshelligkeit bei Vögeln. *Zeitschrift zur vergleichenden Physiologie*, **52**, 370–9.

Harrison, P. J. (1979). Nitrate metabolism of the red tide dinoflagellate *Gonyaulax polyedra* Stein. *Journal of Experimental Marine Biology and Ecology*, **21**, 199–209.

Hastings, J. W. & Sweeney, B. M. (1960). The action spectrum for shifting the phase of the rhythm of luminescence in *Gonyaulax polyedra*. *Journal of General Physiology*, **43**, 697–706.

Hastings, J. W., Rusak, B. & Boulos, Z. (1991). Circadian rhythms: the physiology of biological timing. In *Neural and Integrative Animal Physiology*, ed. C. L. Prosser, pp. 435–546. Cambridge: Wiley-Liss, Inc.

Hau, M. & Gwinner, E. (1992). Circadian entrainment by feeding cycles in house sparrows, *Passer domesticus*. *Journal of Comparative Physiology*, **A170**, 403–9.

Hoffman, K. (1971). Splitting of the circadian rhythm as a function of light intensity. In *Biochemistry*, ed. M. Menaker, pp. 134–51. Washington, DC: National Academy of Sciences.

Honma, K-I., Honma, S. & Hiroshige, T. (1986). Disorganization of the rat activity rhythm by chronic treatment with methamphetamine. *Physiology and Behavior*, **38**, 687–95.

Honma, S., Honma, K-I. & Hiroshige, T. (1991). Methamphetamine effects of rat circadian clock depend on actograph. *Physiology and Behavior*, **49**, 787–95.

Jacobshagen, S. & Johnson, C. H. (1994). Circadian rhythms of gene expression in *Chlamydomonas reinhardtii*: circadian cycling of mRNA abundance of cab II, and possibly of β-tubulin and cytochrome. *European Journal of Cell Biology*, **64**, 142–52.

Jilge, B. (1991). Restricted feeding: a nonphotic Zeitgeber in the rabbit. *Physiology and Behavior*, **51**, 157–66.

Johnson, C. H. & Hastings, J. W. (1989). Circadian phototransduction: phase resetting and frequency of the circadian clock of *Gonyaulax* cells in red light. *Journal of Biological Rhythms*, **4**, 417–37.

Johnson, C. H., Roeber, J. F. & Hastings, J. W. (1984). Circadian changes in enzyme concentration account for rhythm of enzyme activity in *Gonyaulax*. *Science*, **223**, 1428–30.

Johnson, C. H., Golden, S. S., Ishiura, M. & Kondo, T. (1996). Circadian clocks in prokaryotes. *Molecular Microbiology*, **21**, 5–11.

Kennedy, G. A., Coleman, G. J. & Armstrong, S. M. (1991). Restricted feeding entrains circadian wheel-running activity rhythms of the kowari. *American Journal of Physiology*, **261**, R819–27.

Kleinhofs, A. & Warner, R. L. (1990). Advances in nitrate assimilation. *The Biochemistry of Plants*, **16**, 89–120.

Kondo, T., Strayer, C. A., Kulkarni, R. D. *et al.* (1993). Circadian rhythms in prokaryotes: luciferase as a reporter of circadian gene expression in cyanobacteria. *Proceedings of the National Academy of Sciences, USA*, **90**, 5672–6.

Liu, Y., Tsinoremas, N. F., Johnson, C. H. *et al.* (1995). Circadian orchestration of gene expression in cyanobacteria. *Genes and Development*, **9**, 1469–78.

Loros, J. J. & Dunlap, J. C. (1991). *Neurospora crassa* clock-controlled genes are regulated at the level of transcription. *Molecular and Cellular Biology*, **11**, 558–63.

Lythgoe, J. N. (1979). *The Ecology of Vision*. Oxford: Clarendon Press.

McMurry, L. & Hastings, J. W. (1972). No desynchronization among the four circadian rhythms in the unicellular alga, *Gonyaulax polyedra*. *Science*, **175**, 1137–9.

Margulis, L. (1981). *Symbiosis in Cell Evolution*. New York: WH Freeman.

Meijer, J. H., Daan, S., Overkamp, G. J. F. & Hermann, P. M. (1990). The two-oscillator circadian system of tree shrews (*tupaia belangeri*) and its response to light and dark pulses. *Journal of Biological Rhythms*, **5**, 1–16.

Meyer, H., Thienel, U. & Piechulla, B. (1989). Molecular characterization of the diurnal/circadian expression of the chlorophyll *a/b*-binding proteins in leaves of tomato and other dicotyledonous and monocotyledonous plant species. *Planta*, **180**, 5–15.

Michel, S., Geusz, M. E., Zaritsky, J. J. & Block, G. D. (1993). Circadian rhythm in membrane conductance expressed in isolated neurons. *Science*, **259**, 239–41.

Millar, A. J. & Kay, S. A. (1991). Circadian control of *cab* gene transcription and mRNA accumulation in *Arabidopsis*. *The Plant Cell*, **3**, 541–50.

Mittag, M. & Hastings, J. W. (1996). Exploring the signalling pathway of circadian bioluminescence. *Physiologia Plantarum*, **96**, 727–32.

Mittag, M., Lee, D-H. & Hastings, J. W. (1994). Circadian expression of the luciferin-binding protein correlates with the binding of a protein to the 3′ untranslated region of its mRNA. *Proceedings of the National Academy of Sciences, USA*, **91**, 5257–61.

Mittag, M., Lee, D-H. & Hastings, J. W. (1995). Are mRNA binding proteins involved in the translational control of the circadian regulated synthesis of luminescence proteins in *Gonyaulax polyedra*. In *Evolution of Circadian Clocks from Cell to Human*, ed. T. Horoshige & K. Honma, pp. 97–140. Sapporo: Hokkaido University Press.

Morse, D., Fritz, L., Pappenheimer, A. M. J. & Hastings, J. W. (1989*a*). Properties and cellular localization of a luciferin binding protein in the bioluminescence reaction of *Gonyaulax polyedra*. *Journal of Bioluminescence and Chemilumines-cence*, **3**, 79–83.

Morse, D., Milos, P. M., Roux, E. & Hastings, J. W. (1989*b*). Circadian regulation of bioluminescence in *Gonyaulax* involves translational control. *Proceedings of the National Academy of Sciences, USA*, **86**, 172–6.

Morse, D., Pappenheimer, A. M. J. & Hastings, J. W. (1989*c*). Role of a luciferin-binding protein in the circadian bioluminescent reaction of *Gonyaulax polyedra*. *Journal of Biological Chemistry*, **264**, 11822–6.

Morse, D., Fritz, L. & Hastings, J. W. (1990). What is the clock? Translational regulation of the circadian bioluminescence. *Trends in Biochemical Sciences*, **15**, 262–5.

Morse, D., Hastings, J. W. & Roenneberg, T. (1994). Different phase responses of two circadian oscillators in *Gonyaulax*. *Journal of Biological Rhythms*, **9**, 263–74.

Mrosovsky, N., Salmon, P. A., Menaker, M. & Ralph, M. R. (1992). Nonphotic phase shifting in hamster clock mutants. *Journal of Biological Rhythms*, **7**, 41–50.

Nicolas, M. T., Nicolas, G., Johnson, C. H., Bassot, J-M. & Hastings, J. W. (1987). Characterization of the bioluminescent organelles in *Gonyaulax polyedra* after fast freeze-freeze fixation and antiluciferase immunogold staining. *Journal of Cell Biology*, **105**, 723–35.

Nicolas, M. T., Morse, D., Bassot, J. M. & Hastings, J. W. (1991). Colocalization of luciferin binding protein and luciferase to the scintillons of *Gonyaulax polyedra* revealed by double immunolabeling after fast freeze fixation. *Protoplasma*, **160**, 159–66.

Oberschmidt, O., Hücking, C. & Piechulla, B. (1995). Diurnal *Lhc* gene expression is present in many but not all species of the plant kingdom. *Plant Molecular Biology*, **27**, 147–53.

Pickard, G. & Tang, W. (1993). Individual pineal cells exhibit a circadian rhythm in melatonin secretion. *Brain Research*, **627**, 141–6.

Pickard, G. & Tang, W. (1994). Pineal photoreceptors rhythmically secrete melatonin. *Neuroscience Letters*, **171**, 109–12.

Piechulla, B. & Gruissem, W. (1987). Diurnal mRNA fluctuations of nuclear and plastid genes in developing tomato fruits. *EMBO Journal*, **6**, 3593–9.

Pittendrigh, C. S. (1981). Circadian systems: entrainment in *Biological Rhythms*, ed. J. Aschoff, pp. 95–124. New York: Plenum Publ. Corp.

Pittendrigh, C. S. (1993). Temporal organization: reflections of a Darwinian clockwatcher. *Annual Review of Physiology*, **55**, 17–54.

Ramalho, C. B., Hastings, J. W. & Colepicolo, P. (1995). Circadian oscillation of nitrate reductase activity in *Gonyaulax polyedra* is due to changes in cellular protein levels. *Plant Physiology*, **107**, 225–31.

Riley, J. P. & Skirrow, G. (1965). *Chemical Oceanography*. London: Academic Press.

Roenneberg, T. (1992). Spatial and temporal environment. The chrono-ecology of biological rhythms. *Universitas*, **34**, 202–10.

Roenneberg, T. (1994). The *Gonyaulax* circadian system: evidence for two input pathways and two oscillators. In *Evolution of Circadian Clock*, ed. T. Hiroshige & K-I. Honma, pp. 3–20. Sapporo: Hokkaido University Press.

Roenneberg, T. (1996). The complex circadian system of *Gonyaulax polyedra*. *Physiologia Plantarum*, **96**, 733–7.

Roenneberg, T. & Deng, T-S. (1997). Photobiology of the *Gonyaulax* circadian system I: different phase response curves for red and blue light. *Planta*, **202**, 494–501.

Roenneberg, T. & Foster, R. G. (1997). Twilight times – light and the circadian system. *Photochemistry and Photobiology*, in press.

Roenneberg, T. & Hastings, J. W. (1988). Two photoreceptors influence the circadian clock of a unicellular alga. *Naturwissenschaften*, **75**, 206–7.

Roenneberg, T. & Hastings, J. W. (1992). Cell movement and pattern formation in *Gonyaulax polyedra*. In *Oscillations and Morphogenesis*, ed. L. Rensing, pp. 399–412. New York: Marcel Dekker.

Roenneberg, T. & Mittag, M. (1996). The circadian program of algae. *Seminars in Cell and Developmental Biology*, **7**, 753–63.

Roenneberg, T. & Morse, D. (1993). Two circadian oscillators in one cell. *Nature*, **362**, 362–4.

Roenneberg, T. & Morse, D. (1994). Zelluläre Mechanismen der inneren Uhr eines Einzellers. *Naturwissenschaften*, **81**, 343–9.

Roenneberg, T. & Rehman, J. (1996). Nitrate, a nonphotic signal for the circadian system. *Journal of the Federation of American Societies of Experimental Biology*, **10**, 1443–7.

Roenneberg, T. & Taylor, W. (1994). Light induced phase responses in *Gonyaulax* are drastically altered by creatine. *Journal of Biological Rhythms*, **9**, 1–12.

Roenneberg, T., Nakamura, H. & Hastings, J. W. (1988). Creatine accelerates the circadian clock in a unicellular algae. *Nature*, **334**, 432–4.

Roenneberg, T., Nakamura, H., Cranmer, L. D. I., Kishi, Y. & Hastings, J. W.

(1991). Gonyaulin: a novel endogenous period-shortening substance controlling the circadian clock of a unicellular alga. *Experientia*, **47**, 103–6.

Roenneberg, T., Merrow, M. & Eisensamer, B. (1997). Cellular mechanisms of circadian systems. *Proceedings of the German Zoological Society*, in press.

Rosbash, M. (1995). Molecular control of circadian rhythms. *Current Opinion in Genetics and Development*, **5**, 662–8.

Schardt, U., Wilhelm, I. & Erkert, H. G. (1989). Splitting of the circadian activity rhythm in common marmosets (*Callithrix j. jacchus*; Primates). *Experientia*, **45**, 1112–15.

Shinohara, K., Honma, S., Katsuno, Y., Abe, K. & Honma, K. (1995). Two distinct oscillators in the rat suprachiasmatic nucleus *in vitro*. *Proceedings of the National Academy of Sciences, USA*, **92**, 7396–400.

Stephan, F. K. (1992*a*). Resetting of a circadian clock by food pulses. *Physiology and Behavior*, **52**, 997–1008.

Stephan, F. K. (1992*b*). Resetting of a feeding-entrainable circadian clock in the rat. *Physiology and Behavior*, **52**, 985–95.

Sweeney, B. M. & Hastings, J. W. (1960). Effects of temperature upon diurnal rhythms. *Cold Spring Harbor Symposia on Quantitative Biology*, **25**, 87–104.

Takahashi, J. S. (1987). Cellular basis of circadian rhythms in the avian pineal. In *Comparative Aspects of Circadian Clocks*, ed. T. Hiroshige & K. Honma, pp. 3–15. Sapporo: Hokkaido University Press.

Tosini, G. & Menaker, M. (1995). Circadian rhythm of body temperature in Ectotherm (*Iguana iguana*). *Journal of Biological Rhythms*, **10**, 248–55.

Tosini, G. & Menaker, M. (1996). Circadian rhythms in cultured mammalian retina. *Science*, **272**, 419–21.

Van Reeth, O., Byrne, M., Blackman, J. D. *et al.* (1994). Nocturnal exercise phase delays circadian rhythms of melatonin and thyrotropin secretion in normal men. *American Journal of Physiology*, **266**, E964–74.

Welsh, D. K., Logothetis, D. E., Meister, M. & Reppert, S. M. (1995). Individual neurons dissociated from rat suprachiasmatic nucleus express independently phased circadian firing rhythms. *Neuron*, **14**, 697–706.

Wever, R. (1979). *The Circadian System of Man*. Berlin, Heidelberg, New York: Springer.

TEMPORAL ORGANIZATION OF THE CELL DIVISION CYCLE IN EUKARYOTIC MICROBES

DAVID LLOYD[1] AND DON A. GILBERT[2]

[1]*Microbiology Group (PABIO), University of Wales, Cardiff, PO Box 915, Cardiff CF1 3TL, UK*
[2]*11 Mornington Avenue, Ilford, Essex IG1 3QT, UK*

It is the pattern maintained by this homeostasis, which is the touchstone of our personal identity. Our tissues change as we live: the food we eat and the air we breathe become flesh of our flesh, and bone of our bone, and the momentary elements of our flesh and bone pass out of our body every day with our excreta. We are but whirlpools in a river of ever-flowing water. We are not the stuff that abides, but patterns that perpetuate themselves.

Wiener, 1954

What are called structures are slow processes of long duration, functions are quick processes of short duration.

von Bertalanffy, 1952

INTRODUCTION

The smooth and correctly ordered progress of growth and division of a microorganism or a cell is a highly integrated process of great complexity. It involves all the processes of nutrient uptake, the biochemical steps for energy generation, the control of ionic composition, the network of interconversions of intermediary metabolites, the biosynthesis of macromolecules, the assembly of membranes, organelles and cell walls, membrane trafficking, genome duplication and segregation. A web of metabolic, genetic and developmental controls and checkpoints co-ordinate these processes. Essential for normal growth and division is the establishment at an appropriate time of intricate structure, form and function (Nurse, 1990). Abnormal development in the metazoa and metaphyta results from dysfunction of control and co-ordination at the level of the single cell. The detailed dissection of the events and processes that determine cell division, quiescence or cell death are of the greatest importance to understanding; these are central questions of biology. Spectacular progress has been made since 1980 in the identification of some of the control mechanisms that regulate the growth and division of cells. Most accounts deal only with the processes whereby the evident structural elements are developed and duplicated in preparation for cytokinesis. However, these sequences are but the tip of the iceberg; underneath, on more rapid time-scales (Fig. 1) a myriad of intramolecular motions, fast

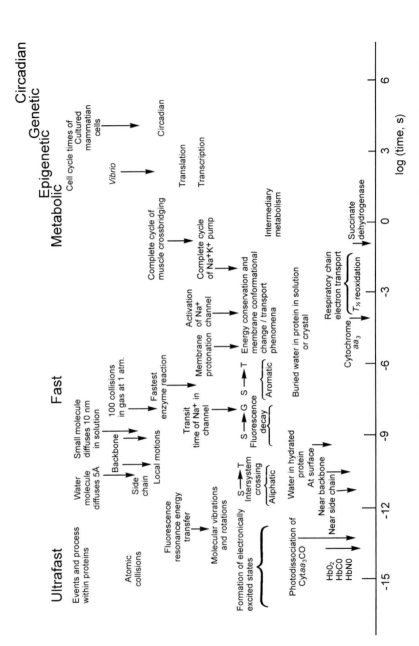

Fig. 1. The time domains of living systems. Approximate relaxation times (to a return of $1/e$ of a small perturbative effect, Goodwin, 1963) are indicated.

electron, proton and ion transport reactions, as well as the web of catabolic and biosynthetic pathways, unceasingly enmesh.

This chapter attempts to integrate a vision of this miraculous panoply. The non-equilibrium driving forces of life produce an exquisite complexity (Aon & Cortassa, 1997). Energy dissipation in an open system gives rise to unexpected order, as well as emergent behaviour beyond the understanding of present-day chemistry and physics. Much of this organization can be ascribed to non-linear oscillatory behaviour (Lloyd, 1998); cells make full use of every aspect of their rhythmic potential on every spatial scale and in every temporal domain (Lloyd, 1987). Homeostasis is an over-simplification of the last century. Except in cryptobiotic organisms, stationary states are vanishingly rare in living systems, and the quality of life depends on its auto-dynamic nature.

THE ULTRAFAST AND FAST TIME DOMAINS

Experiments using femtosecond lasers for activation, and streak-cameras for detection are now routine, not only in investigations of chemical reaction mechanisms, but also in biochemistry. The chemical physics of vision and of the quantum absorption and conversion steps of photosynthetic light harvesting (Martin & Vos, 1992) are examples of studies that require ultrafast methods. Structural intermediates in biological reactions can be very short-lived, with lifetimes spanning the time-scale from femtoseconds to milliseconds or longer (Moffat, 1989; Cruickshank, Helliwell & Johnson, 1992). As well as developing specialized rapid-reaction techniques for real-time studies (Chance et al., 1964), means of artificially prolonging the lifetime of intermediates (for example, by cryogenic techniques, Douzou, 1977) have been used. A splendid example of the elegance and detailed (nanosecond) time resolution now available is provided by the pulsed, Laue X-ray diffraction data on the photolysis and rebinding of the CO complex of myoglobin (Srajer et al., 1996). Numerous examples of nanosecond relaxation times for ligand binding to receptors and enzymes and for membrane-associated events are recognized. It is on these time-scales that a full understanding of biochemical mechanisms must be sought.

Electron transport within and between redox centres, as well as proton translocation across membranes and the ionic pumping that drives energy conservation, are all processes in the fast time domain (ms to s). Techniques used to explore the ultrafast and fast time domains are listed in Lloyd, Poole & Edwards (1982b).

THE METABOLIC TIME DOMAIN

Continuous motion within, and through, the metabolic network has never been described more vividly than by Reich & Sel'kov (1981).

A whole factory indeed. The number of metabolites partaking is counted by thousands. They are all present, from a few up to several million molecules in number, in that tiny volume of 10^{-12} cm^3 which we call a (bacterial) cell. They form more or less a mixture, in aqueous solution, though not a perfect one in the physicochemical sense. The organization is mysterious ... [The metabolic map] ... looks like a cobweb. The first task, obviously, is to unravel the tangle. To make the network more transparent while retaining its principal multicomponent structure. ... In the cell, nearly all the routes are in fact frequented very heavily and simultaneously. If reactions buzzed, we would hear an infernal concert. Everything crooning as in a beehive. We would observe individual molecules bustling to and fro, changing their binding state, or the compartment where they are, becoming engaged in chemical conversions, leaving or entering the cell. If one puts a handy label into it, for example a radioactive molecule, then seconds to minutes later everything carries a mote. The label spreads instantaneously into the most remote corner, and quite unexpected substances will be tagged a few moments later. ... Apparently of all the billions of molecules, very few survive a dozen quiet seconds without undergoing chemical or physical conversion. Yet law and order appears to reign in that seemingly chaotic rush. The average content of metabolites remains unchanged for a long time. Input and output rates are stationary. The whole motion pattern is steady, re-established after perturbation and obstinately defended. The experimenter is often perplexed by the precise status maintenance of a system that may undergo instantaneous and dramatic irreversible change if handled unskillfully. This leads us to the second obvious task of analysis to explain how a system in that bustle can know, find and defend its 'normal' state and maintain homeostasis. To clarify how continuous motion retains coincidence and simultaneity.

On the map, we can discern very many different possible routes. Some of them short-circuit each other to apparent futility. Every function is a cross-way where several enzymes tussle for the common substrate or cofactor. Yet if the motion could be made visible, it would look like a fish shoal: masterly coordinated and capable of instantaneous reorientation being responsive to the slightest specific signal.

Apparently there is co-ordination without confusion and appropriateness of response without error. In some cases the result is close to perfection – amazing, since everything goes on more or less in a broth of thousands of interacting molecules and enzymes. Is there a highly efficient set of commands, collection of interactive regulatory feedback signals which can be demonstrated – literally a second cobweb stretched over the first one of chemical pathways, or just a spontaneous co-ordination and expediency of motion like that of droplets in a stream? ... Thus a third objective is to describe the kinetic functioning of well coordinated and stable flow in a system of potentially chaotic diffuseness.

The years since these words were written have, by and large, seen the successful unravelling of much of this complexity, especially in terms of the mechanisms of metabolic control. The task now relates to the sub-microscopic intracellular organization of all this: on, near, through and around the intricately sculptured but continuously changing latticework of protoplasmic ultrastructure; Sherrington's (1955):

many-chambered gobbet of foam ... a shifting labyrinth of dissolving walls and floors which form melt and reform as the work of the chemical factory requires.

Solution chemistry, pertaining to highly diluted extracts of the macerated and homogenized living substance was never more than a structurally bereft and simplified model, a first look. Now there are new tools and it is no longer necessary to blow the structure away to get at the biochemistry. Ever-increasing sensitivities of charge-coupled device camera imaging, of confocal laser scanning microscopy, and NMR imaging combined with the enormous range of exquisite specificities of molecular and fluorescence probes, and of targeted bioluminescent photoproteins enable anew the re-examination of old questions. It can (almost) all be done now *in situ*, in real time, and in full colour!

The highly dynamic nature of the mammalian cell surface and of cellular redox state has been demonstrated by continuous monitoring of light scattering, intrinsic flavin and NADH fluorescence and differential absorption (Visser *et al.*, 1990; Gilbert & Visser, 1993). Both continuous periodically modulated cycles and repetitive spikes were observed with periods of between 10 s and 20 min (Fig. 2). Sensitivity to transferrin and insulin suggests functional roles for these rhythms. These fast light scattering changes probably correspond to the rapid changes in number, kind, sizes and distributions of cytoplasmic protrusions and localized undulations of the surface membrane ('ruffling' and 'flickering').

The slower rhythms must arise from gross changes in cellular morphology and possibly from the incessant movements of internal components so dramatically evident in phase contrast microscopy. All these shape changes are potentially able to modify and modulate the metabolic processes within these cells (Gilbert & Visser, 1993), a view endorsed by the observation that insulin stimulates the dynamics. Cell–cell interactions must occur in these stirred suspensions so as to provide coherent outputs; the mechanisms by which cells communicate are still largely unfathomed. The power spectrum variations presented here are from direct recordings of cellular sensing processes at the environmental interface.

THE ULTRADIAN TIME DOMAIN

The ultradian time domain includes all those biological processes with time constants of less than 1 day. Many hundreds of different kinds of periodic outputs have been described (Lloyd & Stupfel, 1991; Lloyd & Rossi, 1992, 1993) and include the yeast glycolytic oscillator, the cyclic AMP oscillator in *Dictyostelium discoideum* and the mammalian cytosolic Ca^{2+} oscillator, but it now seems that none of these has timing functions. Indeed, the characteristics of many individual rhythms are sensitive to disturbances and therefore too variable for accurate time keeping although that limitation might be overcome if a number of periodicities are involved (*cf.* Enright, 1980; Klevecz, 1992). However, a special class of ultradian oscillations, sometimes referred to as epigenetic, and with periods that range between 30 min and 4 h

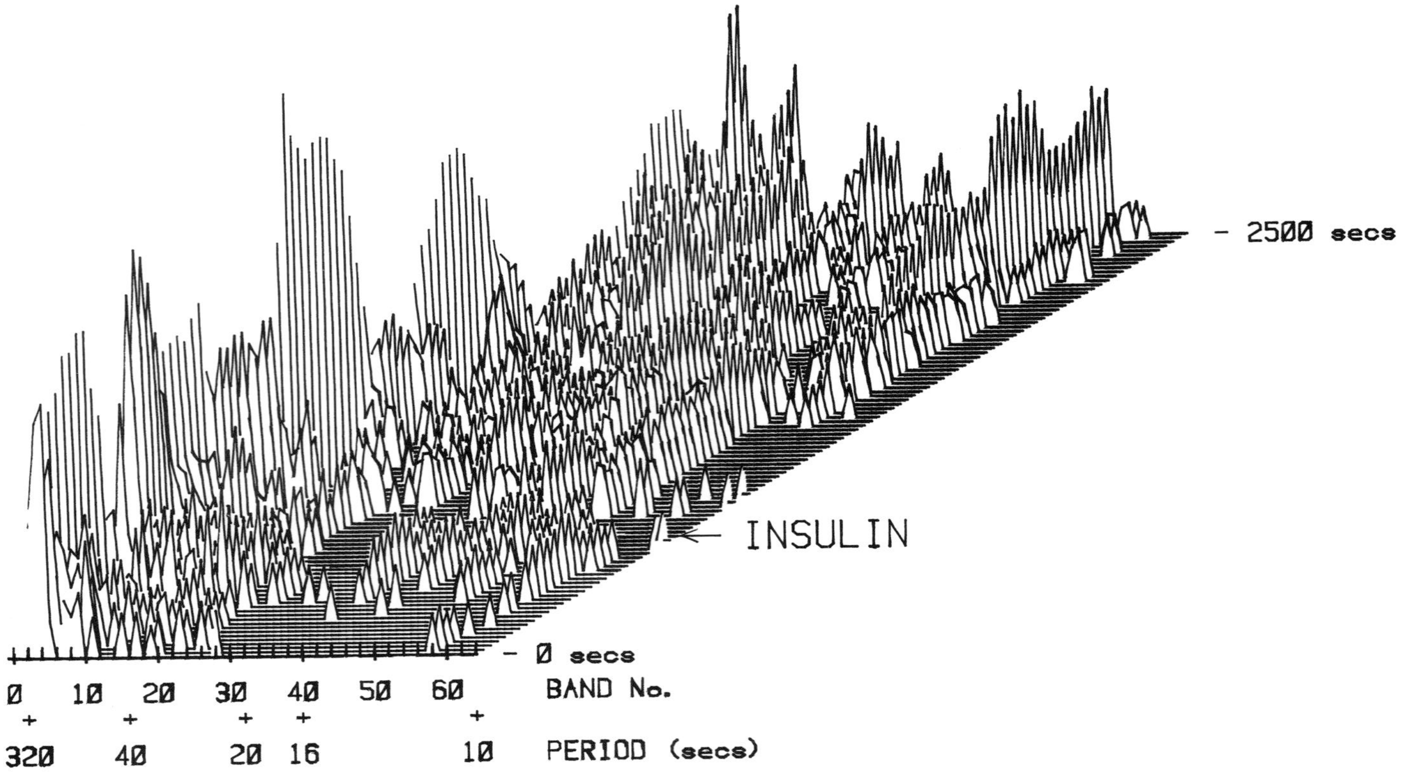

Fig. 2. (caption opposite).

in different species, has been identified as playing a central timekeeping role in embryos (Mano, 1970) and in lower eukaryotes under conditions of rapid growth (Edwards & Lloyd, 1978, 1980; Lloyd, Edwards & Fry, 1982a; Lloyd et al., 1982b; Lloyd & Edwards, 1984). These temperature-compensated ultradian clock-coupled outputs provide the predominant time-dependent changes in energy production, biosynthetic rates and motility; and furthermore they provide a time-frame for the cell division cycle (Lloyd & Kippert, 1987, 1993). Phases when massive decomposition of constituents (while the organism is in a de-energized state) alternate with phases of rapid synthesis with energy generation at full capacity. This rhythm involves the majority of the ribosomes and proteins of the organisms. Thus growth itself is a periodic process that, of necessity, requires phases of destruction (Brodsky, 1975; Edwards & Lloyd, 1980; see also Tsilimigras & Gilbert, 1977; Ferreira, Hammond & Gilbert, 1994a). Available evidence leads us to believe that all proteins are varying periodically at high frequency and in relatively independent fashion (cf. also, Ferreira, Hammond & Gilbert, 1994b; Ferreira, Hammond & Gilbert, 1996b,c).

Caution should be exercised in the interpretation of results obtained by discontinuous observation of a periodic phenomenon, for example, the apparent frequency can be dependent on the sampling frequency. Thus it seems likely that our understanding of all these processes is being hampered by this 'aliasing problem' (Gilbert, 1974b; Gilbert & Tsilimigras, 1981; Ferreira et al., 1994b) which can lead to misinterpretation of the frequency of an oscillation, due to the sampling interval being too long. But dynamic problems demand dynamic solutions which must therefore be rate-based, for they are the only process-determining factors which contain the dimension of time (Gilbert, 1981).

Fig. 2. Burst activity in scatter dynamics. Temporal variations in the power spectrum values for light scattered by murine erythroleukaemic cells in suspension showing the stimulating effect of insulin (Gilbert & Visser, 1993). The plot shows the changes in the power (the contribution of the oscillation of the given frequency to the total fluctuation) for different oscillations before and after addition of the hormone. The cells were circulated through a Hitachi 850 spectrofluorimeter at 0.6×10^6/ml using scatter wavelength of 478 nm, bandwidths of 5 nm and with a 1 s sampling interval being integrated over 5 s at the computer. Further details are as given in Gilbert & Visser (1993) and Visser et al. (1990). High frequency components (right) are taken to reflect periodic movements and shape of the cell surface whereas the slower ones (left) are thought to be due to gross changes in morphology. As described in those articles and in Gilbert & MacKinnon (1992), plots of the time variations for a given individual frequency with time, indicate periodic modulation of each oscillation with some apparently enhanced by the insulin and others depressed: this apparent contradictory behaviour is believed to be due to the hormone altering the frequencies of the oscillations. However, the dominant effect is that of stimulation of the overall dynamics (Gilbert & Visser, 1993; see also Ferreira et al., 1996c).

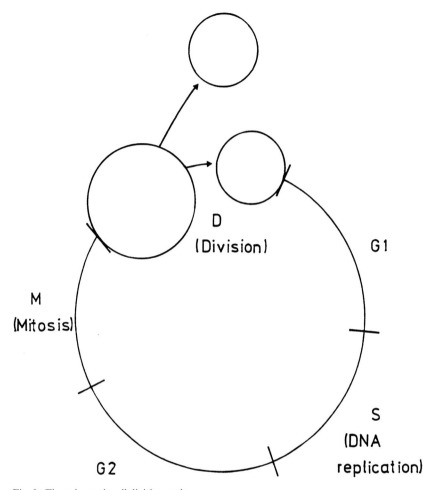

Fig. 3. The eukaryotic cell division cycle.

THE CELL DIVISION CYCLE DOMAIN

The eukaryotic cell division cycle is the most fundamental of all developmental processes. Its outcome includes the proper replication and segregation of the genetic material during the formation of genetically identical daughter cells (Fig. 3). The doubling in cell mass, and on average, of each type of cellular constituent, membrane component and organelle also relies on the efficient co-ordination and completion of this set of events.

There are two basic approaches to the understanding of this phenomenon; (i) the popular one is essentially descriptive and is concerned with the elucidation of a sequence of events and the identification of participating components, (ii) the long-standing heretical view is based on the argument that dynamic problems can only be adequately and quantitatively understood in terms of the kinetics of the reactions and that this can help reveal the compounds involved. We start with the former.

Current views of mitosis see the phenomenon as a sequence of reactions, and the distinct stages of the overall process still bear the designations G_1, S, G_2 and M based on early cytological observations: S (for synthesis) is the phase of DNA replication; M is mitosis, in which segregation of cellular constituents precedes cell division; G_1 and G_2 are the 'gaps' between these two microscopically observable stages. As well as identification of the biochemical steps of this complex dynamic process, both spatial and temporal controls need to be understood (Gilbert, 1981; MacKinnon & Gilbert, 1992). Only then will it become possible to fathom the sources and early consequences of derangement to normal cellular growth and division. Again, it must be stressed that dynamic problems necessitate dynamic solutions, and this demands that explanations include time dependence.

The operationally convenient phases of the cell cycle, G_1, S, G_2 and M are best considered as 'seasons' (Nasmyth, 1996). Seasons are not state variables determining the weather; they are merely descriptive of time intervals; it is the sun's varying inclination that drives the climatic conditions.

It may be asked why it is that, in cultures of simple unicellular organisms, cell division depends on the presence of every necessary nutrient? What of the more complex question of why under similar nutrient status, stem cells (for epidermis or of bone marrow) continue to divide, whereas neurons do not? It has become clear that there are two major control 'sites' that occur universally in the cell division cycles of many organisms from amoeba to humans. These seem to regulate the progress of cells from stage G_1 to S and from G_2 to M, although Gilbert (1974a) has interpreted these in terms of the control system kinetics, as two non-autodynamic (that is static) states.

DNA replication and mitosis are induced by the activation of S-phase and M-phase specific cyclin-dependent protein kinases (CDKs). The catalytic subunits of these kinases are only active when bound to unstable regulators, the cyclins (Evans *et al.*, 1983). Activation of M-phase CDKs, regulated by cyclin synthesis and phosphorylation, precedes the partial resolution of sister chromatids. Duplication of chromosomes is followed by the generation of a bipolar mitotic spindle. Sister kinetochores are attached of microtubules; these become associated with opposite poles of the spindle prior to becoming aligned on the metaphase cell plate. In anaphase, sister chromatids are separated to opposite poles of the cell. Inactivation of M-phase CDK through proteolysis occurs simultaneously with chromosome segregation, but is not a necessary part of the process.

The time spent by any individual cell in each stage depends on many factors. As well as the necessity for all the nutrients normally required for growth (Lu & Means, 1993), the last two decades have seen the discovery of a host of growth factors specific for different cell types, for example epidermal growth factor, and tumour growth promoting factor (Wang, 1992). As is the case for a hormone, each of these has a specific cell surface receptor, and interaction with growth factor initiates a signal transduction pathway involving secondary messenger substances and eventually triggering gene transcription within the nucleus. As well as a requirement for extracellular information, progression through the cell cycle also needs positive inputs from an intracellular control network governed by the products of the genes providing the cell division cycle control system components.

The importance of these enquiries can hardly be overstressed. Cancer cells frequently show genetic instabilities, and many of these lead to gene amplification and chromosomal rearrangements; tumorigenesis is a likely outcome (Hartwell, 1978, 1991, 1992). Although malignancy does not depend directly on such derangements, it may be due to disturbances in the temporal co-ordination of the cell, rather than any molecular instability. Once this occurs, it can drastically affect cell population heterogeneity, and hence the outcome, and treatment aspects, of the disease.

But, is this the only way cancer can arise and what is the mechanism by which oncogenic effects on cell proliferation are achieved? Mitosis and apoptosis (programmed cell death) are also closely related phenomena (Bowen, Bowen & Jones, 1997). Great strides have been made since 1980 in the demonstration of the fundamental uniformity that underlies the apparent diversity of dividing cell types. This comes about because of a conserved network of control processes that operates in all organisms from yeast to humans (Nurse, 1990), as well as in plants (Francis, 1992; Francis & Halford, 1995). This is in spite of the fact that, once embryonic mitoses are complete, the developmental functions of cell division are very different in animals and plants. Whereas continued mitotic activity is largely restricted in the latter of apical meristems, in adult vertebrates the maintenance and regeneration of tissues and circulating cells are reliant on mitoses in stem cells (Jacobs, 1992). The discovery that many, if not all, of the sites and agencies of cell cycle control are universal has enabled a highly successful search in 'simpler' model systems such as yeast to be extended to the identification of homologous processes in higher organisms.

Cyclin-dependent kinases (CDKs) determine several cell cycle transitions, and inhibition of these kinases provides a mechanism by which so-called 'checkpoint pathways' generate cell cycle arrest (Murray, 1992). A kinase subunit CDK, together with an activating subunit, cyclin, give a complex subject to many levels of regulation (Fig. 4).

Direction, order and proper timing of the cell division cycle are provided by the 'awesome power of proteolytic processes' (Fig. 5; King et al., 1996). The eukaryotic cell cycle does not have overlapping phases of DNA synthesis and chromosome separation as happens in prokaryotes. The insertion of the process of chromosome alignment between the two stages of replication and

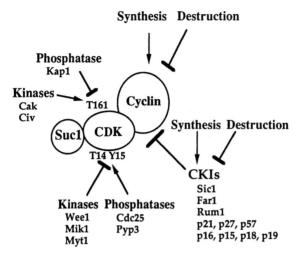

Fig. 4. Regulation of cyclin-dependent kinases (CDK). Arrowheads represent activating events and perpendicular ends represent inhibitory events. Genes known to encode the indicated functions are listed below. Both cyclins and some cyclin-dependent kinase inhibitors (CKIs) are regulated by synthesis and ubiquitin-mediated proteolysis. Checkpoint pathways could act to promote inhibitory pathways or inhibit activating pathways to cause cell cycle arrest. Reproduced with permission from Elledge (1996).

chromosome segregation is characteristic of the eukaryotic cycle. Pivotal in the initiation of this process of chromosome segregation and exit from the mitosis is the active proteolytic destruction of anaphase inhibitors and mitotic cyclins. A large anaphase-promoting complex (APC), the cyclosome, mediates this process.

More generally, cells cannot exist in two phases of the cell cycle at the same time as seems possible in a sequence system (but not according to the limit-cycle oscillator concept of Gilbert, 1981). CDC34 in *Saccharomyces cerevisiae* is required for the G_1–S transition. Here too, proteolysis provides directionality. The CDC34 gene product is a ubiquitin-conjugating enzyme. CDK activity serves as a trigger for destruction for both the CDC34 and APC proteolytic pathways either directly or indirectly. In turn these processes of destruction regulate the level of CDK activity by degrading a CDK activator or inhibitor.

Phosphorylation renders substrates susceptible to the CDC34 pathway; the mechanism by which this occurs is still the subject of speculation.

Besides their roles in late G_1 and mitosis, the CDC34 and APC pathways are implicated at other cell division cycle steps, for example CDC34 at the G_2–M interface (Schwob *et al.*, 1994) and interphase functions for APC, although no substrates have been found at these stages.

Proteolysis directly controls a step in the chromosome cycle as well as in the cell cycle oscillator. Cell duplication is therefore more than a kinase

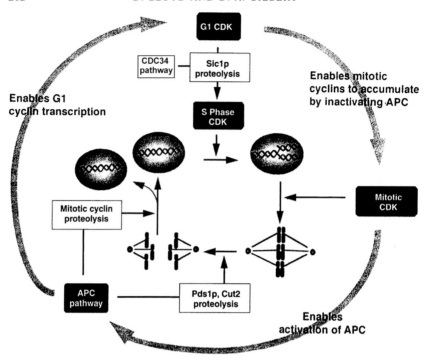

Fig. 5. How proteolysis drives the cell cycle. The model depicts a composite eukaryotic cell cycle and incorporates observations made in several different organisms. The chromosome cycle is depicted in the centre of the figure, with interphase nuclei above and mitotic spindles below. The regulatory states of the cell cycle are interconnected by a series of dependencies. Each regulatory state has two functions: to trigger a chromosomal event such as replication, chromosome alignment, or segregation, and to enable the transition to a subsequent regulatory state (grey arrows). For example, G_1 cyclin-dependent kinases (CDKs) trigger DNA replication by activating S-phase CDKs by proteolysis and also enable mitotic cyclins to accumulate by inactivating the anaphase-promoting complex (APC). Mitotic CDKs trigger chromosome condensation and spindle assembly, and also enable the activation of the APC. The active APC initiates anaphase by ubiquitinating anaphase inhibitors such as CUT2 and PDS1. The APC also catalyses destruction of cyclin B, resulting in exit from mitosis, and enabling G_1 cyclins to be resynthesized in the next cell cycle. The destruction of mitotic CDK activity is also required to allow formation of prereplication complexes, a prerequisite for DNA replication. Reproduced with permission from King et al. (1996).

cascade; phosphorylation and proteolysis are interdependent partners collaborating to effect cell division. But, if there is equilibrium between these features, then stasis would exist; that is there would be no forward 'movement' of the overall process.

Surveillance mechanisms in S. cerevisiae are responsible for the cell division cycle arrest of many DNA replication mutants and irradiated organisms in a G_2-like state (Weinert & Hartwell, 1988). Similarly, improper alignment of chromosomes on the mitotic spindle leads to a delay in the attainment of anaphase (Hoyt, Totis & Roberts, 1991; Li & Murray, 1991). In *Schizosaccharomyces pombe*, delay of mitosis until organisms reach a critical size is dependent on the Wee1 protein kinase (Nurse, 1981). In *Escherichia coli*, inhibition of DNA replication or DNA damage signals activation of RecA protein which proteolyses the LexA repressor. This induces many SOS genes, some of which produce DNA repair factors, and others that inhibit cell division (Walker, 1984).

Accurate duplication of DNA must occur before segregation and transmission to the next generation occur in alternative oscillating cycles (Nurse, 1994; Stillman, 1996). DNA replication requires the ordered assembly of many proteins (the pre-replication complex, pre-RC) at the origins of DNA replication to form a functional pre-replicative chromosomal state. In addition, at least two cell cycle regulated protein kinase pathways are necessary to effect the transition to a post-replicative chromosomal state. Re-replication of DNA is prevented by the protein kinases required to establish mitosis.

The cyclins and CDKs thus have multiple roles in the control of the cell cycle by ensuring co-ordination of S-phase with other cell cycle events. They not only activate DNA replication but prevent re-replication by blocking the formation of the pre-RC in G_2. They also allow establishment of metaphases. Thus, as the cells exit mitosis, the cell cycle is reset to allow a new competent replication state. The ordering of cell cycle events can be achieved in principle by ensuring that the initiation of an event or the commencement of a process cannot occur until the previous event has been completed (Elledge, 1996); this has been likened to a precursor–product relationship (Hartwell & Weinert, 1989). Otherwise, dependent pathways can be established by positive or negative regulatory circuits, and this is the most frequently discussed mechanism (Fig. 6). These regulatory circuits are supposed to monitor the completion of critical steps and allow subsequent cell transitions to proceed. A 'checkpoint' is considered to be a biochemical reaction pathway that ensures dependency, but the way that is achieved may not be via the 'obvious' mechanism: it is not sufficient to invoke reactions without showing that the relevant rates are consistent with the observed temporal features. Things do not happen haphazardly: they must have a cause and that can only be by coincidence or because they form part of the system. The more complex the set of reactions, the less likely it is to be universal and conserved. That a simple substrate–product relationship does not explain the ordered succession of cell cycle events was made explicit when cell fusion experiments using *Physarum polycephalum* were performed (Rusch *et al.*, 1966). These suggested that the ratio of nuclear to cytoplasmic volume of syncytia determined the timing of entry into the mitotic phase. Similar experiments with HeLa cells showed that a G_2 nucleus delayed mitotic entry until the S-phase nucleus completed DNA replication (Rao & Johnson, 1970). This was taken to indicate that S-phase nuclei produced a mitotic inhibitor. In

A

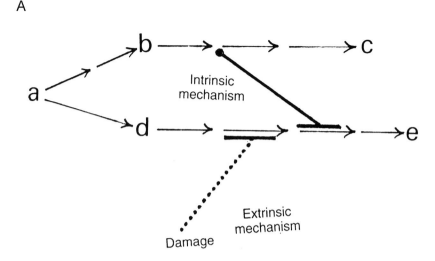

B

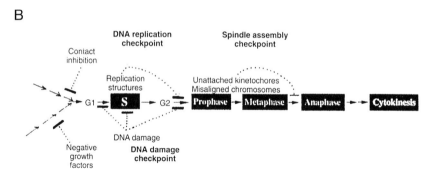

Fig. 6. Checkpoint pathway (A). A genetic pathway illustrating intrinsic and extrinsic check-point mechanisms. Letters represent cell cycle processes. The dot at the beginning of the inhibitory step indicates that, once activated, it maintains its function without the need for upstream signals. The pathway shown with a continuous line indicates an intrinsic checkpoint mechanism that operates to ensure that event c is completed before event e. After event b is completed, an inhibitory signal is activated that blocks completion of event e. After event c is completed, a signal is sent to turn off the inhibitory signal from b, thereby allowing completion of e. The dotted lines represent an extrinsic mechanism that is activated when defects such as DNA damage or spindle errors are detected. It is arbitrarily located on the d to e pathway but could also function by inhibiting a later step in the b to c pathway. In that case, the extrinsic pathway would utilize the intrinsic mechanism for cell cycle arrest. Mutations in either the intrinsic or extrinsic mechanisms would result in a checkpoint-defective phenotype. (B). Schematic repre-sentation of several cell cycle checkpoints. The arrows depict complex signalling pathways that operate in G_1 to transmit information regarding cell proliferation. The dotted lines connecting particular events and cell cycle transitions represent the inhibitory signals generated by checkpoint pathways in response to those events. The points of contact of the negative growth factor and contact inhibition pathways with the cell cycle are arbitrary and meant to indicate arrest in G_1. Reproduced with permission from Elledge (1996).

S. cerevisiae the observation of the effects of the *rad 9* mutation on cell cycle progression after DNA damage (Weinert & Hartwell, 1988), laid the foundations for the concept of the relationship between checkpoint failure, genomic instability and abnormal processes related to tumourigenesis (Hartwell & Kastan, 1994).

Cells in one cell phase appear to actively inhibit processes of another phase through checkpoints (Rao & Johnson, 1970). Thus DNA replication and mitosis are states that are biochemically incompatible; that during a transition the previous state is turned off while the future state is promoted. This is a basic characteristic of cell cycle transitions (Schwob *et al.*, 1994).

A number of other controls are coupled to those already described; it seems that mechanisms for monitoring time as well as cell mass and growth rate must operate in order to ensure orderly progress of cellular growth and division (Murray & Kirschner, 1989) as well as development in metazoans (Edgar & Lehner, 1996). Various models have been proposed; some consist of fixed series of reactions that take a definite time for completion, whereas others involve a self-sustained oscillatory mechanism (Sel'kov, 1970; Gilbert, 1974*a*, 1981) of the type provided by a limit-cycle (Fig. 7). The idea of a redox cycle as an auto-oscillating system can be traced back to Rapkine (1931). The state of the limit-cycle oscillator depends on two variables A and B which must regulate the levels of each other, thereby forming a control loop. In the example shown, auto-oscillation occurs when four requirements hold: (i) that A is converted to B; (ii) negative feedback occurs (that B inactivates or destabilizes A); (iii) that the attainment of a critical threshold of A provides the system's signal output; and (iv) that B has a definite half-life. Thus molecule A increases in concentration, giving rise to an unstable molecule B after a time delay. Molecule B, which itself has a defined half-life, in turn destabilizes molecule A. The time required for the accumulation of A, the delay in the formation of B, the rate of destabilization of A, and the half-life of B are all factors controlling the kinetics of the oscillator, that is its periodicity, amplitude and phase.

There have been many candidates for the role of mitotic oscillator. The currently favoured one involves cyclin–*cdc2* kinase interaction (Hyuver & Le Guyader, 1990; Tyson, 1991; Goldbeter, 1991, 1996; Norel & Agur, 1991; Novak & Tyson, 1993*a,b*, 1995; Tyson, Novak & Val, 1995; Tyson *et al.*, 1996). In the latest version, Tyson *et al.* (1996), used the same approach adopted 25 years earlier with the cytokinetic model (Sel'kov, 1970; Gilbert, 1974*a*) and obtained similar results but they relate the static states to interphase and metaphase arrest rather than G_1 and G_2 quiescence. These states occur when the two relevant isoclines (Tyson's 'balance' curves) intersect outside the turning points on the one isocline. However, the cyclin periodicity is apparently continuous and does not 'stop' at any precise 'checkpoint' reactions. If this view is not as envisaged by the cyclin group, then another oscillation, having a period of about the cell generation time,

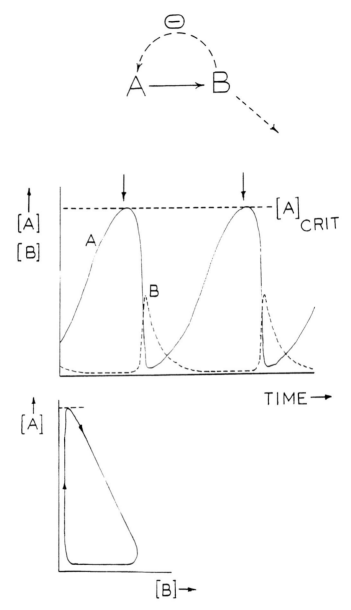

Fig. 7. The cell division cycle as a continuous limit cycle oscillator: the simplest case with a single critical threshold, that is the case where division occurs when the concentration of A reaches its maximal concentration. A threshold concentration for B could also be important or, as first suggested by Sel'kov (1970), a critical value of the ratio A/B (for example, intracellular redox potential). See also Gilbert (1982a).

must cause the reversible switching between the static states described by Tyson *et al.* (1996). Periodic behaviour must exist (i) even when all constituents are being duplicated, and, (ii) despite the fact that many, if not most, cellular constituents are oscillating at very high frequency and high amplitude (see later).

Although it is sometimes referred to as the cell cycle clock, the mitotic oscillator is not a clock in the strict sense. This is because a clock has to be continuous and regular in its characteristics and temperature compensated, whereas the cell division cycle is often irregular and always highly temperature dependent (Lloyd, 1992).

As indicated earlier, in view of the highly dynamic nature of the cell, there is good reason to question a linear-dependent sequence view for mitosis and division (*cf.* Gilbert, 1981): they can only take place when the reactants are available in the right place, in the right amounts and at the right times. The critical issue is what constitutes the driving force for these reactions and how is their co-ordination in time achieved? A reaction scheme is conceptually easy to construct, but do the actual rates and kinetics involved permit the observed pattern of dynamic behaviour? If, as many believe, the cyclin system provides the necessary actions by oscillating, then it has to not only initiate all the other reactions associated with cell replication (including division) but it must be able to ensure that all the constituents needed for mitosis are available under the right circumstances. However, it is not evident that this is so, and there is evidence that ATP and other phosphorylated compounds oscillate at high amplitudes and at high frequencies. These requirements seem extremely complex and yet, mitosis has survived many years of drastically selective evolution. The cyclin system, at least as currently perceived, is a mitosis-specific component, and yet cell replication is essentially an autocatalytic process; this implies that each cellular constituent has some effect on the net rate of its own synthesis. Such an approach suggests that the primary oscillator should be closely involved with cellular processes in general, and is thus able to co-ordinate the supply of the components, not only for mitosis, but also other processes: a coenzyme would seem to fit such a view (Gilbert, 1974a, 1988; MacKinnon & Gilbert, 1992). A coenzyme oscillator model for replication was presented many years before the cyclin studies. This was based on the cellular thiol/disulphide system (Sel'kov, 1970). It was then pointed out that other coenzyme schemes, including a phosphorylation and dephosphorylation system, could be expected to behave similarly (Gilbert, 1974a). Although this view was considered heretical, it was shown, initially, that density-dependent proliferation in culture can be explained on the basis that it only occurs when the coenzyme control system oscillates at low frequency and that quiescence is due to the attainment of a non-oscillatory state (Gilbert, 1974b, 1977, 1978; MacKinnon & Gilbert, 1992). This concept requires that a threshold exists with respect to mitogens (Gilbert, 1977, 1978a; MacKinnon & Gilbert, 1992). This molecular threshold is a measure of the sensitivity of the cell cycle switch and the number of molecules of mitogen required to initiate proliferation. This hypothesis can also provide a simple explanation for the abnormal cellular proliferation characteristics of cancer in terms of a decrease in the magnitude of this threshold. This explains *in vitro* characteristics (Gilbert, 1977; see also Eagle, Piez & Levy, 1961 and Basilico *et al.*, 1974 for evidence for the existence of a threshold). There is also some evidence that the *in vivo* growth properties are inversely related to the maximum density to which cells grow in culture, that is the magnitude of the threshold. The problem of understanding the transition to the uncontrolled proliferative state stems from the fact that the threshold change can be produced in a number of ways and that the pathway of change determines the effects of different mitogens. At this point it has to be stressed that cancer is the result of unwanted cell divisions and not repeated mitoses.

Furthermore, it has been shown that this limit cycle coenzyme oscillator model can account for all the major facets of cell replication (Gilbert, 1974*a*, 1977, 1978*a,b*, 1980, 1982*a,b*; MacKinnon & Gilbert, 1992; see also Gilbert & MacKinnon, 1992). Being largely based on experimental data involving changes in cell numbers, this concept suggests that the cell cycle reflects a cytokinetic oscillator rather than a mitotic one. The question is thus whether these are extremely independent rhythms (in which case it is necessary to consider their co-ordination in more detail) or, whether one drives the other: in view of the above comments, the possibility that a coenzyme system is the fundamental oscillator is favoured, and that it normally triggers the cyclin system (*cf.* MacKinnon & Gilbert, 1992). Exceptions are the multinucleate syncytia and plasmodia that provide examples for the dissociability of division and mitosis.

THE CIRCADIAN TIME DOMAIN

Almost all life in the natural environment experiences the daily alternation of light and dark, of higher and lower temperatures and of relative humidity. Quantitative and temporal features are determined by reaction rates and, hence, by sets of (usually non-linear) differential equations reflecting, for example, the net rates of synthesis and degradation (or utilisation) of components. The only long-term dynamic solutions of such equations are rhythmic. As a result, cellular control systems (like even non-biological processes) have the ability to oscillate when the set of reaction rates lie within certain boundary values (see, for example, Gilbert, 1978*b*; Mac-Kinnon & Gilbert, 1992). Thereby, cells have evolved rhythmic behaviour, show periods which match the daily environmental clues and can be reset daily, for example at dawn by a photic input (Roenneberg, this volume). Though entrainable by light or temperature cycles, circadian rhythms persist for many cycles ('free-run') under constant conditions of complete darkness or continuous light. Subject to intense scrutiny, their genetic control and influence is rapidly being elucidated at the molecular level (Dunlap, this volume), although it is stressed that, in themselves, genes cannot do or determine anything: they must obey the same laws as other constituents; the extent and timing of their expression are governed by other cellular constituents.

It has recently become evident that, as was shown to be the case for the ultradian clock in rapidly growing cells almost two decades ago (Edwards & Lloyd, 1978, 1980; Lloyd *et al.*, 1982*a,b*), the circadian clock-control dominates the entire functioning of the organism in slowly growing lower eukaryotes as well as in some prokaryotes. In both the bioluminescent marine dinoflagellate, *Gonyaulax polyedra* (Morse, Fritz & Hastings, 1990), and in the cyanobacterium *Synecococcus* (Golden, this volume) very many, and perhaps the majority of proteins, show circadian cycles of synthesis and active destruction. In *Gonyaulax polyedra* wholesale disintegration and disappearance of scintillons (the light-emitting organelles) can be observed as they decrease from a peak of about 300 per cell during the night-phase to only 50 per cell in day-phase organisms (Fritz, Morse & Hastings, 1990). Again, the astonishing power of proteolytic digestion as an essential step in a cyclic process of demolition and reconstitution central to cellular growth is evident. However, there is evidence in mammalian cells, that the amount of extractable protein varies at much higher frequency (that is with a period of two min or even less) (Ferreira *et al.*, 1996*c*). This raises the possibility that, due to the problem of aliasing (see above), estimates of much slower rhythms are either incorrect or they result from slow modulation of the amplitude of the rapid changes (as seen in studies on the redox state and cell morphology, Visser *et al.* (1990) and in many enzyme activity measurements). The existence of this protein rhythm also throws into doubt the value of many studies where measurements are presented in terms of specific activities (Tsilimigras & Gilbert, 1977; Ferreira *et al.*, 1994*a*).

INTERACTIONS BETWEEN PROCESSES IN DIFFERENT TIME DOMAINS

The visualization of the various time domains as an hierarchical order is useful but simplistic. The whole edifice of biological time is actually a heterarchy (Yates, 1993), with dynamic interactions at and between every level. Two well-known examples are illustrated here.

The best known cases of strict timekeeping of cell division are those of lower eukaryotes (for example, planktonic algae) that divide each day at dawn. In these organisms cell division is controlled (or 'gated') by a circadian rhythm with a period of about 24 h.

Circadian gating of the cell division cycle is well documented for *G. polyedra* (Vicker *et al.*, 1984; Homma & Hastings, 1988), *Paramecium bursaria* (Volm, 1964) and *Pyrocystis fusiformis* (Sweeney, 1982), in diatoms

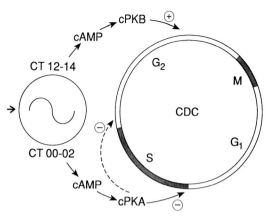

Fig. 8. Gating of the cell division cycle by the circadian oscillator. In *Euglena gracilis*, Ca^{2+} may be involved in the input (entrainment) pathway, in the circadian clock itself or in the output pathways downstream from the oscillator coupling the clock to its overt outputs (the cell division cycle). The cell division cycle is controlled at least at two points by circadian clock outputs. (CT, circadian time where CT00 indicates the phase-point of a free-running rhythm that has been normalized to 24 h and that corresponds to the identical phase-point that occurs at the onset of light in a reference LD 12:12 cycle; the onset of cell division in a dividing culture occurs at approximately CT12. cPKA, cPKB are two different cyclic AMP-dependent kinases. CDC, cell division cycle), modified from Carré & Edmunds (1993).

(Chisholm & Brand, 1981), in *Chlamydomonas reinhardii* (Goto & Johnson, 1995), and in *Euglena gracilis* where it has been studied in great detail (Edmunds, 1966; Carré & Edmunds, 1993). Control of progress through the cell division cycle in *E. gracilis* by the circadian clock involves Ca^{2+}, cyclic AMP and two distinct protein kinases (Fig. 8). A similar subservience to circadian time is seen in the gated cell division of some cyanobacteria (Golden, this volume) and in the generative tissues of higher plants and animals.

A short period (ultradian) clock exerts dominant control in both lower eukaryotic organisms and in higher animal cells in culture growing and dividing more rapidly than in the natural environment (for example, in laboratory cultures close to optimum growth conditions of temperature and nutrient supply) (Klevecz, 1976; Edwards & Lloyd, 1978; Gardner & Feldman, 1980; Gilbert, 1980; Stutzmann, Petrovic & George, 1980; Lloyd & Kippert, 1987; Readey, 1987). Here, cell division times are primarily determined by the mitotic or cytokinetic oscillator, but interaction with regular signals from the ultradian clock when growth conditions are sub-optimum give a series of permitted division times with fixed increments over and above the fundamental cell cycle duration (Lloyd & Kippert, 1993). These "quantal" increases in cell cycle time correspond to the period of the ultradian clock (Lloyd & Volkov, 1990) (Fig. 9). Thus the ultradian clock has an output which can act to control cycle progression. Mathematical models

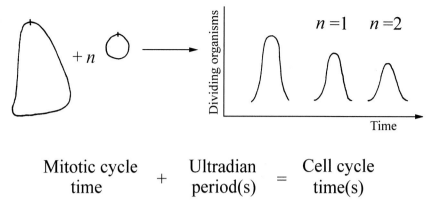

$$\text{Mitotic cycle time} + \text{Ultradian period(s)} = \text{Cell cycle time(s)}$$

Fig. 9. Interaction of the mitotic oscillator with the ultradian clock restricts increases of cell division times to fixed increments: each of these quanta equals one period of the ultradian clock.

of this interaction have been elaborated and can give an explanation for chaotic cell division cycle time dependence (Lloyd, Lloyd & Olsen, 1992). Work with cell division cycle mutants of *Schiz. pombe* indicates that interaction with events of the mitotic cycle occurs at the key cdc^{2+} control network. Thus the $wee1^{+}$ protein kinase is an essential component of the pathway coupling mitosis to the ultradian clock. On the other hand, such high frequency rhythms can act by periodically lowering the threshold (the separatrix) associated with excitability, thereby enabling triggering (Gilbert, 1982a,b).

Finally, two points are raised which stress the significance of dynamic processes and the need to view problems in the correct way if they are to be solved. Theoretical studies long ago indicated that glycolysis can be more efficient at maintaining a high ATP/ADP ratio when oscillating (Termonia & Ross, 1981). Optimization occurs at the natural (resonant) frequency. This fact has recently been used to provide a new concept of ageing. As cell duplication must require adequate supplies of ATP, it seems reasonable to expect the glycolytic efficiency to be a regulatory factor for the cell cycle. *Inter alia*, this could account for the decrease in proliferative ability with age.

Even longer-standing observations on high frequency oscillations in mammalian cells have shown that different phase relationships can exist, for example, between rhythms in normal and virally transformed fibroblasts (Gilbert, 1968, 1969, 1974b, 1980, 1984a). These observations provide support for the view that changes in the pattern of temporal organisation underlie the processes of differentiation, and can therefore also cause cancer (Gilbert, 1966, 1968, 1984b): theoretical studies describe how these phenomena can affect cell replication characteristics, while experimental work has shown that mitogens, like serum and insulin, profoundly affect cell dynamics in many different ways (Gilbert, 1984a; Gilbert & MacKinnon, 1992; Gilbert

& Visser, 1993; Visser *et al.*, 1990). The last observation emphasizes the necessity for simultaneous monitoring of many variables (that is the need to examine more than one rhythm at a time), although it is not possible to predict those systems and phase relationships that may be of significance in a particular situation.

CONCLUSIONS

Life is a whirlpool of organized change. It is the basic property of living matter to exhibit rapid cyclic alternations. Dynamic orchestration extends through a range that encompasses the inconceivably fast (femtosecond) intramolecular vibrations to the slow evolutionary modification of organisms over geological time (Gyr). It is the business of biophysics, biochemistry and cell biology to consider time constants between these extremes. Ultrafast, fast, metabolic, epigenetic, cell division cycle and circadian time scales are convenient concepts which provide not only a framework for ordering sampling routines, but also for giving insights into the temporal order of the living state and its significance. Life is an ensemble of cyclic processes in which energization and de-energization, synthesis and decay, assembly and disassembly, construction and destruction, of necessity alternate (Lloyd & Edwards, 1986; Marques *et al.*, 1987).

Replication is a highly dynamic cellular attribute involving changes in the levels of constituents and behaviour. Quantitative and temporal aspects of cell biology can only be adequately understood in terms of reaction kinetics (Gilbert, 1981). With this in mind, it is found that various dynamic features of the processes underlying cell replication, including the effects of malignant transformation on proliferation characteristics, can be explained in terms of: (i) the rhythmic behaviour of an intracellular control system determining the levels of co-enzyme (like) components, and (ii) the general properties of the oscillatory state.

ACKNOWLEDGEMENT

This work was supported by a Royal Society grant to DAG.

REFERENCES

Aon, M. A. & Cortassa, S. (1997). *Dynamic Biological Organization*, pp. 44–72. London: Chapman & Hall.

Basilico, C., Renger, H. G., Burstin, S. J. & Toniolo, D. (1974). Host cell control of viral transformation. In *Control of Proliferation in Animal Cells*, eds B. Clarkson & R. Baserga, pp. 167–76. New York: Cold Spring Harbor Laboratory.

Bowen, I. D., Bowen, M. & Jones, A. (1997). *Mitosis and Apoptosis*. London: Chapman & Hall.

Brodsky, W. Y. (1975). Protein synthesis rhythm. *Journal of Theoretical Biology*, **55**, 167–200.

Carré, I. A. & Edmunds, L. N. Jr. (1993). Oscillator control of cell division in *Euglena*: cyclic AMP oscillations mediate the phasing of the cell division cycle by the circadian clock. *Journal of Cell Science*, **104**, 1163–73.

Chance, B., Gibson, Q. H., Eisenhart, R. E. & Lonberg-Holm, K. K. (1964). *Rapid Mixing and Sampling Techniques in Biochemistry*. New York: Academic Press.

Chisholm, S. W. & Brand, L. E. (1981). Persistence of cell division, phasing in marine phytoplankton in continuous light after entrainment to light–dark cycles. *Journal of Experimental Marine Biology and Ecology*, **51**, 107–18.

Cruickshank, D. W. J., Helliwell, J. R. & Johnson, L. N. (eds) (1992). *Time Resolved Macromolecular Crystallography*. Oxford University Press.

Douzou, P. (1977). *Cryobiochemistry, An Introduction*. London: Academic Press.

Eagle, H., Piez, K. A. & Levy, M. (1961). The intracellular amino acid concentrations required for protein synthesis in cultured human cells. *Journal of Biological Chemistry*, **236**, 2039–42.

Edgar, B. A. & Lehner, C. F. (1996). Developmental control of cell cycle regulators: a fly's perspective. *Science*, **274**, 1646–52.

Edmunds, L. N. Jr. (1966). Studies on synchronous dividing *Euglena gracilis* Klebs (strain Z). *Journal of Cellular Physiology*, **67**, 35–44.

Edwards, S. W. & Lloyd, D. (1978). Oscillations of respiration and adenine nucleotides in synchronous cultures of *Acanthamoeba castellanii*: mitochondrial respiratory control *in vivo*. *Journal of General Microbiology*, **108**, 197–204.

Edwards, S. W. & Lloyd, D. (1980). Oscillations in protein and RNA content during synchronous growth of *Acanthamoeba castellanii*: evidence for periodic turnover of macromolecules during the cell cycle. *Federation of European Biochemical Societies Letters*, **109**, 21–6.

Elledge, S. J. (1996). Cell cycle checkpoints: preventing an identity crisis. *Science*, **274**, 1664–72.

Enright, J. T. (1980). Temporal precision in circadian systems: a reliable neuronal clock from unreliable components? *Science*, **209**, 1542–5.

Evans, T., Rosenthal, E. T., Youngbloom, J., Distel, D. & Hunt, T. (1983). Cyclin: a protein specified by maternal RNA that is destroyed at each cleavage division. *Cell*, **33**, 389–96.

Ferreira, G. M. N., Hammond, K. D. & Gilbert, D. A. (1994*a*). Oscillatory variation in the amount of protein extractable from marine erythroleukemia cells: stimulation by insulin. *BioSystems*, **32**, 183–90.

Ferreira, G. M. N., Hammond, K. D. & Gilbert, D. A. (1994*b*). Insulin stimulation of high frequency phosphorylation dynamics in murine erythroleukemic cells. *BioSystems*, **33**, 31–43.

Ferreira, G. M. N., Wolfe, H., Hammond, K. D. & Gilbert, D. A. (1996*a*). High frequency oscillations in the activity of phosphotyrosine phosphatase in murine erythroleukemic cells: action of insulin and hexamethylene bisacetamide. *Cell Biology International*, **20**, 599–605.

Ferreira, G. M. N., Hammond, K. D. & Gilbert, D. A. (1996*b*). Independent high-frequency oscillations in the amounts of individual isoenzymes of lactate dehydrogenase in HL60 cells. *Cell Biology International*, **20**, 607–11.

Ferreira, G. M. N., Hammond, K. D. & Gilbert, D. A. (1996*c*). Distinct, very high frequency oscillations in the activity and amount of active isoenzyme of lactate dehydrogenase in murine erythroleukemic cells and a cell-free system. *Cell Biology International*, **20**, 625–33.

Francis, D. (1992). The cell cycle in plant development. *New Phytologist*, **122**, 1–20.

Francis, D. & Halford, N. G. (1995). The plant cell cycle. *Physiologia Plantarum*, **93**, 365–74.

Fritz, L., Morse, D. & Hastings, J. W. (1990). The circadian bioluminescence rhythm of *Gonyaulax* is related to the daily variations in number of light emitting organelles. *Journal of Cell Science*, **95**, 321–8.

Gardner, G. F. & Feldman, J. F. (1980). The frq locus in *Neurospora crassa*: a key element in circadian clock organization. *Genetics*, **96**, 877–86.

Gilbert, D. A. (1966). Isoenzymes and cell regulation. *Discovery*, **27**, 23–6.

Gilbert, D. A. (1968). Differentiation, oncogenesis and cellular periodicities. *Journal of Theoretical Biology*, **21**, 113–33.

Gilbert, D. A. (1969). Phase plane analysis of periodic isozyme pattern changes in cultured cells. *Biochemical and Biophysical Research Communications*, **37**, 860–6.

Gilbert, D. A. (1974*a*). The temporal response of the dynamic cell to disturbance and its possible relationship to differentiation and cancer. *South African Journal of Science*, **70**, 234–44.

Gilbert, D. A. (1974*b*). The nature of the cell cycle and the control of cell proliferation. *BioSystems*, **5**, 197–206.

Gilbert, D. A. (1977). Density-dependent limitation of growth and the regulation of cell replication by changes in the triggering level of the cell cycle switch. *BioSystems*, **9**, 215–28.

Gilbert, D. A. (1978*a*). The mechanism of action and interaction of regulators of cell replication. *BioSystems*, **10**, 227–33.

Gilbert, D. A. (1978*b*). The relationship between the transition probability and oscillator concepts of the cell cycle and the nature of the commitment to replication. *BioSystems*, **10**, 234–40.

Gilbert, D. A. (1980). Mathematics and cancer. In *Mathematical Modelling in Biology and Ecology*, ed. W. Getz, pp. 97–115. Berlin: Springer (Lecture notes in biomathematics, vol. 33).

Gilbert, D. A. (1981). The cell cycle 1981: one or more limit cycle oscillation. *South African Journal of Science*, **77**, 541–6.

Gilbert, D. A. (1982*a*). Cell cycle variability: the oscillator model of the cell cycle yields transition probability alpha and beta type curves. *BioSystems*, **15**, 317–30.

Gilbert, D. A. (1982*b*). The oscillator cell cycle model needs no first or second chance event. *BioSystems*, **15**, 331–9.

Gilbert, D. A. (1984*a*). Temporal organisation, re-organisation and disorganisation in cells. In *Cell Cycle Clocks*, ed. L. N. Jr. Edmunds, pp. 5–25. New York: Marcel Dekker Inc.

Gilbert, D. A. (1984*b*). The nature of tumour cell proliferation. *Nature*, **311**, 160.

Gilbert, D. A. (1988). On G_0 and cell cycle controls. *BioEssays*, **9**, 135–6.

Gilbert, D. A. & MacKinnon, H. (1992). Oscillations and cancer. In *Ultradian Rhythms in the Life Sciences*, ed. D. Lloyd & E. Rossi, pp. 71–87. London: Springer-Verlag.

Gilbert, D. A. & Tsilimigras, C. W. A. (1981). Cellular oscillations: the relative independence of enzyme activity rhythms and periodic variations in the amount of extractable protein. *South African Journal of Science*, **77**, 66–72.

Gilbert, D. A. & Visser, G. R. (1993). Insulin-induced enhancement of cell morphological dynamics: non-specific biophysical mechanisms for the generalized stimulation of metabolism? *BioSystems*, **29**, 143–9.

Goldbeter, A. (1991). A minimal cascade model for the mitotic oscillator involving cyclin and *cdc2* kinase. *Proceedings of the National Academy of Sciences, USA*, **88**, 9107–11.

Goldbeter, A. (1996). *Biochemical Oscillations and Cellular Rhythms*. Cambridge: Cambridge University Press.

Goodwin, B. C. (1963). *Temporal Organization in Cells*. London: Academic Press.

Goto, K. & Johnson, C. H. (1995). Is the cell division cycle gated by a circadian clock? The case of *Chlamydomonas reinhardtii*. *Journal of Cell Biology*, **129**, 1061–9.

Hartwell, L. H. (1978). Cell division from a genetic perspective. *Journal of Cell Biology*, **77**, 627–37.

Hartwell, L. H. (1991). Twenty-five years of cell cycle genetics. *Genetics*, **129**, 975–80.

Hartwell, L. H. (1992). Defects in cell cycle checkpoint may be responsible for the genomic instability of cancer cells. *Cell*, **71**, 543–6.

Hartwell, L. H. & Kastan, M. B. (1994). Cell cycle control and cancer. *Science*, **266**, 1821–8.

Hartwell, L. H. & Weinert, T. A. (1989). Checkpoints: control that ensures the order of cell cycle events. *Science*, **246**, 629–34.

Homma, K. & Hastings, J. W. (1988). Cell cycle synchronization of *Gonyaulax polyedra* by filtration: quantized generation times. *Journal of Biological Rhythms*, **3**, 49–58.

Hoyt, M. A., Totis, L. & Roberts, B. T. (1991). *S. cerevisiae* genes required for cell cycle arrest in response to loss of microtubule function. *Cell*, **66**, 507–17.

Hyuver, C. & Le Guyader, H. (1990). MPF and cyclin: modelling of the cell cycle minimum oscillator. *BioSystems*, **24**, 85–90.

Jacobs, T. (1992). Control of the cell cycle. *Developmental Biology*, **153**, 1–15.

King, R. W., Deshaies, R. J., Peters, J-M. & Kirschner, M. W. (1996). How proteolysis drives the cell cycle. *Science*, **274**, 1652–9.

Klevecz, R. R. (1976). Quantized generation time in mammalian cells as an expression of the cellular clock. *Proceedings of the National Academy of Sciences, USA*, **73**, 4012–16.

Klevecz, R. R. (1992). A precise circadian clock from chaotic cell cycle oscillations. In *Ultradian Rhythms in Life Processes*, ed. D. Lloyd & E. R. Rossi, pp. 41–70. London: Springer-Verlag.

Li, R. & Murray, A. W. (1991). Feedback control of mitosis in budding yeast. *Cell*, **66**, 519–31.

Lloyd, D. (1987). The cell division cycle. *Biochemical Journal*, **242**, 313–21.

Lloyd, D. (1992). Intracellular timekeeping: epigenetic oscillations reveal the functions of an ultradian clock. In *Ultradian Rhythms in Life Processes*, ed. D. Lloyd & E. R. Rossi, pp. 5–22. London: Springer-Verlag.

Lloyd, D. (1998). Circadian and ultradian clock-controlled rhythms in unicellular microorganisms. *Advances in Microbial Physiology and Biochemistry*, **39**, 291–338.

Lloyd, D. & Edwards, S. W. (1984). Epigenetic oscillations during the cell cycles of lower eukaryotes are coupled to a clock. Life's slow dance to the music of time. In *Cell Cycle Clocks*, ed. L. N. Jr. Edmunds, pp. 27–46. New York: Marcel Dekker.

Lloyd, D. & Edwards, S. W. (1986). Temperature-compensated ultradian rhythms in lower eukaryotes: periodic turnover coupled to a timer for cell division. *Journal of Interdisciplinary Cycle Research*, **17**, 321–26.

Lloyd, D. & Kippert, F. (1987). A temperature-compensated ultradian clock explains temperature-dependent quantal cell cycle times. In *Temperature and Animal Cells* (Soc. Explt. Biol. Symp. no. 41), ed. K. Bowler & J. B. Fuller, pp. 135–55. Cambridge: Society of Biologists.

Lloyd, D. & Kippert, F. (1993). Intracellular coordination by the ultradian clock. *Cell Biology International*, **17**, 1047–52.

Lloyd, D. & Rossi, E. R. (1992). *Ultradian Rhythms in Life Processes*. London: Springer.

Lloyd, D. & Rossi, E. R. (1993). Biological rhythms as organization and information. *Biological Reviews*, **68**, 563–77.

Lloyd, D. & Stupfel, M. (1991). The occurrence and functions of ultradian rhythms. *Biological Reviews*, **66**, 275–99.

Lloyd, D. & Volkov, E. (1990). Quantized cell cycle times: interaction between a relaxation oscillator and ultradian clock pulses. *BioSystems*, **23**, 305–10.

Lloyd, D., Edwards, S. W. & Fry, J. C. (1982*a*). Temperature-compensated oscillations is respiration and cellular protein content in synchronous cultures of *Acanthamoeba castellanii*. *Proceedings of the National Academy of Sciences, USA*, **79**, 3785–8.

Lloyd, D., Poole, R. K. & Edwards, S. W. (1982*b*). *The Cell Division Cycle: Temporal Organisation of Cellular Growth and Reproduction*, pp. 523. London: Academic Press.

Lloyd, D., Lloyd, A. L. & Olsen, L. F. (1992). The cell division: a physiologically plausible dynamic model can exhibit chaotic solutions. *BioSystems*, **27**, 17–24.

Lu, K. P. & Means, A. R. (1993). Regulation of the cell cycle by calcium and calmodulin. *Endocrine Reviews*, **14**, 40–58.

MacKinnon, H. & Gilbert, D. A. (1992). To divide or not to divide? That is the question. In *Fundamentals of Medical Cell Biology*, vol. 7, Developmental Biology, ed. E. Bittar, pp. 1–14. Greenwich: JAI Press.

Mano, Y. (1970). Cytoplasmic regulation and cyclic variation in protein synthesis in the early cleavage stage of the sea urchin embryo. *Developmental Biology*, **22**, 433–60.

Marques, N., Edwards, S. W., Fry, J. C., Halberg, F. & Lloyd, D. (1987). Temperature-compensated ultradian variations in cellular protein content of *Acanthamoeba castellanii* revisited. In *Advances in Chronobiology*, Part A, ed. J. Pauly & L. E. Scheving, pp. 105–119. New York: Alan R. Liss.

Martin, J. L. & Vos, M. H. (1992). Femtosecond biology. *Annual Reviews of Biophysics and Biomolecular Structure*, **21**, 199–222.

Moffat, K. (1989). Time resolved macro-molecular crystallography. *Annual Reviews of Biophysics and Biophysical Chemistry*, **18**, 309–22.

Morse, D., Fritz, L. & Hastings, J. W. (1990). What is the clock? Translational regulation of the circadian bioluminescence. *Trends in Biochemical Science*, **15**, 262–5.

Murray, A. W. (1992). Creative blocks: cell cycle checkpoints and feedback controls. *Nature*, **159**, 599–604.

Murray, A. M. & Kirschner, M. W. (1989). Dominoes and clocks: the union of two views of the cell cycle. *Science*, **246**, 614–21.

Nasmyth, K. (1996). Viewpoint: putting the cell cycle in order. *Science*, **274**, 1643–5.

Norel, R. & Agur, Z. (1991). A model for the adjustment of the mitotic clock by cyclin and MPF levels. *Science*, **251**, 1076–8.

Novak, B. & Tyson, J. J. (1993*a*). Modelling the cell division cycle: M-phase trigger, oscillations and size control. *Journal of Theoretical Biology*, **165**, 101–34.

Novak, B. & Tyson, J. J. (1993*b*). Numerical analysis of a comprehensive model of M-phase control in *Xenopus* cycles, extracts and intact embryos. *Journal of Cell Science*, **106**, 1158–68.

Novak, B. & Tyson, J. J. (1995). Quantitative analysis of a molecular model of mitotic control in fission yeast. *Journal of Theoretical Biology*, **173**, 283–305.

Nurse, P. (1981). Genetic analysis of the cell cycle. In *Genetics as a Tool in Microbiology*, SGM Symposium 31, ed. S. W. Glover & D. A. Hopwood, pp. 291–316. Cambridge: Cambridge University Press.

Nurse, P. (1990). Universal control mechanism regulating onset of M-phase. *Nature*, **344**, 503–8.

Nurse, P. (1994). Ordering S-phase and M-phase in the cell cycle. *Cell*, **79**, 547–50.

Rao, P. N. & Johnson, R. T. (1970). Mammalian cell fusion: studies on the regulation of DNA synthesis and mitosis. *Nature*, **225**, 159–64.

Rapkine, L. (1931). Sur les processes chimiques au course de la division cellulaire. *Annals of Physiology and Physicochemistry*, **7**, 382–418.

Readey, M. A. (1987). Ultradian photosynchronization in *Tetrahymena pyriformis* GLC is related to modal cell generation time: further evidence for a common timer model. *Chronobiology International*, **4**, 195–208.

Reich, J. G. & Sel'kov, E. E. (1981). *Energy Metabolism of the Cell*. London: Academic Press.

Rusch, H. P., Sachsenmaier, W., Behrens, K. & Gruter, V. (1966). Synchronization of mitosis by the fusion of the plasmodia of *Physarum polycephalum*. *Journal of Cell Biology*, **31**, 204–9.

Schwob, E., Bohm, T., Mendenhall, M. D. & Nasmyth, K. (1994). The b-type cyclin kinase inhibitor p40 controls G_1 to S transition in *S. cerevisiae*. *Cell*, **79**, 233–44.

Sel'kov, E. E. (1970). Two alternative, self oscillating stationary states in thiol metabolism – two alternative types of cell division: normal and malignant ones. *Biophysika*, **15**, 1065–73.

Sherr, C. J. (1996). Cancer cell cycles. *Science*, **274**, 1672–7.

Sherrington, C. S. (1955). *Man on his Nature*, 2nd edn. Hermonsworth: Penguin Books.

Srajer, V., Teng, T., Ursby, T., Pradervand, C., Ren, Z., Adachi, S., Schildkamp, W., Bourgeois, D., Wulff, M. & Moffat, K. (1996). Photolysis of the carbon monoxide complex of myoglobin: nanosecond time resolved crystallography. *Science*, **274**, 1726–9.

Stillman, B. (1996). Cell cycle control of DNA replication. *Science*, **274**, 1659–64.

Stutzmann, J., Petrovic, A. & George, D. (1980). Life cycle length, number of cell generations, mitotic index and model chromosome number as estimated in tissue culture of normal and sarcomatous bone cells. In *Bone and Tumours*, 3rd Symp. CEMO, ed. A. Donath & B. Courvoisier, pp. 188–94. Geneva: Editions Médicine et Hygiene.

Sweeney, B. M. (1982). Interaction of the circadian cycle with the cell cycle in *Pyrocystis fusiformis*. *Plant Physiology*, **70**, 272–6.

Termonia, Y. & Ross, J. (1981). Oscillations and control features in glycolysis: analysis of resonance effects. *Proceedings of the National Academy of Sciences, USA*, **78**, 3563–6.

Tsilimigras, C. W. A. & Gilbert, D. A. (1977). High frequency, high amplitude oscillations in the amount of protein extractable from cultured cells. *South African Journal of Science*, **77**, 123–5.

Tyson, J. J. (1991). Modeling the cell division cycle: *cdc2* and cyclin interactions. *Proceedings of the National Academy of Sciences, USA*, **88**, 7328–32.

Tyson, J. J., Novak, B. & Val, J. (1995). Checkpoints in the cell cycle from a modeler's perspective. In *Progress in Cell Cycle Research*, Vol. 1, ed. L. Meijer, S. Guidet & H. Y. L. Tung, pp. 1–8. New York: Plenum Press.

Tyson, J. J., Novak, B., Odell, G. M., Chen, K. & Thron, C. D. (1996). Chemical kinetic theory: understanding cell-cycle regulation. *Trends in Biochemical Sciences*, **21**, 89–96.

Vicker, M. G., Becker, J., Gebauer, G., Schill, W. & Rensing, L. (1984). Circadian rhythms of cell cycle processes in the marine dinoflagellate, *Gonyaulax polyedra*. *Chronobiology International*, **5**, 5–17.

Visser, G., Reinten, C., Coplan, P., Gilbert, D. A. & Hammond, K. (1990). Oscillations in cell morphology and redox state. *Biophysical Chemistry*, **37**, 383–94.

Volm, M. (1964). Die tagesperiodikder zeltielung von *Paramecium bursaria*, *Zell Vergl Physiologie*, **48**, 157–80.

Von Bertalanffy, L. (1952). *Problems of Life*, pp. 216. New York: Harper & Brothers.

Walker, G. C. (1984). Mutagenesis and inducible response to DNA damage in *Escherichia coli. Microbiological Reviews*, **48**, 60–93.

Wang, J. Y. J. (1992). Oncoprotein phosphorylation and cell cycle control. *Biochemica et Biophysica Acta*, **1114**, 179–92.

Weiner, N. (1954). *The Human Use of Human Beings*, p. 96. New York: Doubleday.

Weinert, T. A. & Hartwell, L. H. (1988). The *RAD9* gene controls the cell cycle. Response to DNA damage in *Saccharomyces cerevisiae. Science*, **241**, 317–22.

Yates, F. E. (1993). Self-organizing systems. In *The Logic of Life: The Challenge of Integrative Physiology*, ed. C. A. R. Boyd & D. Noble, pp. 189–218. Oxford: University Press.

THE CIRCADIAN REGULATORY SYSTEM IN
NEUROSPORA

JAY C. DUNLAP, JENNIFER J. LOROS,
SUSAN K. CROSTHWAITE, YI LIU,
NORMAN GARCEAU, DEBORAH BELL-PEDERSEN,
MARI SHINOHARA, CHENGHUA LUO,
MICHAEL COLLETT, ANNE B. COLE
AND CHRISTIAN HEINTZEN

*Department of Biochemistry, Dartmouth Medical School, Hanover,
NH 03755-3844, USA*

INTRODUCTION

Nearly all eukaryotic organisms exhibiting photobiological responses, and even some photoresponsive prokaryotes, possess the capacity for endogenous temporal control and organization over their activities. Because of the dominant influence of the daily light–dark and corresponding daily temperature cycle on living things, a pre-eminent aspect of temporal regulation is seen as 24-hour cycles – diurnal rhythms in behaviour, activity, metabolism and development. It has been known now for several hundred years that such rhythms are endogenous (that is, they arise from biochemistry within the organism) but can be synchronized with external cycles. In the absence of external time cues, the rhythms continue with period lengths of about a day – *circadian* – whence they derive their common name of circadian rhythms. The catch-all phrase for everything in an organism involved with rhythmic control and expression of rhythmic behaviour is the circadian system, and the cellular and biochemical machinery underlying this ability is known as the biological clock. All circadian clocks share a few common characteristics: a period length of about 24 h, the ability to be reset by light and a temperature compensation mechanism, whereby period length is kept nearly constant despite changes in ambient temperature. Circadian clocks dictate the timing of cell division in a variety of microbes and vertebrate tissues, regulate mating in *Paramecium* and reproductive timing in mammals (including people), photosynthesis in plants, algae and cyanobacteria, development in plants (through the clock's central role in photoperiodic timing) and in fungi, and contribute to the regulation of almost every physiological parameter measurable in people. In humans, clock malfunction is linked to a variety of psychiatric disorders including displaced sleep-phase syndromes, seasonal affective disorders and some types of manic-depressive illness. A well-adapted clock contributes to our physical and mental health in a variety of ways. Whether microbe or mammal, all levels of cellular and organismal organization display temporal control.

Big problems such as dissection of a circadian system are made more digestible by breaking them into smaller parts. In this context, there are three aspects to all circadian systems, corresponding to the three great questions in circadian biology; (i) input: how do environmental signals reset the clock? (ii) the oscillator: what are the gears and cogs in the feedback loop generating time and how is the loop assembled? (iii) output: how is time information generated by the clock transduced within the cell to effect changes in the behaviour of the oscillator cell and thereby (in a multicellular organism) to change the behaviour of the organism? Understanding how each of these works, and how they all work together, is a truly great research problem, perhaps one of the last great problems in cellular regulation and moreover, a research problem that touches all of biology.

A WELL-UNDERSTOOD MODEL ORGANISM FOR UNDERSTANDING THE CLOCK AT THE LEVEL OF THE CELL

All circadian oscillators are assembled at the level of the cell; intercellular communication plays no role in the basic feedback loop oscillator in any organism. Given the cellular basis of the clock, the ubiquitous nature of these rhythms and their shared characteristics, any student of modern molecular biology would be comfortable in asserting that common elements and mechanisms will be found working in all circadian clocks. For this reason, microbial systems, and systems amenable to genetic analysis, have long been appreciated for the insights they can provide into the mechanism of the clock and the means by which clocks act to regulate the metabolism and behaviour of cells. The microbial clock best understood at the molecular level, and arguably the best-understood circadian system in any context, is that of the ascomycete fungus *Neurospora crassa*, in which a circadian clock controls several aspects of growth and development.

The cartoon shown in Fig. 1 is used to keep sight of the overall picture and of each of the separate research approaches. One avenue has been to identify and characterize clock components through the identification, cloning and characterization of genes that, when mutated, affect the expression of rhythmicity. To this end, seven genes have been identified in genetic screens for clock-affecting genes, two other genes required for rhythmicity have been found, and the involvement of three of these (*frequency, wc-1*, and *wc-2*) have been characterized in some detail (Dunlap, 1993, 1996; Loros, 1995). A second approach has been aimed at understanding how environmental cues such as light and temperature act to synchronize internal cycles with external time cues. In this work, existing strains defective in blue-light photoresponses in *Neurospora* have yielded insights into how light acts to reset the pacemaker mechanism and also how the mechanism itself is assembled. Thirdly, the

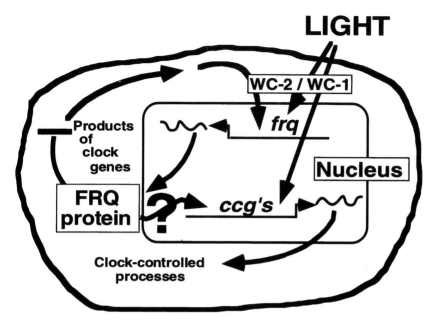

Fig. 1. A schematic view of the rhythmic cell provides a framework for understanding the nature of circadian biology. The clock mechanism is represented by a feedback oscillator possessing the ability to keep time with a reliable period length regardless of changes in temperature or the metabolic environment. Identified components of the oscillator are the *frq* gene and FRQ proteins, which may have several functions, including down-regulation of *frq* mRNA, as part of the negative feedback loop that creates the circadian oscillator. Another role of FRQ is to transcriptionally regulate, either directly or indirectly, the rhythmic activation of clock controlled gene (*ccg*) expression. Identification of components in the light entrainment pathway, such as the *wc* genes, gives clues to how clocks perceive environmental stimuli and how and with what kinetics that stimuli is effective in resetting the subjective time of the cell. The *wc* genes are also required for sustained rhythmicity and *frq* expression in the dark. Figure adapted from Aronson *et al.* (1994*a*).

pathways through which clocks act to control cellular metabolism and behaviour are being dissected. Initially, genes whose transcript levels are controlled by the clock were identified. Utilizing a subtractive hybridization procedure two such genes, designated *clock controlled gene-1* (*ccg-1*) and *clock-controlled gene-2* (*ccg-2*) were cloned (Loros *et al.*, 1989). Six additional genes have now been identified through the use of differential hybridization (Bell-Pedersen *et al.*, 1996*b*).

THE CIRCADIAN OSCILLATOR IS A NEGATIVE AUTOREGULATORY FEEDBACK LOOP INVOLVING THE PRODUCTS OF THE *FREQUENCY* GENE

A number of genes have been identified in *Neurospora* having the characteristic that, when these genes are mutated, the result is an organism in which the period length and sometimes the temperature compensation properties of the pacemaker are altered. Many of these genes were identified by Jerry Feldman in screens for genes which when mutated affected the operation of the clock (Feldman, 1982; Feldman, Gardner & Dennison, 1979; Morgan & Feldman, 1997). Seven of them, *cla-1*, *chr*, and *prd-1,2,3,4*, and 6, are identified by single alleles while an eighth locus, called *frequency* (*frq*) is identified by 11 independently isolated alleles. There are a number of additional genes isolated in other contexts that might be classified as clock-affecting (see, for instance, Dunlap, 1993), since mutations in these genes also result in real, albeit generally small, effects on the clock.

How does the clock itself work? Nearly all of what is known about how the circadian oscillator actually works in *Neurospora* is based on genetic and molecular biological studies that have shed light on the actions and regulation of *frq*, a gene that encodes a central component of a feedback loop that is essential for normal operation of this circadian clock (Aronson *et al.*, 1994*a*; Aronson, Johnson & Dunlap, 1994*b*; Dunlap, 1996). The circadian oscillator in *Neurospora* is a negative autoregulatory feedback cycle (Aronson *et al.*, 1994*a*), wherein rhythmic transcription of the *frq* gene yields a transcript that encodes two forms of the protein FRQ, a long form of 989 amino acids and a shorter form of 890 amino acids, which result from alternative initiation of translation (Garceau *et al.*, 1997). Both forms of FRQ contain a functional nuclear localization signal (C. Luo, J. Loros & J. Dunlap, unpublished observations), serine-rich and hyperacidic regions (Aronson *et al.*, 1994*b*), and a region predicted to have structural similarity to a helix-turn-helix DNA binding domain (Lewis & Feldman, 1993), all of which are phylogenetically conserved (Merrow & Dunlap, 1994; Lewis & Feldman, 1997; Lewis, Morgan & Feldman, 1997). The levels of *frq* mRNA and FRQ proteins cycle in amount (Aronson *et al.*, 1994*a*; Garceau *et al.*, 1997), the latter acting, at least indirectly, to depress the level of the *frq* transcript (Aronson *et al.*, 1991*a*; Merrow, Garceau & Dunlap, 1997). No distinct functions have been assigned to either form of FRQ, so FRQ is applied to connote both forms, unless otherwise specified. Importantly, rhythmic change in the amount of *frq* transcript is required for the overt circadian rhythm as no level of constant *frq* expression supports the overt rhythm and step changes in the expression of *frq* reset the clock (Aronson *et al.*, 1994*a*). Since rhythmic, autoregulated expression and not merely the unregulated expression of *frq* is essential for the circadian clock, the levels of *frq* and FRQ are therefore not just components but state variables of the circadian oscillator, elements whose oscillation defines the clock.

A journey through molecular time

The *Neurospora* clock cycle can be imagined starting from subjective midnight in a cell kept in constant darkness and constant temperature where only internal cellular processes will affect its timecourse. At this time, *frq* and FRQ levels are low but *frq* transcript is beginning to rise, a process that requires the WC-2 (and maybe WC-1) proteins (Crosthwaite, Dunlap & Loros, 1997) (see also below) and that takes about 10–12 h to reach peak. Following an approximately 4 h lag, FRQ protein begins to appear (Garceau *et al.*, 1997). FRQ expressed following light induction (Garceau, 1996) or from a heterologous promoter (Merrow *et al.*, 1997) appears with little delay, suggesting that the 4 h lag seen in the normal cycle is a regulated part of the circadian cycle (Merrow *et al.*, 1997). Soon after its synthesis FRQ enters the nucleus (Luo, C., Loros, J. & Dunlap, J. unpublished observations; Garceau, 1996) where it acts to repress its own expression (Aronson *et al.*, 1994*a*; Merrow *et al.*, 1997) and probably other genes (including perhaps some of the eight known *ccg*s, Bell-Pedersen *et al.*, 1996*b*; Loros, 1995). *frq* mRNA levels peak in the mid-morning (Aronson *et al.*, 1994*a*; Crosthwaite *et al.*, 1995) around 4 h before the peak of FRQ which occurs in the early afternoon (Garceau *et al.*, 1997). As soon as either form of FRQ can be seen, it is partially phosphorylated and both forms become progressively more phosphorylated over the course of the day. At midday *frq* mRNA is at its peak and during the afternoon *frq* transcript levels fall whilst FRQ declines through the early night after having become extensively phosphorylated, consistent with a model in which the phosphorylation triggers its turnover. By midnight, the levels of both RNA and protein are low and the cycle can begin again.

ENVIRONMENTAL INPUT ENTRAINS THE CLOCK BY AFFECTING THE
LEVELS OF *frq* MRNA AND FRQ PROTEIN

Light

Visible light can reset the clock in every rhythmic organism examined to date. Light is perceived via a photoreceptor, the molecular entity that most directly interfaces with the environment. In different organisms photoreceptors sensitive to different wavelengths of light are utilized. Photoreceptor pigments may vary widely among organisms, and in fact more than one photoreceptor may be involved in the entrainment of different circadian rhythms within a single organism (for instance, Engelman & Mack, 1978; Roenneberg & Rehman, this volume). Among the photoreceptors suggested as being involved in clock resetting are phytochrome (widespread in plants but also known in fungi, Kaufman, 1993), rhodopsin, and blue light photoreceptors (clearly involved in insects, some plants, and in the fungi); among these phytochrome (Quail, 1991), a flavin-mediated blue light

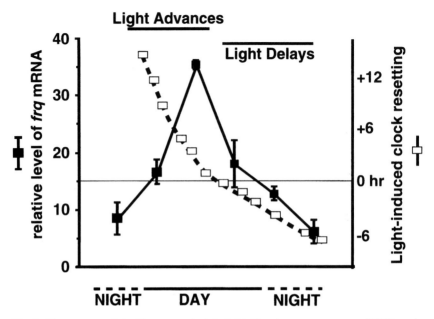

Fig. 2. The response of the *Neurospora* clock to light. The phase response curve (PRC) can be predicted from the response of *frq* transcript levels to light. The molecular cycling in the level of *frq* mRNA (thin solid line), plotted as mean ± SE (*n* = 3), is correlated in time with the response of the circadian oscillator to light, the *Neurospora* PRC (thick broken line) as determined by Dharmananda (1980). In broad outline delays occur from late day through to early evening, whilst advances are seen from late night through to early morning. The times when light causes an advance or a delay can be predicted simply from the known response of the *frq* gene to light, and the knowledge that the clock is reset by changes in the level of *frq* expression. Figure adapted from Crosthwaite *et al.* (1995).

photoreceptor (Ahmad & Cashmore, 1993), and an opsin (Max *et al.*, 1995) have been cloned. The *Neurospora* photoreceptor has yet to be identified, but some are now hot on its trail (see Macino *et al.*, this volume).

Light resets the clock by acting very rapidly (within minutes) to increase the level of the *frq* transcript; the dose response for this light induction matches the dose response for clock resetting. For light to synchronize an internal clock with the environment successfully, the same light signal must elicit different responses at different times of day: light in the early subjective evening should be interpreted as a long day and delay the clock by moving it back into the previous daytime, whereas light in the early subjective morning should be interpreted as an early dawn and advance the clock into the next day (Crosthwaite *et al.*, 1995; Pittendrigh, 1976, 1993). In broad outline, the response of *frq* to light explains this organismal resetting response – light in the late night/early morning acts rapidly to bring *frq* transcript and FRQ levels up to their peak, a process that would normally take several hours (Fig. 2). This precocious rise causes an advance in the rhythm. Conversely, light in the late day and early evening delays the rhythm. This results from the light-induction of *frq* which restores *frq* transcript levels to the peak amounts from which they have declined over the past hours. Newly synthesized *frq* mRNA gives rise to new FRQ protein which begins the slow process of phosphorylation and eventual turnover. Light appears to act directly on the *frq* promoter, independently from the negative feedback loop, rather than just acting to interrupt the negative regulation by the clock (Crosthwaite, Loros & Dunlap, 1995).

Efforts to dissect the signal transduction pathway, whereby light triggers the activation of *frq*, surprisingly led to the identification of two more proteins required for the execution of the feedback loop (Crosthwaite, Dunlap & Loros, 1997). These are the global regulatory genes, *white collar-1* (*wc-1*) and *wc-2*, that govern nearly all known light responses in *Neurospora*, including the light induction of *frq* RNA (Harding & Turner, 1981; Harding & Melles, 1983; Russo, 1988; Ballario *et al.*, 1996; Linden & Macino, 1997; Macino *et al.*, this volume). Although no functions had been assigned to these proteins in the dark, *frq* levels were surprisingly found to be vanishingly low in the dark in these strains, the mutant strains display no overt rhythms, and rhythmicity could not be initiated by temperature treatments which act independently of the light response pathway (see below) (Crosthwaite *et al.*, 1997). In short, *wc-1* and *wc-2* appear to encode novel clock components and/or clock-associated functions, and for this reason are included both in the feedback loop and within the light input pathway in Fig. 1. Both *wc-1* and *wc-2* have recently been cloned and appear to encode members of the GATA family of transcription factors (Ballario *et al.*, 1996; Linden & Macino, 1997). Each contains a zinc-finger DNA binding region, a transcriptional activation domain, and PAS domains putatively associated with protein–protein interactions. Although the two molecularly characterized circadian clocks, in *Neurospora* and in *Drosophila*, are both believed to be transcription/translation based negative feedback loops, these WC proteins are the first clock proteins having any known biochemical function, and specifically the first transcription factors to be clearly associated with oscillator function. The identification of PAS domains in both WC proteins was exciting and unanticipated. These are functional domains that had turned up in two different classes of proteins, one involving photoresponse proteins including bacterial photoresponse proteins like PYP (Ahmad & Cashmore, 1993), and the plant photoreceptor phytochrome (Lagarias, Shu-Hsing & Lagarias, 1995; Quail *et al.*, 1995) and the other in the *Drosophila* clock protein PERIOD (Huang, Edery & Rosbash, 1993). The identification of PAS domains in the WC proteins associated both with photoresponsivity and with clock function thus suggested that clock proteins may have evolved from ancient proteins involved with photoresponses (Crosthwaite *et al.*, 1997). Finally, the similarity in the PAS region between PERIOD and the WC proteins is the first example of sequence similarity among clock-associated proteins and, given the distant evolutionary connections between flies and fungi, suggested that other putative clock components might be identified through these conserved structures (Crosthwaite *et al.*, 1997). A recently identified clock-associated protein in the mouse contains sequence motifs found in the WC proteins, having a PAS domain, DNA binding region and putative transcriptional activation domains (Antoch *et al.*, 1997; King *et al.*, 1997). These data have provided convincing confirmation that the molecular mechanisms of many, if not all, circadian oscillators will be similar and that *Neurospora* is an excellent model system for describing the underpinnings of vertebrate and mammalian clocks as well as microbial clocks. The data suggest a working model in which *frq* is normally activated at night by one or both WC proteins acting perhaps in concert (Crosthwaite *et al.*, 1997), although this has yet to be shown. Since *frq* can always be light induced, WC-1 is probably always there, but not necessarily always active: perhaps it can be slowly activated as a part of the clock cycle, or rapidly activated in response to light. The identification of these positive elements (*wc-1* and *wc-2*) effectively closes the clock loop, conceptually, for the first time and provides a concrete molecular connection between effects of light on the clock and the progress of the clock in the dark.

Temperature influences the clock

Along with light, temperature is the other major universally active clock resetting agent. As noted above, *frq* gives rise to two forms of FRQ through alternative translation initiation sites. When mutations were induced in *frq* that eliminated the use of either the long or short form of FRQ, it was found that either form alone was sufficient to complete a functional clock at some temperatures but that under all conditions both forms of FRQ were necessary for robust overt rhythmicity (Liu *et al.*, 1997). This is due, apparently, to the facts that, as the temperature rises, more of both forms of FRQ are made and that temperature regulates the ratio of the two FRQ forms, different temperatures favoring different initiation codons. When either initiation codon is eliminated, the temperature range permissive for rhythmicity is demonstrably reduced. This temperature-influenced choice of translation initiation site represents a novel adaptive mechanism that extends the physiological temperature range over which the clock can function. Data further indicated that there is a threshold level of FRQ required to establish the feedback loop, which comprises the oscillator, and that this threshold level increases as the temperature increases (Liu *et al.*, 1997). Altogether, these data may explain how the temperature limits permissive for rhythmicity are established, thus providing a molecular understanding for one of the basic characteristics of circadian clocks (Liu *et al.*, 1997; Pittendrigh, 1993).

Light

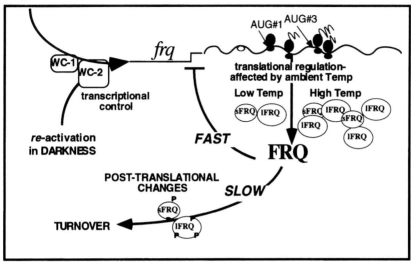

Fig. 3. Temporal dynamics of *frq* gene expression and the effect of temperature on FRQ protein production. The findings and the summary figures of Garceau *et al.* (1997) and Liu *et al.* (1997) have been assimilated in this figure. Light onset in natural conditions activates the transcription of *frq* (horizontal line with arrow). When the level of FRQ is low and the response to light is fast, *frq* mRNA can increase. Hence, a DELAY occurs between the rise of the transcript's level and that of FRQ. Post-translational components of FRQ's temporal dynamics occur as slow processes over the course of most of the daytime. These processes include progressive phosphorylation as well as turnover. Light onset restarts the cycle, but progression into a new cycle (involving WC-1 and WC-2) can also occur in complete darkness (Crosthwaite *et al.*, 1997). *frq* mRNA (wavy line) contains an open reading frame (ORF) having both upstream and downstream AUGs in a good context for translation initiation. Ribosomes (ovals, with the small subunit by itself being open) are believed to scan the *frq* transcript, beginning upstream of the 5'-most (upstream) AUG, continuing and eventually reaching the downstream one. Ambient temperature influences (a) the absolute amount of total FRQ protein; and (b) the efficiency of the ribosomes' two initiation events. As the temperature rises, ribosomes successfully initiate at the first AUG more often so more of the large form of FRQ is made. Figure adapted from Hall (1997).

A summary cartoon describing just the oscillator-specific aspects of the clock cycle is shown in Fig. 3. The general path and characteristics of the oscillator cycle are now known with some confidence, although there are a large number of issues still under study. In particular, we still do not know what FRQ actually does or how its activity and stability are regulated. Furthermore, we do not know what the immediate regulators of the WC proteins are nor their immediate targets. With the identification of the obligate role of WC-2 within the feedback loop and the requirement of both WC-1 and WC-2 for sustained rhythmicity in the dark (Crosthwaite *et al.*, 1997), there are now more clock genes described in *Neurospora* than in any

Table 1. *A summary of the* ccgs *identified in* Neurospora

Gene	Peak	Amplitude	Location	Light	Develop.	Identity
ccg-1	CT3	8X	LGVR	+	+	stress
ccg-2	CT22	10X	LGIIR	+	+	hygrophobin
ccg-4	CT5	22X	LGIL	+	+	?
ccg-6	CT19	18X	LGVR	+	+	regulator?
ccg-7	CT21	9X	LGIIR	−	−	GAPDH
ccg-8	CT20	7X	LGVC	−	−	?
ccg-9	CT19	9X	LGVIIL	+	+	regulator?
ccg-12	CT18	4X	LGVIL	−	−	CuMT

Other clock-regulated genes have also been identified in other contexts; see text for details. The circadian time when each of the clock-controlled transcripts peak and the fold difference between peak and trough levels is shown. Circadian time (CT) is a formalism whereby the period of an organism is divided in to 24 equal parts. By convention CT 0 is subjective dawn and CT 12 subjective dusk. The chromosomal location (linkage group, LG) of each gene determined by RFLP analysis and regulation of transcript level by light and during development is noted: no change (−) or increased levels (+) of transcript.

other system and the roles of these proteins in the clock are better understood.

DESCRIBING CLOCK OUTPUT AT THE LEVEL OF MOLECULES

During the 1980s subtractive and differential hybridization screens were developed in order to identify classes of genes that were co-ordinately regulated by environmental or endogenous stimuli. Utilizing subtractive hybridization, the first global screen for clock regulated genes was executed in *Neurospora* in the mid-1980s and resulted in the identification of two 'clock-controlled genes' or *ccgs* (Loros, Denome & Dunlap, 1989). The term '*ccg*' is now used generically in the field of rhythms to refer to components of the output pathway, or to terminally regulated products, as distinct from elements within the oscillator. The initial genes identified in this study, *ccg-1* and *ccg-2* (Loros *et al.*, 1989) have now both been shown to be controlled by the clock at the level of transcription (Loros & Dunlap, 1991). Differential hybridization has been used more recently to identify six additional *ccgs* (Bell-Pedersen *et al.*, 1996*b*) (Table 1). An important finding to come out of this latest round of screens is that, although daily regulation of development by the clock is the major overt rhythm in *Neurospora*, not all of the *ccg* genes are required for, or even associated with, the execution of the daily developmental program. For instance, CCG-7 was identified as the single *Neurospora* glyceraldehyde-3-phosphate dehydrogenase (GAPDH) (Shinohara, Loros & Dunlap, 1997), a pivotal control enzyme in the glycolytic pathway of all living things. This provides the first evidence for circadian regulation of both a gene and protein found in all living things. Moreover, GAPDH is not induced by light, stress, or development, suggesting that circadian regulation may affect the core of metabolism and cannot be thought of as simply an epiphenomenon that lies at the periphery of the life of the cell. With other genes now known to be clock-regulated, including several *con* genes (for instance, Lauter & Yanofsky, 1993) and *al-3* (Arpaia, Carottoli & Macino, 1995*a*), nearly a dozen different circadianly regulated output genes are known in *Neurospora*. They display a diverse set of functions involved in development, stress response and intermediary metabolism. As the largest collection of clock-regulated genes known in any organism, they provide a window into the role of the clock in the life of the organism. Additionally, they provide an excellent context in which to study the mechanisms through which clock regulation is achieved.

Genes that are important to the life of the organism and that interface with the environment will be acutely regulated by a variety of factors in addition to the clock; some of these may be independent of the clock or co-dependent with the clock and all aspects of regulation need to be integrated at the level of the target controlled gene. A major effort in recent years has been devoted to developing an understanding of this complex regulatory network, the first steps of which have involved the dissection of a clock regulated promoter with the goal of identifying each of the constituent parts of the promoter (Bell-Pedersen, Dunlap & Loros, 1996*a*). The cartoon in Fig. 4 shows the results of this work for *eas* (*ccg-2*) which, with the *Arabidopsis* CAB2 gene (Anderson & Kay, 1995), is the best described circadianly regulated promoter. Moving in a 5′ to 3′ direction from the more distal regions of the promoter towards the start site of transcription, there is (i) a general transcriptional activator (enhancer) between 1900 and 625 bp from the start of transcription, which is disrupted by the canonical *eas* mutation (Bell-Pedersen *et al.*, 1992). This enhancer lies within (ii) a mycelia specific repressor that acts to keep *ccg-2* repressed prior to terminal differentiation. More proximal to the promoter lies (iii) the positive acting light element that is required to mediate the acute activation of the gene by light. Still closer to the start site of transcription is (iv) a light responsive element that acts to modify the kinetics of the light response but which is not absolutely required for the light response. Finally, closest to the transcription start site lies (v) the activating circadian element (ACE) or positive clock element, now localized to a small region between -57 and -102 (Bell-Pedersen *et al.*, 1996*a*). Sequences within this region are similar to sequences found in other clock-regulated and light-regulated genes. The ACE is both necessary and sufficient for conferring circadian regulation: removal of the ACE eliminates circadian regulation (reducing *ccg-2* transcript to the level characteristic of the low points in the cycle) but leaves other aspects of regulation intact. Furthermore, insertion of the ACE sequence upstream of a basal non-clock regulated promoter confers circadian clock regulation on

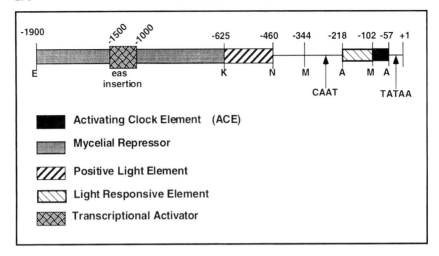

Fig. 4. Diverse regulatory sequences govern the expression of the clock-controlled gene *eas* (*ccg-2*). To determine the identity and importance of various parts of the promoter, promoter DNA was removed and the effects of the changes on the regulation of the gene subsequently examined. When the activating circadian element (ACE) was localized in this way through resection analysis, its identity was confirmed further by placing it in front of a constitutively expressed (non-clock-regulated) promoter, and showing that just the 45 base pairs of the ACE are necessary and sufficient to confer morning specific clock regulation (Bell-Pedersen *et al.*, 1996*a*). K (*KpnI*), N (*NdeI*), M (*MscI*) and A (*AvaII*) refer to the positions of restriction endonuclease cleavage sites within the promoter of the *ccg-2* gene. Figure adapted from Bell-Pedersen *et al.* (1996*a*).

a reporter gene (β-galactosidase) driven by the naive promoter (Bell-Pedersen *et al.*, 1996*a*). A protein factor from nuclear extracts that binds to the ACE region has now been identified (Bell-Pedersen, Loros & Dunlap, unpublished observations); this is a candidate for the clock-regulated *trans*-acting factor that confers circadian regulation on *ccg-2* and potentially on other clock-regulated genes in *Neurospora*.

Separate collaborative work has focused on the light regulation of the *ccg* genes, a number of which are rapidly induced following brief exposure to blue light (Arpaia *et al.*, 1993, 1995*b*). In a *frq* null (clock disrupted) strain, these genes remained photoinducible, suggesting that the signal for this particular light induction is not obligately transduced through the clock but instead acts on the genes directly. Light regulation of the genes was blocked in strains bearing single gene mutations in either *wc-1* or *wc-2*, which are known to result in photo-blindness as discussed above for *frq*. This *ccg* data shows that clock-regulated morning-specific genes can be acutely regulated by light; in this way external (e.g. environmental) and internal (e.g. circadian) signals reinforce each other in driving the expression of downstream target genes at appropriate times of day.

CONCLUSIONS FROM A MODEL SYSTEM

Of the various approaches taken to understanding the molecular basis of circadian rhythmicity, the greatest advances in understanding have come from genetic and molecular genetic analysis. The *Neurospora* system has already provided an excellent model system in which to pursue this and the recent identification of a clock-associated gene from the mouse that is a potential sequence and functional homologue of the *wc* genes has added convincing proof of the validity of the *Neurospora* model, supporting this as an approach for understanding circadian oscillations in more complex organisms. Genetics originally identified the *frequency* locus as encoding a possible clock component and molecular analysis has shown the two products of this locus to be clock components. The *Neurospora* circadian oscillator is now understood to be an autoregulatory feedback loop wherein the products of the *frq* locus act to regulate their own expression and thereby help to regulate the behaviour of the cell as a function of time. It is now clear that the products of the *wc-1* and *wc-2* genes also participate in this feedback loop, helping in some way to promote the expression of *frq* at an appropriate time. Light resets the clock by acting rapidly to promote transcription and thus to increase the level of *frq* mRNA; here again the *wc* gene products play an essential role. Temperature acts principally at the level of translation to influence both the total amount of FRQ made and also the relative amount of the two forms of FRQ. In this way the physiological temperature limits for rhythmicity are set and temperature steps can act to reset the clock. Nearly a dozen output genes are now known in *Neurospora* and they are involved in a variety of cellular functions including development, stress responses and the core of intermediary metabolism. The *cis*-acting elements mediating circadian regulation of target genes are now being elucidated.

Within the past few years the identity of several components of circadian oscillators have been revealed as well as the overall scheme governing the assembly of these components to create this and perhaps all circadian clocks. This knowledge in the context of existing theory led to the discovery of the means through which these oscillators are reset by light and temperature. Since the behavioural characteristics of all clocks are so similar, it is expected that this knowledge of clock mechanism and resetting will pave the way for understanding other microbial, as well as animal and plant, clocks.

ACKNOWLEDGEMENTS

This work was supported by grants from the National Institute of Health (GM34985 and MH01186 to J.C.D and MH44651 to J.C.D and J.J.L.), the AFOSR (F49620-94-1-0260 to J.J.L.), the National Science Foundation (MCB-9307299 to J.J.L.), and the Norris Cotton Cancer Center core grant at Dartmouth Medical School.

REFERENCES

Ahmad, M. & Cashmore, A. R. (1993). HY4 gene of *A. thaliana* encodes a protein with characteristics of a blue-light photoreceptor. *Nature*, **366**, 162–6.

Anderson, S. L. & Kay, S. A. (1995). Functional dissection of circadian clock and phytochrome-regulated transcription of the *Arabidopsis CAB2* gene. *Proceedings of the National Academy of Sciences, USA*, **92**, 1500–4.

Antoch, M., Soog, E., Chang, A., Vitaterna, M., Zhao, Y., Wilsbacher, L., Sangoram, A., King, D., Pinto, L. & Takahashi, J. (1997). Functional identification of the mouse circadian *CLOCK* gene by transgenic BAC rescue. *Cell*, **89**, 655–67.

Aronson, B., Johnson, K., Loros, J. J. & Dunlap, J. C. (1994*a*). Negative feedback defining a circadian clock: Autoregulation in the clock gene *frequency*. *Science*, **263**, 1578–84.

Aronson, B. D., Johnson, K. A. & Dunlap, J. C. (1994*b*). The circadian clock locus *frequency*: A single ORF defines period length and temperature compensation. *Proceedings of the National Academy of Sciences, USA*, **91**, 7683–7.

Arpaia, G., Loros, J. J., Dunlap, J. C., Morelli, G. & Macino, G. (1993). The interplay of light and the circadian clock: independent dual regulation of clock-controlled gene *ccg-2 (eas)*. *Plant Physiology*, **102**, 1299–305.

Arpaia, G., Carattoli, A. & Macino, G. (1995*a*). Light and development regulate the expression of the *albino-3* gene in *Neurospora crassa*. *Developmental Biology*, **170**, 626–35.

Arpaia, G., Loros, J. J., Dunlap, J. C., Morelli, G. & Macino, G. (1995*b*). The circadian clock-controlled gene *ccg-1* is induced by light. *Molecular and General Genetics*, **247**, 157–63.

Ballario, P., Vittorioso, P., Magrelli, A., Talora, C., Cabibbo, A. & Macino, G. (1996). *White collar-1*, a central regulator of blue-light responses in *Neurospora crassa*, is a zinc-finger protein. *EMBO Journal*, **15**, 1650–7.

Bell-Pedersen, D., Dunlap, J. C. & Loros, J. J. (1992). The *Neurospora* circadian clock-controlled gene, *ccg-2*, is allelic to *eas* and encodes a fungal hydrophobin required for the formation of the conidial rodlet layer. *Genes and Development*, **6**, 2382–94.

Bell-Pedersen, D., Dunlap, J. C. & Loros, J. J. (1996*a*). Distinct cis-acting elements mediate clock, light, and developmental regulation of the *Neurospora crassa eas (ccg-2)* gene. *Molecular and Cellular Biology*, **16**, 513–21.

Bell-Pedersen, D., Shinohara, M., Loros, J. J. & Dunlap, J. C. (1996*b*). Circadian clock-controlled genes isolated from *Neurospora crassa* are late night to early morning specific. *Proceedings of the National Academy of Sciences, USA*, **93**, 13096–101.

Crosthwaite, S. K., Loros, J. J. & Dunlap, J. C. (1995). Light-induced resetting of a circadian clock is mediated by a rapid increase in *frequency* transcript. *Cell*, **81**, 1003–12.

Crosthwaite, S. K., Dunlap, J. C. & Loros, J. J. (1997). *Neurospora wc-1* and *wc-2*: Transcription, photoresponses, and the origins of circadian rhythmicity. *Science*, **276**, 763–9.

Dharmananda, S. (1980). Studies of the circadian clock of *Neurospora crassa*: Light-induced phase shifting. PhD thesis, University of California, Santa Cruz.

Dunlap, J. C. (1993). Genetic analysis of circadian clocks. *Annual Review of Physiology*, **55**, 683–728.

Dunlap, J. C. (1996). Genetic and molecular analysis of circadian rhythms. *Annual Review of Genetics*, **30**, 579–601.

Engelman, W. & Mack, J. (1978). Different oscillators control the circadian rhythms of eclosion and activity in *Drosophila*. *Journal of Comparative Physiology*, **127**, 229–37.

Feldman, J. F. (1982). Genetic approaches to circadian clocks. *Annual Review of Plant Physiology*, **33**, 583–608.

Feldman, J. F., Gardner, G. F. & Dennison, R. A. (1979). Genetic analysis of the circadian clock of *Neurospora*. In *Biological Rhythms and Their Central Mechanism*, ed. M. Suda, pp. 57–66. Amsterdam: Elsevier.

Garceau, N. (1996). Molecular and genetic studies on the *frq* and *ccg-1* loci of *Neurospora*. PhD thesis, Dartmouth.

Garceau, N., Liu, Y., Loros, J. J. & Dunlap, J. C. (1997). Alternative initiation of translation and time-specific phosphorylation yield multiple forms of the essential clock protein FREQUENCY. *Cell*, **89**, 469–76.

Hall, J. (1997). Circadian pacemakers blowing hot and cold – but they're clocks, not thermometers. *Cell*, **90**, 9–12.

Harding, R. W. & Melles, S. (1983). Genetic phototropism of *Neurospora crassa* perithecial beaks using white collar and albino mutants. *Plant Physiology*, **72**, 996–1000.

Harding, R. W. & Turner, R. V. (1981). Photoregulation of the carotenoid biosynthetic pathway in albino and white collar mutants of *Neurospora crassa*. *Plant Physiology*, **68**, 745–9.

Huang, Z. J., Edery, I. & Rosbash, M. (1993). PAS is a dimerization domain common to *Drosophila* Period and several transcription factors. *Nature*, **364**, 259–62.

Kaufman, L. (1993). Transduction of blue-light signals. *Plant Physiology*, **102**, 333–7.

King, D., Zhao, Y., Sangoram, A., Wilsbacher, L., Tanaka, M., Antoch, M., Steeves, T., Vitaterna, M., Kornhauser, J., Lowrey, P., Turek, F. & Takahashi, J. (1997). Positional cloning of the mouse circadian *CLOCK* gene. *Cell*, **89**, 641–53.

Lagarias, D. M., Shu-Hsing, W. & Lagarias, J. C. (1995). Atypical phytochrome gene structure in the green alga *mesotaenium caldariorum*. *Plant Molecular Biology*, **29**, 1127–42.

Lauter, F. R. & Yanofsky, C. (1993). Day/night and circadian rhythm control of *con* gene expression in *Neurospora*. *Proceedings of the National Academy of Sciences, USA*, **90**, 8249–53.

Lewis, M. & Feldman, J. F. (1997). Evolution of the *frequency* clock locus in ascomycete fungi. *Molecular and Biological Evolution*, **13**, 1233–41.

Lewis, M. T. & Feldman, J. F. (1993). The putative frq clock protein of *Neurospora crassa* contains sequence elements that suggest a nuclear transcriptional regulatory role. *Protein Sequencing and Data Analysis*, **5**, 315–23.

Lewis, M., Morgan, L. & Feldman, J. F. (1997). Cloning of (*frq*) clock gene homologs from the *Neurospora sitophila* and *Neurospora tetrasperma*, *Chromocrea spinulosa* and *Leptosphaeria australiensis*. *Molecular and General Genetics*, **253**, 401–14.

Linden, H. & Macino, G. (1997). White collar-2, a partner in the blue-light signal transduction, controlling expression of light-regulated genes in *Neurospora crassa*. *EMBO Journal*, **16**, 98–109.

Liu, Y., Garceau, N., Loros, J. J. & Dunlap, J. C. (1997). Thermally regulated translational control mediates an aspect of temperature compensation in the *Neurospora* circadian clock. *Cell*, **89**, 477–86.

Loros, J. (1995). The molecular basis of the Neurospora clock. *Seminars in Neuroscience*, **7**, 3–13.

Loros, J. & Dunlap, J. C. (1991). *Neurospora crassa* clock-controlled genes are regulated at the level of transcription. *Molecular and Cellular Biology*, **11**, 558–63.

Loros, J. J., Denome, S. A. & Dunlap, J. C. (1989). Molecular cloning of genes under the control of the circadian clock in *Neurospora*. *Science*, **243**, 385–8.

Max, M., McKinnon, P., Seidenman, K., Barrett, R., Applebury, M., Takahashi, J. & Margolskee, R. (1995). Pineal opsin: a nonvisual opsin expressed in chick pineal. *Science*, **267**, 1502–6.

Merrow, M. & Dunlap, J. C. (1994). Intergeneric complementation of a circadian rhythmicity defect: phylogenetic conservation of the 989 amino acid open reading frame in the clock gene *frequency*. *EMBO Journal*, **13**, 2257–66.

Merrow, M., Garceau, N. & Dunlap, J. C. (1997). Dissection of a circadian oscillation into discrete domains. *Proceedings of the National Academy of Sciences, USA*, **94**, 3877–82.

Morgan, L. & Feldman, J. (1997). Isolation and characterization of a temperature-sensitive circadian clock mutant in *Neurospora crassa*. *Genetics*, **146**, 525–30.

Pittendrigh, C. S. (1976). Circadian clocks: What are they? In *The Molecular Basis of Circadian Rhythms*, eds J. W. Hastings and H. G. Schweiger, pp. 11–48. Berlin: Abakon Verlagsgesellschaft.

Pittendrigh, C. S. (1993). Temporal organization: reflections of a Darwinian clock-watcher. *Annual Review of Physiology*, **55**, 17–54.

Quail, P. (1991). Phytochrome: a light-activated molecular switch that regulates plant gene expression. *Annual Review of Genetics*, **25**, 389–409.

Quail, P., Boylan, M. T., Parks, B. M., Short, T. W., Xu, Y. & Wagner, D. (1995). Phytochromes: photosensory perception and signal transduction. *Science*, **268**, 675–80.

Russo, V. E. A. (1988). Blue light induces circadian rhythms in the *bd* mutant of Neurospora; double mutants *bd,wc-1* and *bd,wc-2* are blind. *Journal of Photochemistry and Photobiology, B; Biology*, **2**, 59–65.

Shinohara, M., Loros, J. J. & Dunlap, J. C. (1997). Glyceraldehyde-3-phosphate dehydrogenase is regulated on a daily basis by the circadian clock. *Journal of Biological Chemistry* submitted.

THE MOLECULAR GENETICS OF *DROSOPHILA* CLOCKS

CHARALAMBOS P. KYRIACOU, ALBERTO PICCIN AND EZIO ROSATO

Department of Genetics, University of Leicester, Leicester LE1 7RH, UK

Most eukaryotes and some prokaryotes show biological cycles in behaviour and physiology, which are controlled by endogenous biological clocks. A layman-friendly topic such as biorhythm research, gets its fair share of reviews. In a characteristically whimsical recent report on the molecular biology of clock genes, Jeff Hall (Hall, 1996), considered something of a sage in these matters, mused that the highest frequency biological rhythm was the appearance rate of clock reviews '...every 15 s'. As if to prove his point, here is another review...

Biological cycles can be circadian, ultradian or infradian, with periods of about 24 h, considerably less or considerably more than 24 h, respectively. The possibility that similar molecular mechanisms are operating in these different time-scales has recently been reviewed (Iwasaki & Thomas, 1997), and there are tantalizing glimpses in the literature of common control elements that span the different temporal domains (Hall & Kyriacou, 1990; Hall, 1995). The best-studied biological oscillators, however, are the circadian clocks, which generate phenotypes that are relatively easy to monitor and thereby amenable to mutagenesis. The organisms of choice here are Cyanobacteria (Kondo *et al.*, 1994), *Drosophila* (Konopka & Benzer, 1971; Sehgal *et al.*, 1994), *Neurospora* (Loros, 1995) and *Arabidopsis* (Millar & Kay, 1997), and more recently, the mouse (Vitaterna *et al.*, 1994). Furthermore, the forward genetic approach of mutate, map and clone, has been enhanced by reverse genetic methods, where, using various criteria, molecules that might be important for clock function are first identified, before the corresponding gene is isolated. This has also had its successes in *Neurospora* (Loros, 1995), *Drosophila* (van Gelder & Krasnow, 1995) and *Xenopus* (Green & Besharse, 1996). However, for the purposes of this review, the focus will be almost exclusively on insect clocks, and particularly on *Drosophila*.

A MODEL FOR THE CIRCADIAN CLOCK

Circadian time is not a linear but a circular dimension, characterized by a period of 24 h, and various models involving transcriptional/translational,

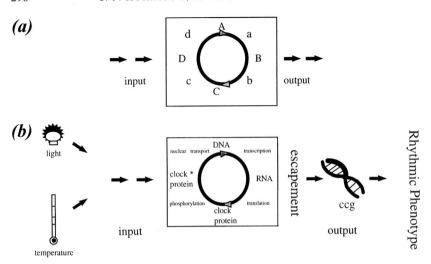

Fig. 1. (*a*) Model of a circadian clock with a number of state variables, A, B, C and D. a, b, c and d are the parameters of the feedback loop. Input and output signals are shown as arrows. (*b*) The model above, translated into a molecular feedback loop. The negative feedback loop involves the transcriptional and translational control of a clock protein, whereby both the mRNA and protein oscillate out of phase with each other. The protein is translated and undergoes a number of steps, for example, phosphorylation generating a lag (see text), before nuclear entry and feedback on to the gene. The parameters of transcription, translation, phosphorylation and nuclear entry are shown, as is the 'escapement' from the loop by which the state variables control the clock-controlled genes (*ccg*s) which generate the rhythmic phenotypes. * denotes phosphorylated protein.

biochemical and membrane loops have been proposed (Edmunds, 1988), which rely on some kind of negative feedback loop. It might be imagined *a priori*, that a circadian oscillation will be determined by dynamic elements called 'state variables' which describe the state of the system at various points (or phases) in the cycle, and these will be continually changed by the 'parameters' of the system. In other words, a clock molecule that represents a state variable, might be expected to show a circadian cycle in some aspect of its activity. Figure 1 (top) shows a general mechanism for a circadian clock. A state variable evolves from one state to another, by the action of a number of parameters, around a negative loop. To this is added an input pathway which keeps the endogenous clock in phase with the external world, and an output pathway which represents the 'hands' of the clock and is reflected in rhythmic phenotypes.

How can this general model be translated into a biological model? Possible state variables are DNA (genes), mRNA, clock proteins. Possible parameters are transcription, translation, post-translational modifications, nuclear transport, etc. (Fig. 1 bottom). Although to date there is no universally agreed set of criteria by which to identify a state variable, some requirements are particularly relevant, for example:

(i) the state variable should display circadian periodicity in some aspect of its activity;
(ii) induced variations in the state variable should change (by negative feedback) its own activity;
(iii) modification of a state variable should affect the clock;
(iv) external stimuli able to influence the clock (i.e. light, temperature) must be able to change the activity of the state variable. Conversely, changes in the state variable must mimic the effects of changes produced by these external stimuli;
(v) abolishing the oscillation of a state variable should also abolish the rhythmicity of the clock.

Thus whenever a 'clock' molecule is identified, it is important to discover whether it represents a component of the input to the clock, the pacemaker itself, or the output. In practice, this is not always very easy.

THE *DROSOPHILA* CLOCK

In *D. melanogaster* there are two genes, *period* (*per*) and *timeless* (*tim*), whose products have been identified as putative state variables of the clock. The two loci control the circadian pupal–adult eclosion rhythm, and the adult locomotor activity cycle, and both genes were originally defined by mutation (Konopka & Benzer, 1971; Sehgal et al., 1994). Both the eclosion and the locomotor activity rhythms represent true circadian phenotypes in that they show periods of about 24 h under constant conditions (constant darkness, DD, constant temperature), they can be entrained (they respond to external stimuli, light and temperature), and they are temperature compensated (the period does not significantly change over a broad range of temperatures). The classic *per* mutations, per^{L1}, per^s and per^{01} lengthened, shortened and obliterated respectively, both the eclosion and the locomotor activity cycles (Konopka & Benzer, 1971), whereas the first *tim* mutant, tim^0, like per^{01}, showed an arrhythmic phenotype (Sehgal et al., 1994).

The PER and TIM products satisfy a number of the state variable criteria outlined above. Both transcripts exhibit circadian cycles in abundance in the head, with a peak early in the night phase (Hardin, Hall & Rosbash, 1990: Sehgal et al., 1995). PER (Edery et al., 1994) and TIM (Hunter-Ensor, Ousley & Sehgal, 1996; Lee et al., 1996; Myers et al., 1996; Zeng et al., 1996) also show circadian rhythms in abundance, both in light/dark (LD) cycles (see Fig. 2) and under constant darkness (DD), but their peak abundance is about 6 h later than their peak mRNA levels, a lag suggesting some post-transcriptional delay. Both proteins are also progressively phosphorylated during the night phase (Edery et al., 1994), but the function of this cycle of post-translational modification is unclear (Fig. 2). Differences between PER and TIM cycles include an approximate 2 h anticipation of TIM

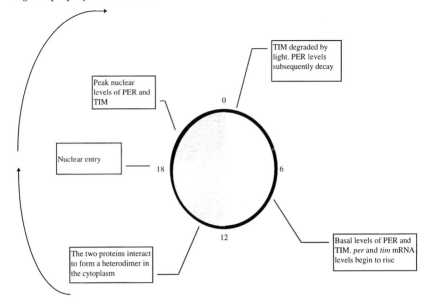

Progressive phosphorylation in the nucleus

Progressive phosphorylation in the cytoplasm

Fig. 2. The PER and TIM negative feedback loop (see text). The circadian cycle is shown as a circle with a day (Zeitgeber Time, ZT 0–12) and night phase (ZT 12–24, white at night, shaded during day). PER and TIM start to accumulate in the cytoplasm at the beginning of the night. At this time the mRNA levels are high and the protein levels can rise. Both proteins start to be progressively phosphorylated (TIM less extensively than PER). TIM is more stable than PER in the cytoplasm so it can drive PER accumulation by associating with it until the middle of the night, when the complex enters the nucleus. The two proteins stay together for the next 6 h or so. The nuclear entry of PER and TIM approximately coincides with the decline of *per* and *tim* mRNA levels, in agreement with the hypothesis of the feedback loop. In nuclei, both proteins are phosphorylated further. PER now is more stable than TIM which disappears more quickly. Thus in the last part of the night PER is found both as a complex with TIM and as a monomer. By the beginning of the day, TIM has disappeared. PER instead persists for few more hours, until ZT4–ZT6. This difference can, in part, be explained by the finding that TIM is degraded after light exposure, whereas the direct effect of light on PER is minimal. The fact that *per* and *tim* mRNA levels start to rise only in the middle of the day, after PER has completely disappeared, might indicate that PER could control transcription by associating with another protein in the absence of TIM (Hunter-Ensor, Ousley & Sehgal, 1996: Lee *et al.*, 1996; Myers *et al.*, 1996; Zeng *et al.*, 1996).

accumulation with respect to PER, and a much quicker decrease of TIM after its peak level (Lee *et al.*, 1996; Zeng *et al.*, 1996). Because the PER and TIM protein cycles go up as their transcripts go down, this implies a negative autoregulatory feedback loop of transcription and translation. In *per*[s] mutants, which have short 19 h cycles due to a single amino acid substitution (Yu *et al.*, 1987), the *per* mRNA cycles with a correspondingly truncated period in DD, further suggesting a feedback of the PER protein on its own mRNA (Hardin *et al.*, 1990).

How PER or TIM regulate their own mRNA cycles in DD is unknown, as no DNA-binding domains have been identified in either product. Using reporter constructs, it has been demonstrated that the *per* mRNA cycle is transcriptionally regulated (Hardin, Hall & Rosbash, 1992). Furthermore, within the *per* promoter, a sequence which corresponds to an E-box (to which bHLH proteins can bind) has been identified, which, if mutated, significantly reduces the amplitude of *per* mRNA cycling (Hao, Allen & Hardin, 1997). A second regulatory element in the transcribed region of the mRNA is also required, in addition to the promotor sequences, to give the normal wild-type *per* mRNA cycling dynamics (Stanewsky *et al.*, 1997).

Certain neurons found in lateral regions of the fly brain have been identified as putative pacemakers for the locomotor activity cycle (Ewer *et al.*, 1992). In these cells (termed 'lateral neurons' – LNs, Siwicki *et al.*, 1988; see also Vosshall & Young, 1995), in photoreceptor neurons (e.g. Curtin, Huang & Rosbash, 1995; Myers *et al.*, 1996; Hunter-Ensor *et al.*, 1996), and in the Malpighian tubules (Giebultowicz & Hege, 1997) dramatic cycles of PER and TIM abundance have been visualized. Immunohistochemical studies focusing on the LNs and visual system neurons in LD12:12 cycles, have revealed that, during the early part of the night, the two products are seen to accumulate in the cytoplasm (for review see Helfrich-Foerster, 1996). The mRNA is at peak levels early at night, so if there was no lag in translation of this pool of mRNA, high protein levels would be expected soon afterwards. However, peak levels of PER and TIM are seen several hours later in the dark phase, so translation appears to be delayed (Helfrich-Foerster, 1996). When both proteins reach their peak levels late at night, they are observed to move into the nucleus. Consequently there are two lags, one between peak transcription and peak protein levels, and the second, which represents the delay between the proteins appearing in the cytoplasm until nuclear translocation (Curtin *et al.*, 1995). Current negative feedback models (Goldbeter, 1995; Ruoff & Rensing, 1996), require a lag between the appearance of the transcript and protein, otherwise any attempt at oscillation grinds to an immediate halt. One of the unsolved problems for the PER/TIM negative feedback model is how these lags between peak levels of mRNA protein and nuclear entry, are generated.

In *tim*^{0} mutants, a PER-reporter protein fails to translocate to the nucleus (Vosshall *et al.*, 1994), and so the implication is that the *tim* gene encodes a nuclear translocator for PER. The two products would therefore associate to form a heterodimer which passes through the nuclear membrane. This dimerization has been demonstrated both *in vivo* and *in vitro* (Gekakis *et al.*, 1995; Lee *et al.*, 1996; Zeng *et al.*, 1996; Saez & Young, 1996). If nuclear entry and heterodimer formation are key components of the feedback loop, then a mutation in either of the two products which affects these processes would be expected to affect the expression of the partner, as well as changing the circadian phenotype. In fact, the *per* mRNA cycle is obliterated in *tim*^{0}

mutants (Sehgal *et al.*, 1994), and the *tim* mRNA cycle is similarly affected in *per^{01}* mutants (Sehgal *et al.*, 1995), confirming that each product is involved in the other's feedback loop. In addition, the *per^{L1}* mutation, which changes the affinity of PER to bind to TIM (Huang, Edery & Rosbash, 1993; Huang, Curtin & Rosbash, 1995; Gekakis *et al.*, 1995), produces a significant delay in nuclear translocation of the PER^L product compared to the wild-type (Curtin *et al.*, 1995), an observation which fits nicely with its extended long circadian period of 29 h.

Thus, the biochemical basis of this nuclear entry seems to rest on the association of the two proteins into a PER-TIM heterodimer (Zeng *et al.*, 1996; Lee *et al.*, 1996). When this complex enters the nucleus late at night, it is presumably involved in repressing the transcription of the two genes, before the complex is ultimately degraded. With degradation, the block on transcription is lifted, and the mRNAs, and later the PER and TIM products, begin to re-accumulate (Fig. 2). The sequences within PER and TIM that are necessary for nuclear translocation (apart from the nuclear localization signals that are present in both proteins) have been determined (Saez & Young, 1996). In *per^{L1}* mutants the extra lag between cytoplasmic accumulation and nuclear translocation is due to the missense mutation (valine-to-apartate, Baylies *et al.*, 1987) in the PAS domain. PAS is a motif found in several bHLH transcription factors (for review see Hall, 1995), and allows PER to dimerize to TIM (Gekakis *et al.*, 1995). In experiments with *Drosophila* cells, it was observed that the presence of PER and TIM was sufficient for nuclear translocation (Saez & Young, 1996). The importance of the PAS domain, and, in addition, another *per* region encoded downstream of PAS, termed a CLD (cytoplasmic localization domain) was demonstrated. TIM also has a C-terminal CLD, which is required to maintain TIM in the cytoplasm. With both PER and TIM, deletion of their corresponding CLD's, in the presence of the intact partner molecule, results in nuclear translocation (Saez & Young, 1996). Thus these CLDs are sequences that are required to anchor the PER and TIM products to the cytoplasm, until the onset of nuclear translocation.

One prediction of the negative feedback model is that increasing the amount of PER protein should decrease the amount of *per* mRNA. When a constitutive eye specific promoter was used to drive high levels of PER in photoreceptor cells, the endogenous *per* mRNA (*per* is expressed in many anatomical regions, including the photoreceptors, Siwicki *et al.*, 1988) was reduced, consistent with the model (Zeng, Hardin & Rosbash, 1994). Another prediction is that light pulses should reset the PER and TIM molecular cycles. It turns out that TIM is sensitive to light and degrades on exposure to illumination (Hunter-Ensor *et al.*, 1996; Lee *et al.*, 1996; Myers *et al.*, 1996; Zeng *et al.*, 1996). Because the stability of TIM is intimately linked to that of PER (Lee *et al.*, 1996; Zeng *et al.*, 1996), the reduction of TIM levels when lights come on will also subsequently affect PER levels, and indeed the studies cited above show the decreasing levels of TIM anticipating those of PER. Early at night a light pulse will degrade cytoplasmic TIM, but the available *tim* transcript levels will allow TIM translation and recovery to its pre-pulse levels. The time it takes to recover generates a delay. Late at night, nuclear TIM is degraded by the light pulse, but *tim* mRNA is at its trough. TIM levels decrease prematurely, followed by PER, and the clock appears to advance to a phase corresponding with these lower levels.

It is well known in the fly and many other organisms that light pulses cause early evening phase delays and phase advances late at night, generating the phase response curve (PRC, Saunders, 1982; Saunders, Gillanders & Lewis, 1994). Thus the PRC can be readily understood in terms of TIM's light-sensitivity and negative feedback. Light pulses also produced an immediate reduction of *per* mRNA, followed by a more leisurely increase in transcript levels, as well as a behavioural phase shift (Lee *et al.*, 1996; Young *et al.*, 1996). What causes the rapid decrease of *per* mRNA on exposure to light is not known, but the subsequent increase is probably mediated by the light-induced loss of TIM, and consequent destabilization of PER, leading to an increase in *per* transcript levels compared to controls. In other words, these experiments show indirectly and *in vivo*, that as PER goes down, the mRNA goes up, again supporting negative feedback.

Abolishing the oscillation of *per* and *tim* mRNA and proteins should disrupt circadian rhythmicity. Clearly this requirement of the model is met in that *per*^{01} and *tim*^{0} arrhythmic mutants abolish both circadian behaviour and mRNA cycling, not only of their own transcripts, but also of each others, as mentioned earlier. In arrhythmic *per*^{01} mutants, constitutive peak transcript levels would be expected as no PER is detectable in the mutants, but in contrast, *per*^{01} mRNA levels remain intermediate. This could be due to the *per*^{01} transcript being more prone to degradation, as suggested by Brandes *et al.* (1996), because it has a premature translational stop (Yu *et al.*, 1987; Baylies *et al.*, 1987). In addition, the regulatory sequences in the *per* mRNA identified by Stanewsky *et al.* (1997) may also contribute to the less-than-peak levels of *per*^{01} transcript that are observed in the mutants.

Another apparent difficulty for the negative feedback model is that when *per*^{01} flies carrying a *hsp-per* construct are transferred from 18 to 29°C, quite robust circadian cycles in locomotor behaviour were observed (Ewer *et al.*, 1988, 1990). This is unexpected because constitutive high levels of PER should effectively stop the clock at a time corresponding to peak PER levels. This result was explained by showing that PER from the transgene did indeed cycle in cells which normally express PER, but not in those that do not (*hsp-per* expresses PER promiscuously). Furthermore, the endogenous *per*^{01} transcript cycled (Frisch *et al.*, 1994). There are two possible explanations for this result. First, regulatory sequences in the *hsp-per* construct, again, possibly those identified by Stanewsky *et al.* (1997 – see above), were driving a cycling of the *hsp-per* mRNA, which was driving the PER protein rhythm, which in turn was presumably feeding back to generate the cycling of the endogenous per^{01} transcript. Secondly, because the PER partner TIM, is light-sensitive and degrades on exposure to illumination (Hunter-Ensor *et al.*, 1996; Myers *et al.*, 1996; Lee *et al.*, 1996; Zeng *et al.*, 1996; Young *et al.*, 1996), consequently destabilizing PER, the PER cycles seen in LD could represent light driving PER oscillations via TIM cyclical degradation. Another intriguing observation, namely that promoterless *per* genes also manifest mRNA cycling in a few transgenic lines (Frisch *et al.*, 1994), can also be 'explained' along similar lines.

Finally, in LD cycles, a *per-lacZ* fusion gene shows a modest protein cycling of the hybrid protein in a per^{01} background (Dembinska *et al.*, 1997). This differs from other such transgenes which have less *per* coding material and whose products do not cycle (Dembinska *et al.*, 1997). These results imply the presence of a temporal proteolysis (experiments show that this is TIM and LD dependent), which targets particular PER sequences. Interestingly, these sequences (Dembinska *et al.*, 1997) include a Thr-Gly/Ser-Gly repeat region that seems to be under complex forms of natural selection (Costa *et al.*, 1992; Peixoto *et al.*, 1993; Nielsen *et al.*, 1994; Rosato *et al.*, 1994, 1997), and which structurally appears to provide a nicely non-globular degradable module (Ishida, Mitsui & Tsukada, 1994; Castiglioni-Morelli *et al.*, 1995).

Clearly, the role or PER and TIM within a negative feedback model is largely supported. If a case could be found where PER or TIM did not cycle in the key pacemaker tissues, but the fly still retained circadian behaviour, then it would be difficult to defend PER as a state variable. No such example currently exists. So far, there is no evidence that cycling PER does not somehow feedback to its own mRNA, nor is there any evidence that cycling *per* mRNA does not generate cycling PER protein. However, just to be argumentative, there is, as yet, no direct evidence that PER, TIM, the PER-TIM complex, or any other protein, binds dynamically to the *per* or *tim* promoters, as would be expected from a simple autoregulatory negative feedback model. Nevertheless, if, as predicted by the model, PER and TIM directly, or more likely, indirectly, regulate their own transcription, then it is also possible that they may regulate the clock-controlled genes (*ccg*, see Fig. 1(b)) that are required to transduce the signal of the pacemaker to the organs that will express circadian behaviour and physiology.

COMPARATIVE STUDIES

From an evolutionary perspective, there is little doubt that circadian clocks are adaptive, given their ubiquity, and so one might imagine *a priori*, that natural selection would find various ways of maintaining and reinforcing the cycling of state variables. One method is to have cycling transcripts giving rise to cycling proteins after the appropriate lag. A clock-regulated degradation pathway would be required to make sure that the state variable is degraded with 'circadian' kinetics, thereby releasing the negative regulation on the transcript. This feedback of the protein on to the mRNA could be via the promoter to give transcriptional cycling, and via post-transcriptional effects on other regions of the transcript, again safeguarding the mRNA cycle. Perhaps *Drosophila* has just made the best of a variety of mechanisms, suggesting that there may be several other ways of generating a negative feedback loop. This somewhat contrived opening leads us into the giant silkmoth, which has turned up something of a surprise.

Anthereae pernyi, a huge insect which manifests robust circadian phenotypes, is much easier to manipulate than *D. melanogaster* and consequently, a wealth of cellular and physiological information has accumulated over the years. This reveals that the pacemaker for the timing for the photoperiodic termination of pupal diapause, adult eclosion, and adult flight rhythms, is located to the dorsal lateral protocerebrum (Truman, 1972, 1974, 1992; Truman & Riddiford, 1970). The molecular dissection of the moth circadian clock began with the cloning of the *per* homologue (Reppert *et al.*, 1994). As expected, just as in the fly, a prominent circadian oscillation was found for *per* mRNA levels in moth heads and for the PER product in photoreceptor nuclei. In addition, the moth *per* gene can function as an adequate clock when transformed into *Drosophila per^{01}* hosts and show the dynamic nuclear–cytoplasmic features of fly PER in the LNs (Levine *et al.*, 1995). What was not expected was a striking difference in the spatial and temporal pattern of *per* expression between photoreceptors and brain cells (Reppert *et al.*, 1994; Sauman & Reppert, 1996). For the photoreceptors, the results are in complete agreement with those from *Drosophila*. The *per* mRNA cycles with peak levels between early at night, and PER immunoreactivity shows intense nuclear straining late in the photoperiod, giving the characteristic lag between the mRNA and protein cycles (Reppert *et al.*, 1994).

The surprise comes from the brain. Only eight neurosecretory cells express PER, quite unlike the dozens of neurons and hundreds of glial cells in *Drosophila* (Sauman & Reppert, 1996). PER (and TIM) cycle in phase with each other in these cells, but remain largely confined in the cytoplasm at all times, and are even found in the axonal projections. In addition *per* mRNA also cycles, but with the same phase as the PER product. As PER does not enter the nucleus in these cells, it follows that it cannot negatively autoregulate its own transcription directly. However, a cytoplasmic antisense RNA cycles in antiphase with the sense mRNA, suggesting a possible mechanism for producing mRNA rhythms through the formation of RNA–RNA duplexes (Sauman & Reppert, 1996). The cycling antisense RNA has also been found in the photoreceptors, even if in these cells PER is nuclear. Constant light (LL), which abolishes behavioural rhythmicity, disrupts the molecular cycles of PER, TIM and the two *per* mRNAs, consistent with these molecules presumed functional properties (Sauman & Reppert, 1996).

To learn something about the possible function of the PER-staining tissues, *per* antisense mRNA was injected into the embryo. At this developmental stage, a number of tissues express PER and TIM, including eight dorsolateral neurons, fat body cells, and the midgut epithelium (Sauman *et al.*, 1996). Of these, only the gut cells show characteristic PER nuclear–cytoplasmic rhythms, but neither gut nor fat body cells appear to express TIM. The neurons show no evidence of cytoplasmic oscillations in PER or TIM, unlike their adult counterparts. Nevertheless, the circadian emergence of the pharate larva from the egg shell was blocked by the anti-sense injections, suggesting that PER is involved in generating this clock phenotype.

So, do these studies on *Antheraea* put the PER/TIM negative feedback model in mothballs? Probably not, because the eyes and the gut do show nuclear localization of PER as in the fly, but variety appears to be the rule. By this is meant that PER and TIM can be rhythmically or constitutively expressed, in cytoplasm or nucleus, together or alone, depending on the tissue, and the developmental stage. In the adult moth, there is no direct evidence that PER and TIM are involved in any circadian behavioural phenotypes, but if they are, then their neuronal cytoplasmic rhythms are difficult to incorporate into the standard feedback model. Rhythmic egg hatching requires PER, suggesting that the relevant embryonic neurons can function as an oscillator with cytoplasmic non-cycling PER and TIM. It is unlikely that the antisense *per* RNA affected rhythmic egg-hatching in the embryo by blocking PER translation in the fat body cells or in the gut epithelium, because then one would be obliged to build a clock based on cytoplasmic non-cycling PER without TIM, or cycling nuclear PER without TIM, respectively. This is improbable although formally possible. Rather, PER and TIM in all these adult and embryonic tissues may represent the output of a clock whose state variables are as yet unidentified. Might this be why PER and TIM are expressed in axonal processes, and could they be the chemical messengers of the clock, with hormonal functions? Intriguingly, the *per*-expressing neurons and their axons are strategically placed near other cells that contain peptides known to be under circadian control (Sauman & Reppert, 1996).

Perhaps the moth work is not so surprising, in that *Drosophila* also shows tissue-specific regulation of PER, for example, in female ovaries, where the antigen is non-rhythmic and cytoplasmic (Hardin, 1994). It is not known whether TIM is absent in these tissues, which if it was, would explain the absence of nuclear movements. The only clue about what PER might be doing there is that the photoperiodic gonadal regression of *per^{01}* females, in short-day, long-night conditions which mimic winter, differs significantly from that of wild-type (Saunders, 1990). Could PER be providing some aspect of photoperiodic timing?

In addition, ancient *per* history reminds us that transplanting a *per^s* brain into a *per^{01}* abdomen gives hosts with circadian behavioural cycles characteristic of the donor's brain (when this revolting experiment works), in the apparent absence of normal connections, thus providing fly PER with 'hormonal-like' properties (Handler & Konopka, 1979). This is something of an overstatement, as a molecule which works downstream of PER could be the key diffusible factor in these studies, but taken in the context of the moth PER biology, it is an interesting reminder of an almost forgotten experiment.

Furthermore, the 50–60s ultradian male lovesong cycle that is altered by the *per* mutations in parallel with their effects on circadian rhythms, is a long-standing thorn in the side of PER if we believe that PER is *only* a circadian transcriptional regulator (Kyriacou & Hall, 1980, 1989). Clearly, this behavioural oscillation is unlikely to be determined by a transcriptional–translational cycle. The key tissues that generate this cycle are thoracic (Konopka, Kyriacou & Hall, 1997), and in this anatomical area, only glial cells express PER (Siwicki *et al.*, 1988). The effects of the *per* mutations on this cycle have been rationalized recently as reflecting perhaps PER's role as a transcriptional regulator of downstream genes (*ccgs* – Fig. 1) required to build the song oscillator (Konopka *et al.*, 1997). Another group, who have recently independently confirmed the original *per*-mutant song rhythm findings (Alt, Ringo & Dowse, in press), argue that the cell is a bag full of rhythmic processes, some with very high frequencies, which PER couples to generate lower frequency metaoscillators, including the circadian clock (see also Dowse, Hall & Ringo, 1987). Perhaps even a third role for PER can now be hypothesized in *Drosophila*, namely that of a neuromodulatory factor which generates the song cycle via a glial–neuronal pathway. Therefore, even in the fly, there are some data that show that PER has more than just a circadian function.

In the other tissues where PER is expressed in *Drosophila*, in photo-receptors, corpora cardiaca, gut, Malpighian tubules (Siwicki *et al.*, 1988), the nuclear–cytoplasmic oscillations comfort us into thinking that these molecular rhythms reflect an underlying circadian physiology within these organs. Interestingly, the PER/TIM cycles in the Malpighian tubules persist even when the fly is decapitated, and these can be entrained to different LD cycles (Giebultowicz & Hege, 1997). This shows that the brain and peripheral tissue can use the same mechanisms and molecules to programme their circadian clocks. However, the presence of a timer in Malpighian tubules, as well as one for the ultradian song cycle in the thorax (Konopka *et al.*, 1997), also shows how these timers are autonomous from the brain. There is a tendency to use the term 'pacemaker' to refer to the LNs of the fly brain, implying that these neurons represent some kind of master clock which controls other 'slave' oscillators in other tissues. The LNs may be an important pacemaker for the activity cycle, but even here, Helfrich-Foerster (1996) has suggested that other, distinct neurons that express GLASS (Vosshall & Young, 1995), may also contribute to the locomotor rhythm.

Finally, just to emphasize the point about autonomous tissue specific clocks, Emery *et al.* (1997) have managed to culture *Drosophila* pupal ring glands, and cycling of *per* gene expression can be maintained in these tissues for a week. This opens the door for studying the relationship between the state variables *per* and *tim* and cellular physiology.

PERSPECTIVES

So perhaps the silkmoth is simply a somewhat bizarre evolutionary example, and other species will follow the fly model. This can only be answered by initially studying other Diptera, before moving further afield. However, some spectacular forward genetics have identified a potential clock molecule in the mouse, which when mutated gives defective locomotor cycles. This gene, called *Clock* (Vitaterna *et al.*, 1994), encodes a bHLH protein which also has a PAS domain (King *et al.*, 1997), thereby giving it an evolutionary link with PER. The gene rescues the original *Clock* mutation in transgenic mice (Antoch *et al.*, 1997). Its primary sequence suggests it could be a transcriptional regulator, and the clock community will follow this work with great interest. Is this a *per* analogue? Apparently not, as a human and mouse gene called *rigui* (named after an ancient Chinese sundial) has been identified which also has a PAS domain, a bHLH region, and other clusters of similarities to fly *per* (Sun *et al.*, 1997). The overall level of protein similarity between *rigui* and fly *per* is about 44% (Sun *et al.*, 1997). Furthermore, the mouse *rigui* transcript cycles in various regions of the brain that have been implicated with circadian rhythms, whereas the *Clock* mRNA on the other hand, does not appear to oscillate (Sun *et al.*, 1997). The *rigui* gene has also been identified by another group (Tei *et al.*, 1997), who failed to spot the bHLH region, but who also noticed that, as well as the PAS domain similarity with fly *per*, there were also other regions of conservation, including the Thr-Gly/Ser-Gly repeat. The PAS domain has again reared its head in the form of two *Neurospora* genes, *wc1* and *wc2*, which also encode DNA-binding proteins. Mutations in these genes are insensitive to light and are intimately associated with light regulation of the *Neurospora frq* gene (Crosthwaite, Dunlap & Loros, 1997; Dunlap *et al.*, this volume). Consequently, of the seven genes that can be termed 'clock loci', five of them, *per*, *Clock*, *rigui*, *wc1* and *wc2*, have PAS domains, suggesting that these dimerization motifs represent an evolutionary conserved feature of clock proteins. Some light-sensing proteins in bacteria also have PAS domains (e.g. Borgstahl, Williams & Getzoff, 1995), and it has been suggested that the origins of circadian rhythmicity might be traced to the ancient clock proteins that may have originally recruited PAS domains from these microbial proteins (Crosthwaite *et al.*, 1997). Given the association between light and clocks, this hypothesis has some intuitive appeal, but remains to be tested rigorously phylogenetically.

Apart from the silkmoth, other animals, *Aplysia, Bulla*, rats, beetles, and frogs, have also shown labelling of anatomical structures with anti-PER antibodies (for review see Kyriacou *et al.*, 1996). The staining in most of these organisms is usually cytoplasmic, and importantly, sometimes labels structures that may represent clock machinery. However, it could be that these antibodies are simply labelling cross-reacting material that is not related to PER. In spite of a lack of progress in identifying the genes that give rise to these apparent PER-related antigens, the comparative approach will eventually reveal the generality of the negative feedback model, once the relevant clock proteins have been identified. One particularly significant step in this direction has been the reverse genetic approach used in *Xenopus*, by Green & Besharse (1996). A cycling photoreceptor mRNA was isolated whose predicted sequence encoded a transcription factor motif and a leucine-zipper. Whether this molecule represents a clock component or a downstream clock-controlled gene (*ccg*s – see Fig. 1) is not known at present. However, the detection of cycling mRNAs may offer one way of identifying clock molecules in organisms which have little genetics, but it is likely that this approach will tend to identify more of the genes that lie downstream of the oscillator than state variables. For example, a similar approach used in *Drosophila* by van Gelder and Krasnow (1995) identified 20 cycling transcripts in the head, but *per* was not one of them.

To conclude, the molecular biology of circadian clocks is beginning to move at a tremendous pace. No longer are the fruit-fly and *Neurospora* the only molecular model systems, although they still lead the way. Higher vertebrates like *Xenopus*, and particularly the mouse, will contribute significantly to the molecular dissection of the biological clock. How similar the pacemaker mechanisms will be between *Neurospora* and *Cyanobacteria* and the mouse, remains to be seen, but one thing is certain, an exhilarating phase is dawning in research in this area.

REFERENCES

Alt, S., Ringo, J. & Dowse, H. B. (1997). Song rhythms in *per*-altered mutants of *Drosophila melanogaster. Animal Behavior*, In Press.

Antoch, M. P., Song, E-J., Chang, A-M., Vitaterna, M. H., Zhao, Y., Wilsbacher, L. D., Sangoram, A. M., King, D. P., Pinto, L. H. & Takahashi, J. S. (1997). Functional identification of the mouse circadian *Clock* gene by transgenic BAC rescue. *Cell*, **89**, 655–67.

Baylies, M. K., Bargiello, T. A., Jackson, F. R. & Young, M. W. (1987). Changes in abundance or structure of the per gene product can alter periodicity of the *Drosophila* clock. *Nature*, **326**, 390–2.

Borgstahl, G. E. O., Williams, D. R. & Getzoff, E. D. (1995). 1.4 A structure of photoactive yellow protein, a cytosolic photoreceptor: unusual fold, active site and chromophore. *Biochemistry*, **34**, 6278–87.

Brandes, C., Plautz, J. D., Stanewsky, R., Jamison, C. F., Straume, M., Woods, K. V.,

Kay, S. A. & Hall, J. C. (1996). Novel features of *Drosophila per* transcription revealed by real-time luciferase reporting. *Neuron*, **16**, 687–92.

Castiglione-Morelli, M. A., Guantieri, V., Villani, V., Kyriacou, C. P., Costa, R. & Tamburro, A. M. (1995). Conformational study of the Thr-Gly repeat in the *Drosophila* clock protein period. *Proceedings of the Royal Society of London (Biology)*, **260**, 155–63.

Costa, R., Peixoto, A. A., Barbujani, G. & Kyriacou, C. P. (1992). A latitudinal cline in a *Drosophila* clock gene. *Proceedings of the Royal Society of London (Biology)*, **250**, 43–9.

Crosthwaite, S. K., Dunlap, J. C. & Loros, J. J. (1997). *Neurospora wc-1* and *wc-2*: Transcription, photoresponses, and the origins of circadian rhythmicity. *Science*, **276**, 763–9.

Curtin, K. D., Huang, Z. J. & Rosbach, M. (1995). Temporally regulated nuclear entry of the *Drosophila* period protein contributes to the circadian clock. *Neuron*, **14**, 365–72.

Dembinska, M. E., Stanewsky, R., Hall, J. C. & Rosbash, M. (1997). Circadian cycling of a period-lacZ fusion protein in *Drosophila*: evidence for cyclical degradation. *Journal of Biological Rhythms*, **12**, 157–72.

Dowse, H. B., Hall, J. C. & Ringo, J. M. (1987). Circadian and ultradian rhythms in *period* mutants of *Drosophila melanogaster*. *Behavior Genetics*, **17**, 19–35.

Edery, I., Zweibel, L. J., Dembinska, M. E. & Rosbash, M. (1994). Temporal phosphorylation of the *Drosophila* period protein. *Proceedings of the National Academy of Sciences, USA*, **91**, 2260–4.

Edmunds, L. N. Jr (1988). *Cellular and Molecular Bases of Biological Clocks*. New York: Springer.

Emery, I. F., Noveral, J. M., Jamison, C. F. & Siwicki, K. K. (1997). Rhythms of *period* gene expression in culture. *Proceedings of the National Academy of Sciences, USA*, **94**, 4092–6.

Ewer, J., Rosbash, M. & Hall, J. C. (1988). An inducible promoter fused to the *period* gene of *Drosophila* conditionally rescues adult *per*-mutant arrhythmicity. *Nature*, **333**, 82–4.

Ewer, J., Hamblen-Coyle, M., Rosbash, M. & Hall, J. C. (1990). Requirement for *period* gene expression in the adult and not during development for locomotor activity rhythms of imaginal *Drosophila melanogaster*. *Journal of Neurogenetics*, **7**, 31–73.

Ewer, J., Frisch, B., Hamblen-Coyle, M., Rosbash, M. & Hall, J. C. (1992). Expression of the *period* clock gene within different cell types in the brain of *Drosophila* adults and mosaic analysis of these cells' influence on circadian behavioral rhythms. *Journal of Neuroscience*, **12**, 3321–49.

Frisch, B., Hardin, P. E., Hamblen-Coyle, M., Rosbash, M. & Hall, J. C. (1994). A promotorless *period* gene mediates behavioral rhythmicity and cyclical *per* gene expression in a restricted subset of the *Drosophila* nervous system. *Neuron*, **12**, 555–70.

Gekakis, N., Saez, L., Delahaye-Brown, A. M., Myers, M. P., Sehgal, A., Young, M. W. & Weitz, C. J. (1995). Isolation of *timeless* by PER protein interaction: defective interaction between timeless protein and long-period mutant *per*[L]. *Science*, **70**, 811–15.

Giebultowicz, J. M. & Hege, D. M. (1997). Circadian clock in Malpighian tubules. *Nature*, 386, 664.

Goldbeter, A. (1995). A model for circadian oscillations in the *Drosophila* period protein. *Proceedings of the Royal Society of London (Biology)*, **261**, 319–24.

Green, C. B. & Besharse, J. C. (1996). Identification of a novel vertebrate circadian clock-regulated gene encoding the protein nocturnin. *Proceedings of the National Academy of Sciences, USA*, **93**, 14884–8.

Hall, J. C. (1995). Tripping along the trail to the molecular mechanisms of biological clocks. *Trends in Neuroscience*, **18**, 230–40.

Hall, J. C. (1996). Are cycling gene products as internal Zeitgebers no longer the Zeitgeist of chronobiology? *Neuron*, **17**, 799–802.

Hall, J. C. & Kyriacou, C. P. (1990). Genetics of biological rhythms in *Drosophila*. *Advances in Insect Physiology*, **22**, 221–98.

Handler, A. M. & Konopka, R. J. (1979). Transplantation of a circadian pacemaker in *Drosophila*. *Nature*, **279**, 236–8.

Hao, H., Allen, D. L. & Hardin, P. E. (1997). A circadian enhancer mediates PER-dependent mRNA cycling in *Drosophila*. *Molecular and Cellular Biology*, **17**, 3687–93.

Hardin, P. E. (1994). Analysis of *period* mRNA cycling in *Drosophila* head and body tissues indicates that body oscillators behave differently than head oscillators. *Molecular and Cellular Biology*, **14**, 7211–18.

Hardin, P. E., Hall, J. C. & Rosbash, M. (1990). Feedback of the *Drosophila period* gene product on circadian cycling of its messenger RNA. *Nature*, **343**, 536–40.

Hardin, P., Hall, J. C. & Rosbash, M. (1992). Circadian oscillations in *period* gene mRNA levels are transcriptionally regulated. *Proceedings of the National Academy of Sciences, USA*, **89**, 11711–15.

Helfrich-Foerster, C. (1996). *Drosophila* rhythms: from brain to behavior. *Seminars in Cell and Developmental Biology*, **7**, 791–802.

Huang, Z. J., Edery, I. & Rosbash, M. (1993). PAS is a dimerization domain common to *Drosophila* period and several transcription factors. *Nature*, **364**, 259–62.

Huang, Z. J., Curtin, K. D. & Rosbash, M. (1995). PER protein interactions and temperature compensation of a circadian clock in *Drosophila*. *Science*, **267**, 1169–72.

Hunter-Ensor, M., Ousley, A. & Sehgal, A. (1996). Regulation of the *Drosophila* protein timeless suggests a mechanism for resetting the circadian clock by light. *Cell*, **84**, 677–85.

Ishida, N., Mitsui, Y. & Tsukada, M. (1994). Molecular conformation and physical properties of poly(Gly-Thr)n: a compound model for the period repetitive sequences. *Molecular Biology (Life Sciences Advances)*, **13**, 107–11.

Iwasaki, K. & Thomas, J. H. (1997). Genetics in rhythm. *Trends in Genetics*, **13**, 111–15.

King, D. P., Zhao, Y., Sangoram, A. M., Wilsbacher, L. D., Tanaka, M., Antoch, M. P., Steeves, T. D. L., Vitaterna, M. H., Kornhauser, J. M., Lowrey, P. L., Turek, F. W. & Takahashi, J. S. (1997). Positional cloning of the mouse circadian *Clock* gene. *Cell*, **89**, 641–53.

Kondo, T., Tsinoremas, N. F., Golden, S. G., Johnson, C. H., Kutsuna, S. & Ishiura, M. (1994). Circadian clock mutants of *Cyanobacteria*. *Science*, **266**, 1233–6.

Konopka, R. J. & Benzer, S. (1971). Clock mutants of *Drosophila melanogaster*. *Proceedings of the National Academy of Sciences, USA*, **68**, 2112–16.

Konopka, R. J., Kyriacou, C. P. & Hall, J. C. (1997). Mosaic analysis in the *Drosophila* CNS of circadian and courtship song rhythms affected by a *period* clock mutant. *Journal of Neurogenetics*, **11**, 117–39.

Kyriacou, C. P. & Hall, J. C. (1980). Circadian rhythm mutations in *Drosophila* affect short-term fluctuations in the male's courtship song. *Proceedings of the National Academy of Sciences, USA*, **77**, 6729–33.

Kyriacou, C. P. & Hall, J. C. (1989). Spectral analysis of *Drosophila* courtship songs. *Animal Behaviour*, **37**, 850–9.

Kyriacou, C. P., Sawyer, L. A., Piccin, A., Couchman, M. E. & Chalmers, D. (1996). Evolution and population biology of the *period* gene. *Seminars in Cell and Developmental Biology*, **7**, 803–10.

Lee, C., Parikh, V., Itsukaichi, T., Bae, K. & Edery, I. (1996). Resetting the *Drosophila* clock by photic regulation of PER and a PER-TIM complex. *Science*, **271**, 1740–4.

Levine, J. D., Sauman, I., Imbalzano, M., Reppert, S. M. & Jackson, F. R. (1995). Period protein from the giant silkmoth *Antheraea pernyi* functions as a circadian clock element in *Drosophila melanogaster*. *Neuron*, **15**, 147–57.

Loros, J. (1995). The molecular basis of the *Neurospora* clock. *Seminars in the Neurosciences*, **7**, 3–13.

Millar, A. J. & Kay, S. A. (1997). The genetics of phototransduction and circadian rhythms in *Arabidopsis*. *Bioessays*, **19**, 209–14.

Myers, M. P., Wager-Smith, K., Rothenfluh-Hilfiker, A. & Young, M. W. (1996). Light-induced degradation of Timeless and entrainment of the *Drosophila* circadian clock. *Science*, **271**, 1736–40.

Nielsen, J., Peixoto, A. A., Piccin, A., Costa, R., Kyriacou, C. P. & Chalmers, D. (1994). Big flies, small repeats: the 'Thr-Gly' region of the *period* gene in *Diptera*. *Molecular Biology and Evolution*, **11**, 839–53.

Peixoto, A. A., Campesan, S., Costa, R. & Kyriacou, C. P. (1993). Molecular evolution of a repetitive region within the *per* gene of *Drosophila*. *Molecular Biology and Evolution*, **10**, 127–39.

Reppert, S. M., Tsai, T., Roca, A. L. & Sauman, I. (1994). Cloning of a structural and functional homolog of the circadian clock gene *period*, from the Giant Silkmoth *Antheraea pernyi*. *Neuron*, **13**, 1167–76.

Rosato, E., Peixoto, A. A., Barbujani, G., Costa R. & Kyriacou, C. P. (1994). Molecular evolution of the *period* gene in *Drosophila simulans*. *Genetics*, **138**, 693–707.

Rosato, E., Peixoto, A. A., Costa, R. & Kyriacou, C. P. (1997). Mutation rate, linkage disequilibrium, and selection in the repetitive region of the *period* gene in *Drosophila melanogaster*. *Genetic Research*, **69**, 89–99.

Ruoff, P. & Rensing, L. (1996). The temperature-compensated Goodwin model simulates many circadian clock properties. *Journal of Theoretical Biology*, **179**, 275–85.

Saez, L. & Young, M. W. (1996). Regulation of nuclear entry of the *Drosophila* clock proteins period and timeless. *Neuron*, **17**, 911–20.

Sauman, I. & Reppert, S. M. (1996). Circadian clock neurons in the silkmoth *Antheraea pernyi*: novel mechanisms of period protein regulation. *Neuron*, **17**, 889–900.

Sauman, I., Tsai, T., Roca, A. L. & Reppert, S. M. (1996). Period protein is necessary for circadian control of egg hatching behavior in the silkmoth *Antheraea pernyi*. *Neuron*, **17**, 901–9.

Saunders, D. S. (1982). *Insect Clocks*. 2nd edn. Oxford: Pergamon Press.

Saunders, D. S. (1990). The circadian basis of ovarian diapause regulation in *Drosophila melanogaster*. Is the *period* gene causally involved in photoperiodic time measurement? *Journal Biological Rhythms*, **5**, 315–31.

Saunders, D. S., Gillanders, S. W. & Lewis, R. D. (1994). Light-pulse phase response curves for the locomotor activity rhythm in *period* mutants of *Drosophila melanogaster*. *Journal of Insect Physiology*, **40**, 957–68.

Sehgal, A., Price, J. L., Man, B. & Young, M. W. (1994). Loss of circadian behavioral rhythms and *per* mRNA oscillations in the *Drosophila* mutant *timeless*. *Science*, **263**, 1603–9.

Sehgal, A., Rothenfluh-Hilfiker, Hunter-Ensor, M., Chen, Y., Myers, M. P. & Young, M. W. (1995). Rhythmic expression of *timeless*: a basis for promoting circadian cycles in *period* gene autoregulation. *Science*, **270**, 808–10.

Siwicki, K. K., Eastman, C., Petersen, G., Rosbash, M. & Hall, J. C. (1988). Antibodies to the *period* gene product of *Drosophila* reveal diverse distribution and rhythmic changes in the visual system. *Neuron*, **1**, 141–50.

Stanewsky, R., Jamison, C. F., Plautz, J. D., Kay, S. A. & Hall, J. C. (1997). Multiple circadian elements contribute to cycling *period* gene expression in *Drosophila*. *EMBO Journal*, **16**, 5006–18.

Sun, Z. H., Albrecht, U., Zhuchenko, O., Bailey, J., Eichele, G. & Lee, C. C. (1997). *Rigui*, a putative mammalian ortholog of the *Drosophila period* gene. *Cell*, **90**, 1003–11.

Tei, H., Okamura, H., Shigeyoshi, Y., Fukuhara, C., Ozawa, R., Hirose, M. & Sakaki, Y. (1997). Circadian oscillation of a mammalian homologue of the *Drosophila period* gene. *Nature*, **389**, 512–16.

Truman, J. W. (1972). Physiology of insect rhythms. II. The silkmoth brain as the location of the biological clock controlling eclosion. *Journal of Comparative Physiology*, **81**, 99–114.

Truman, J.W. (1974). Physiology of insect rhythms. IV. Role of the brain in regulation of the flight rhythm of the giant silkmoths. *Journal of Comparative Physiology*, **95**, 281–96.

Truman, J. W. (1992). The eclosion hormone system of insects. *Progress in Brain Research*, **92**, 361–74.

Truman, J. W. & Riddiford, L. M. (1970). Neuroendocrine control of ecdysis in silkmoths. *Science*, **167**, 1624–6.

Van Gelder, R. N. & Krasnow, M. A. (1995). Extent and character of circadian gene expression in *Drosophila melanogaster* – identification of 20 oscillating mRNAs in the fly head. *Current Biology*, **5**, 1424–36.

Vitaterna, M. H., King, D. P., Chang, A. M., Kornhauser, J. M., Lowrey, P. L., McDonald, J. D., Dove, W. F., Pinto, L. H., Turek, F. W. & Takahashi, J. S. (1994). Mutagenesis and mapping of a mouse gene, *Clock*, essential for circadian behavior. *Science*, **264**, 719–25.

Vosshall, L. B. & Young, M. W. (1995). Circadian rhythms in *Drosophila* can be driven by *period* expression in a restricted group of central brain cells. *Neuron*, **15**, 345–60.

Vosshall, L. B., Price, J.L., Sehgal, A., Saez, L. & Young, M. W. (1994). Block in nuclear localisation of period protein by a second clock mutation, *timeless*. *Science*, **263**, 1606–9.

Young, M. W., Wager-Smith, K., Vosshall, L, Saez, L. & Myers, M. P. (1996). Molecular anatomy of a light-sensitive circadian pacemaker in *Drosophila*. *Cold Spring Harbor Symposia on Quantitative Biology*, **61**, 279–84.

Yu, Q., Jacquier, A. C., Citri, Y., Hamblen, M., Hall, J. C. & Rosbash, M. (1987). Molecular mapping of point mutations in the *period* gene that stop or speed up biological clocks in *Drosophila melanogaster*. *Proceedings of the National Academy of Sciences, USA*, **84**, 784–8.

Zeng, H., Hardin, P. E. & Rosbash, M. (1994). Constitutive expression of the *Drosophila* period protein inhibits *period* mRNA cycling. *EMBO Journal*, **13**, 3590–8.

Zeng, H., Quian, Z., Myers, M. P. & Rosbash, M. (1996). A light-entrainment mechanism for the *Drosophila* circadian clock. *Nature*, **380**, 129–35.

Index